COMPUTERIZED CONTROL SYSTEMS IN THE FOOD INDUSTRY

FOOD SCIENCE AND TECHNOLOGY

A Series of Monographs, Textbooks, and Reference Books

1. Flavor Research: Principles and Techniques, *R. Teranishi, I. Hornstein, P. Issenberg, and E. L. Wick*
2. Principles of Enzymology for the Food Sciences, *John R. Whitaker*
3. Low-Temperature Preservation of Foods and Living Matter, *Owen R. Fennema, William D. Powrie, and Elmer H. Marth*
4. Principles of Food Science
   Part I: Food Chemistry, *edited by Owen R. Fennema*
   Part II: Physical Methods of Food Preservation, *Marcus Karel, Owen R. Fennema, and Daryl B. Lund*
5. Food Emulsions, *edited by Stig E. Friberg*
6. Nutritional and Safety Aspects of Food Processing, *edited by Steven R. Tannenbaum*
7. Flavor Research: Recent Advances, *edited by R. Teranishi, Robert A. Flath, and Hiroshi Sugisawa*
8. Computer-Aided Techniques in Food Technology, *edited by Israel Saguy*
9. Handbook of Tropical Foods, *edited by Harvey T. Chan*
10. Antimicrobials in Foods, *edited by Alfred Larry Branen and P. Michael Davidson*
11. Food Constituents and Food Residues: Their Chromatographic Determination, *edited by James F. Lawrence*
12. Aspartame: Physiology and Biochemistry, *edited by Lewis D. Steginkand and L. J. Filer, Jr.*
13. Handbook of Vitamins: Nutritional, Biochemical, and Clinical Aspects, *edited by Lawrence J. Machlin*
14. Starch Conversion Technology, *edited by G. M. A. van Beynum and J. A. Roels*

Additional Volumes in Preparation

COMPUTERIZED CONTROL SYSTEMS IN THE FOOD INDUSTRY

edited by
GAURI S. MITTAL

School of Engineering
University of Guelph
Guelph, Ontario, Canada

CRC Press
Taylor & Francis Group
Boca Raton London New York

CRC Press is an imprint of the
Taylor & Francis Group, an **informa** business

CRC Press
Taylor & Francis Group
6000 Broken Sound Parkway NW, Suite 300
Boca Raton, FL 33487-2742

First issued in paperback 2019

© 1997 by Taylor & Francis Group, LLC
CRC Press is an imprint of Taylor & Francis Group, an Informa business

ISBN-13: 978-0-8247-9757-7 (hbk)
ISBN-13: 978-0-367-40121-4 (pbk)

Library of Congress Cataloging–in–Publication Data

Computerized control systems in the food industry / edited by Gauri S.
Mittal.
    p. cm.— (Food science and technology ; 78)
    Includes index.
    ISBN 0-8247-9757-4 (hardcover : alk. paper)
    1. Food industry and trade—Data processing. 2. Automatic
control—Data processing. I. Mittal, G.S. (Gauri S.) II. Series:
Food science and technology (Marcel Dekker, Inc.) ; 78.
TP370.5.C67 1996
664'.024—dc20   96-31578
                                                          CIP

Visit the Taylor & Francis Web site at
http://www.taylorandfrancis.com

and the CRC Press Web site at
http://www.crcpress.com

Preface

The food processing industry is in an era of transition due to competition and a rapidly changing business environment. It is spending billions of dollars every year on modernization including increased computerized or digital automation. Although analog control is adequate under normal operating conditions, it cannot respond quickly enough during significant process deviations. Moreover, when software-based controls are used, it is much easier to implement new control algorithms. Process control technology has advanced in recent years because of the availability of inexpensive digital technology, circuit boards, sensors, programmable logic controllers, and microcomputers.

Computer-based process control is now universally regarded as an essential tool. It is crucial not only for quality control, but for securing economic survival. Better process controls are also needed to convert batch operations to continuous, and to improve plant sanitation. Automation offers to the food industry better profitability, productivity, process efficiency, and product uniformity, and offers to the consumer a higher-quality and safer product. To achieve Total Quality Management (TQM), improved process control is very important. Because food processes are difficult to automate and control, computer-based controls offer significant benefits. The documentation capability of these control systems is virtually unlimited, thus providing integration of controls and information management.

This is the first comprehensive, unified book on this topic. Encyclopedic in scope, it systematically covers the basics, emerging techniques, and advances in process control, control algorithms, and specific applications. This book is directed to the present need of the food industry to improve productivity, food quality, and safety through computerized automation and process control. It offers an introduction to the subject to introduce food engineers and scientists in one convenient single volume. Both established and innovative control schemes are covered. The book provides an important source of information for all those involved in food manufacturing, process control, engineering, and production planning.

A number of books are available on computer-based process controls. However, because food processes have unique control requirements relating to optimal palatability, safety, and the biological nature of foods, books that have been written for chemical and other industries on process control are not directly suitable. No one book is available that treats the needs of the food industry regarding the lack of suitable sensors, inadequate models of process dynamics, complexities and nonhomogeneity of raw materials, changes in the physical properties of raw material during processing, sanitary requirements, etc. Some books have surveyed the controls and automation in food processing but inadequately cover the basics and integration with the equipment. Although this book deals with the control of food processes, the principles involved are applicable to many other processes within related fields such as the biotechnological and pharmaceutical industries.

The book consists of 20 chapters divided into four parts. The first two parts contain seven chapters which explain various tools, sensors, and other related information required to develop computerized process controls. Basic and advanced process modeling, and neuro-fuzzy, sensor, and image processing technologies are included. Part III describes computerized automation for important unit operations used in food manufacturing. These included fermentation, drying, thermal processing, chilling, packaging and warehouse management. The last part explains automation in major food industries such as meat, dairy, fish, cereal/bakery. A discussion of computer-integrated food manufacturing concludes the book. Thus this well-integrated, single-source reference offers comprehensive coverage of the computerized food process control.

Chapter 1 introduces the subject of computerized food process controls, their problems, and solutions. Computer-based instrumentation and sensors for in-line measurements are discussed in Chapter 2. Chapter 3 deals with methods for dynamic modeling of food processes and introduces the use of these models for prediction of process behavior. System modeling concepts for developing computer-based food process controls are discussed. Chapter 4 presents classical approaches to control system analysis and design and extends these to computerized controllers. This includes controller turning, direct digital control implementation, multivariable control, and supervisory control techniques. Chapter 5 describes the basic principles of fuzzy logic and neural networks, and their applications in food process automation. Further discussion of these topics is given in Chapter 6. The emerging areas of digital image processing and applications in the food industry are covered in Chapter 7.

Chapter 8 describes the control of fermentation processes, including problem aspects, procedures, tools, and current trends with practical illustrations. Chapter 9 explains the control of thermal processing operations by illustrating feed-forward control of microbial inactivation, feedback control of retort temperature, and prediction of cumulative lethality. Introduction to aseptic processes is also included. Chapter 10 focuses on the automatic control of drying processes in food industry. On the basis of the corn drying example, different control strategies with increasing complexities are compared. Chapter 11 discusses freezing systems, giving a background on computerized approaches for freezing time forecasting, and a survey of available modern equipment for food freezing and chilling. Chapter 12 illustrates mathematical models describing three selected separation processes: filtration, extraction/leaching, and membrane processes. These process models can be integrated with the suitable control strategies. Chapter 13 examines computerized automation in the food warehouse by reviewing information requirements and computer use in warehouses, discussing the planning process, identifying system design issues, describing system architecture, and presenting applications. Next, Chapter 14 describes sensing equipment used on packaging lines to monitor and evaluate various characteristics of packages and packaging machines. This chapter also discusses the use of computerized control units to interpret the information provided by the sensors and manage line operations.

In Part IV, Chapter 15 presents computer-based automation in dairy processing. It emphasizes essential elements of a dairy processing plant, dairy processing operations, plant sanitation, and manufacturing of important dairy products. Chapter 16 focuses on automation in meat processing, including slaughter and dressing technologies, computer simulation of slaughter floor operations, carcass classification, inspection and grading, fabrication, boning and refrigeration. Automation in various cooking processing is discussed in Chapter 17. This also presents different tools and aids related to computer use for cooking operations. Chapter 18 points out industrial advances and presents results that can be applied in-line, for the control of the baking of cereal products. How computers are used to control and integrate unit operations and improve overall efficiency during harvesting and processing of fish is addressed in Chapter 19. Lastly, Chapter 20 introduces several issues related to the implementation of computer-integrated manufacturing in the food industry. Basic concepts of data and knowledge bases are reviewed. Computer-aided design techniques for food manufacturing processes with applications and practical implementations are also discussed.

Acknowledgment is due to all the contributors to this book, and to the many individuals who reviewed chapters and provided useful suggestions. Assistance provided by the staff of Marcel Dekker, Inc., is highly appreciated. I also acknowledge, on behalf of the contributors, all the organizations and individuals who gave permission to use their published work.

Gauri S. Mittal

Contents

Contributors

C. G. Bublitz Fishery Industrial Technology Center, School of Fisheries and Ocean Sciences, University of Alaska–Fairbanks, Kodiak, Alaska

John W. Buhot Division of Food Science and Technology, Commonwealth Scientific and Industrial Research Organisation (CSIRO), Tingalpa D.C., Queensland, Australia

Gour S. Choudhury Fishery Industrial Technology Center, School of Fisheries and Ocean Sciences, University of Alaska–Fairbanks, Kodiak, Alaska

Lino R. Correia Agricultural Engineering Department, Technical University of Nova Scotia, Halifax, Nova Scotia, Canada

Georges Corrieu Department of Food and Fermentation Process Engineering, National Institute for Agronomical Research, Thiverval-Grignon, France

Francis Courtois Department of Food Process Engineering, High School of Food Science and Technology, National Institute for Agronomical Research, Massy, France

J. Pemberton Cyrus   Industrial Engineering Department, Technical University of Nova Scotia, Halifax, Nova Scotia, Canada

V. J. Davidson   School of Engineering, University of Guelph, Guelph, Ontario, Canada

Sundaram Gunasekaran   Department of Biological Systems, University of Wisconsin–Madison, Madison, Wisconsin

Gordon L. Hayward   School of Engineering, University of Guelph, Guelph, Ontario, Canada

Harold A. Hughes   School of Packaging, Michigan State University, East Lansing, Michigan

D. S. Jayas   Department of Biosystems Engineering, University of Manitoba, Winnipeg, Manitoba, Canada

Eric Latrille   Department of Food and Fermentation Process Engineering, National Institute for Agronomical Research, Thiverval-Grignon, France

X. Luo   Department of Biosystems Engineering, University of Manitoba, Winnipeg, Manitoba, Canada

S. Majumdar   Department of Biosystems Engineering, University of Manitoba, Winnipeg, Manitoba, Canada

Gauri S. Mittal   School of Engineering, University of Guelph, Guelph, Ontario, Canada

Bart M. Nicolaï   Department of Agro-Engineering and Economics, Katholieke Universiteit Leuven, Heverlee, Belgium

K. Niranjan   Department of Food Science and Technology, The University of Reading, Whiteknights, Reading, England

Feng Ou-Yang   Department of Food Science, Purdue University, West Lafayette, Indiana

Bruno Perret   Department of Food and Fermentation Process Engineering, National Institute for Agronomical Research, Thiverval-Grignon, France

Daniel Picque Department of Food and Fermentation Process Engineering, National Institute for Agronomical Research, Thiverval-Grignon, France

Hosahalli S. Ramaswamy Department of Food Science and Agricultural Chemistry, Macdonald Campus, McGill University, Ste. Anne-de-Bellevue, Quebec, Canada

A. Ryniecki Institute of Food Technology, University of Agriculture, Poznan, Poland

Shyam S. Sablani Department of Food Science and Agricultural Chemistry, Macdonald Campus, McGill University, Ste. Anne-de-Bellevue, Quebec, Canada

N. C. Shilton Department of Agriculture and Food Engineering, University College Dublin, Dublin, Ireland

Rakesh K. Singh Department of Food Science, Purdue University, West Lafayette, Indiana

Christina Skjöldebrand SIK—The Swedish Institute for Food and Biotechnology, Göteborg, Sweden

Gilles Trystram Department of Food Processing Engineering, High School of Food Science and Technology, National Institute for Agronomical Research, Massy, France

COMPUTERIZED CONTROL SYSTEMS IN THE FOOD INDUSTRY

1

Process Controls in the Food Industry: Problems and Solutions

Gauri S. Mittal
University of Guelph, Guelph, Ontario, Canada

I. INTRODUCTION

A food plant should convert raw materials into products with desired attributes using available technology and energy resources in the most economical way. For this, the food plant uses various unit operations (evaporation, filtration, drying, mixing, extrusion, etc.) in a systematic, rational manner. These unit operations are run at controlled process conditions (temperature, pressure, flow, concentration, etc.). For safety, the operating conditions should be within allowable limits, and they should follow environment regulations. Operating constraints should also be satisfied. Finally, the process conditions should be controlled at the optimum levels for minimum operating cost, maximum profit, and desired attributes of the product.

Process control technology has advanced with the availability of low-cost microprocessors and their accessories. Advances in programmable logic controllers (PLC) have also greatly assisted process automation. Further, neuro-fuzzy logic has improved operational control in food processing. Statistical process controls (SPC) are used to maintain product quality when continuous monitoring of product quality is feasible [1]. Presently, a computer-based process control system is an essential tool of any industry. Even for the small to medium sized industry, handling the flow of process information with an automated system is crucial not only to control the process, but to ensure optimum product quality and productivity [2]. Computerized controls can be used to upgrade existing equipment as well as for new facilities.

Digital or computer-based process controls optimize process and/or equipment efficiency [3]. Analog controls, used earlier, were adequate under normal operating conditions, but they could not respond quickly enough during significant process deviations to make the required adjustments without overshooting. This was reported [3] for spray dryers used to dry milk. In the past, under these situations, experienced operators were needed for manual process control.

Although pneumatic control systems have long been used in the food industry, the use of computer-based controls provides many additional advantages such as a capability to monitor and control many operations independently and concurrently. Electrical switches, relays, timers, and solenoids controlled processes before programmable controllers and microprocessors evolved. Downtime has been drastically reduced with the replacement of electromechanical controls with programmable controllers. Compared with other industries such as the chemical and metal industries, the food industry has been slow to adopt the advances in computer-based controls. However, several large companies are making significant progress in automation of various processes.

II. NEED AND LIMITATIONS

The food industry is facing increasing global competition, regulation, and consumer demands. These require new technologies and practices for competitive advantages in the market. The need for increased automation in the food industry is due to [4] (1) the elimination of extremely repetitive and monotonous tasks, which resulted in repetitive strain injury to workers, (2) better quality control needed because of consumer sophistication, regulatory labeling requirements, and narrow quality boundaries, (3) the elimination of off-line quality control due to the need for more rapid correction of deviations from process and quality standards/specifications, and (4) the detection of foreign and contaminant material in food.

A. Costs

Automation lowers production cost by reducing the labor requirement, decreasing wastage, and permitting faster, more efficient volume production. Automation generally reduces the number of machines required in a plant. This will provide savings in labor and maintenance costs. With automation, savings in personnel cost is generally significant. The automation system shows none of the human characteristics of fatigue, mood changes, or inconsistent judgement. Computer control and automation provide accurate and repeatable batches. Such controls allow one to reach optimum operating conditions faster and with less adjustment. Using these controls, operators can implement any control strategy from proportional-integral-derivative (PID) to complex multiloop control and sequencing, process models and mathematics, setpoint ramping, alarm logging along with complex interactive control such as lead/lag, dead-time compensation, and adaptive gain [5]. Special control functions can easily be incorporated into the control system. These also provide considerable information on such thing as process variables, outputs, setpoints, and alarm trip points. Inputs to these systems can come from many transducers including from remote field-installed transmitters. Thus these controls are required to create higher efficiencies and reduce overall costs in addition to increasing productivity.

B. Documentation

The *documentation capability* of the system is virtually unlimited. Considerable process and product information can be stored in the computer system. If required, these systems can generate ingredient usage reports for inventory control [6]. These reports can be retrieved on daily, weekly, monthly, and yearly bases. Other quality data can also be stored for each batch and can be used later on for batch-to-batch variation studies. Smart systems also monitor and record periodic and transient variations in product variables. This helps in further product quality improvement. An operator can use these systems to generate reports, view processes in real time, alter setpoints, change system configurations, and perform testing. The collected statistics can be used to study and improve the process.

Computer-based controls can increase productivity. Oil-processing output was increased by 30% [7] by using three discrete microcomputer-based process controllers. These were designed to perform all continuous loops involving complex, integrated algorithms, valve interlocking, and some sequencing. Similar controls can also be used to optimize blend formulations and production scheduling, and process modeling can be incorporated. These systems also increase production, safety, and efficiency in every area of the manufacturing process. The ability to produce the best product at the least cost with employee safety is enhanced [8]. Total quality management (TQM) can be easily implemented with improved process control, quality monitoring of raw material, and statistical process control.

Products can be electronically weighed to precise standards, providing buyers a highly uniform pack [9]. A computerized on-line check-weighing system automatically calculated average, range, standard deviation, and confidence limits in a frozen pizza manufacturing facility [10]. Application of the many ingredients of pizza was precisely controlled to meet net weight requirements and product formulation standards. Data and signals by the controller alerted the technicians or supervisors to take corrective action.

C. Robotics

Computer-controlled *robots* are programmed to perform many operations in food manufacturing. They are useful in material handling, providing flexibility in loading and unloading [11]. Advanced robotics and modern sensor technology help to develop flexible modular solutions to meet the challenges in food industry. In 1992, of 4000 robots sold in the food industry worldwide, 1500 were sold in Japan, mostly for sushi manufacturing [12]. It became possible to use robots in food processing with developments in special gripping capabilities, sensors, vision, and sophisticated information management systems. Modern robots are highly sophisticated programmable manipulators that possess mechanical arms powered by electromechanical devices. These permit multiple combinations of rotational/transitional motion in 3-D space. Robots can pick, place, transport, and orient similarly to a human arm, but with greater power, precision, and repeatability. Carcasses are automatically split by robotic splitters traveling in tandem at a rate of 1106 pork carcasses per hour [13].

D. Machine Vision

Machine vision inspection techniques have been developed to inspect and detect process defects at line speed. Vision systems can examine food products for foreign material, wrong color, bruises, scars, and other flaws [14]. Such systems use cameras to detect flaws by color as well as by size. They use lasers to sense differences in internal structure and color that help separate foreign material similar in color to the product. Thus they are useful in quality control on processing lines. However, there have been problems in grading and/or sorting fruits and vegetables because of their variable nature.

Food-processing machinery manufacturers are now providing equipment diagnostics to avoid breakage. The programmable control with special input/output (I/O) reduces the danger of a process failure and downtime. This improves the process diagnostic capabilities and thus makes error tracing easier.

Due to their simplicity, digital signals are less susceptible to electrical noise than analog signals. Thus these digital I/O devices are particularly needed in the food industry, where heavy electrical equipment provides disturbances in low-voltage signals from sensors [15].

III. PROBLEMS

Problems faced during food process automation include a large number of recipes, hygiene requirements, perishability of raw and processed foods, and the seasonal nature of the production of food products. In food processing, repeatable cycles are needed to assure the production of safe, consistent products batch to batch. Improper control of a batch can require expensive product losses or extensive reprocessing of the product [16].

Process variables such as flow rate, temperature, pressure, level, and agitation, need to be precisely controlled about their desired setpoints. The control system must be flexible enough to provide easy operator interface for the variety of products processed. Precise control of ingredients is crucial in any food processing operation. Therefore food process control is much more complex than other process industries. Many times, ingredients are heterogeneous. For example, in a pizza, too much or too little of one ingredient component (sauce, cheese, meat, vegetables, etc.) can adversely affect the quality, taste, appearance, cook time, and profitability [10]. Excess sauce, say, can bleed over the crust edges. Thus the greatest problem for process automation is variation in size, shape, and homogeneity (or wider variability) of the raw product. The ease of microbial and physical damage to food also creates problems.

Many specialized food processes have unique control requirements. Food process controllers increase in complexity with the need to detect and control product composition and quality. Moreover, satisfactory in-line sensors for most flavors and tastes are not available.

The resources, such as water, steam, energy, and ingredients, in the food industry should be efficiently utilized for maximum profit—especially with rising costs of labor, raw materials, and energy. Most food manufacturers want to automate their plants, but most of the time they end up with small islands of automation that do not solve plantwide problems [17]. Most food plants are not fully automated but are a combination of batch and continuous processing. Generally, beverage and dairy industries, handling principally fluid products, use continuous processes. Others use mostly batch processes, but slowly these processes are being converted to continuous or semicontinuous processes [15]. Different pieces of equipment may not be able to communicate with each other unless automated by the same manufacturer. This often creates problems, as suitable equipment is not all available from the same manufacturer. Moreover, process automation (sensors, hardware, software) is changing too fast to keep pace with. Equipment can become outdated in a few years. This rapidly changing technology complicates the proper choice of control hardware and software.

Management must justify the vast expenditure for automation against the total needs of the business. In many cases, if traditional investment criteria are used, it is difficult to evaluate the potential of computer-aided manufacturing operations.

A. Process

Before getting an automated system for a process, it must be simplified. If there is some problem with the process, you cannot correct it with hardware or software. Process parameters should be optimized before purchasing or designing a control system [18]. Comprehensive information should be collected from the operator about the process before automating it through a computer. Most food processes are dynamic, and to tune a controller properly, the process must be studied carefully. All features essential to product uniformity should be automatically controlled. Many times, however, the skills and intuitive reasoning of an experienced operator are very difficult to define and implement within a computer-based system [15].

Design of a food process controller will require consideration of the type of unit operations, the range of their operating conditions, possible disturbances, manipulations and possible measurements, and problems related to process startup or shutdown [19].

Traditionally, large food industries first used advanced manufacturing technology, including automation. However, medium and small industries need these advanced technologies more because of their limited supply of raw materials or market restrictions. Process automation in medium and small industries will assist in reducing product rejection rate and raw material wastage, and in improving product quality and consistency.

B. Process Models

To analyze the behavior of food processes, the physical and chemical phenomena occurring during the processes are mathematically modeled. These models are needed to design food process controllers. Digital computers make it easy to use the models in process control algorithms. The models assist in making decisions and thus provide more accurate and easier processing, and they predict the behavior of food processes continuously when the external disturbances and manipulated variables are changing with time. Thus understanding of the food process dynamics greatly influences the design of effective controls. This is a major bottleneck in the proper designing of food process controllers: many times food processes are poorly understood, or their physical, chemical, or microbial parameters are poorly known or measured [19]. In these cases, process models based on inadequate knowledge are insufficient to describe the dynamic behavior of an actual process. Due to the uncertain biological nature of the food, food engineers face the most difficult and challenging task of effective process control design. They have to rely heavily on their knowledge of microbiology, biochemistry, food chemistry, and food processing coupled with basic engineering principles.

IV. SOLUTIONS

Industry first needs to identify specific processes to be targeted for automation. Future growth and upgrading of the control system has to be anticipated in process automation. Many processes in plants can be partially automated using small computer-based systems. These can be expanded into larger, plantwide systems later without much problem. For example, batch or continuous ingredient mixing can start using a flow meter and a digital control system. Later on, other instruments such as more sensors, a PLC, or a distributed control system can be added.

A high level of security can be built into these systems. An extensive security system for the control strategies and information is crucial in computer-based controls, as these can be easily modified, copied, and destroyed. A single virus can damage the whole system. Thus access to these systems must be restricted to prevent unauthorized use [20].

Failure of a single instrument should leave unrelated functions unaffected, thus minimizing risk of sudden process shutdown. Thus these controls must provide proper contingencies for emergency shutdown and possible utility failure. The advantage of a distributed, modular architecture is that the malfunctioning device or module can be changed without disrupting the system operation. Thus the failure of one component will not cause system failure. The controls should be programmable on line to allow any required process changes without shutting down the entire operation.

With adequate volume, automated manufacturing will provide the most economical process. To determine the economical feasibility of automation, the company must compare all of the anticipated operation costs with the potential savings and benefits that can result from automated processes. Engineers have to explain to plant management the advantages of automatic control in terms of overall profitability to get their approval. Otherwise plant automation is delayed because of the fear of the complexities involved. However, rapid development of computer technology has reduced this fear, as most people are using this technology at some level.

A. Programmable Logic Controllers

Earlier plant automation systems have used central computer systems. Presently, programmable logic controllers (PLCs) and distributed control systems are used. With these systems, operators can review plant data on a microcomputer screen and control plant processes. Combining PLCs with smart transmitters provides a flexible, low-cost alternative to distributed control systems for better process control. This approach gives increased accuracy, greater control functionality, and lower time and cost for installation and maintenance [21]. For example, a PLC-controlled refrigeration system cooled carcasses from 38 to 1°C in 24 h [13]. Another PLC system boosted cookie cook-time efficiency to 99% [22]. Troubleshooting was simplified, as the status of all inputs and outputs is indicated by LEDs; e.g., if a bag placement fails, a proximity switch sends this to the controller, which will stop the flow of cookies to the bag. Similarly, packaging lines in a bakery were controlled by PLC interfaced to workstations [23]. The distributive control system consisted of cell controllers, workstations, a PLC, a fiber-optic backbone of 100 megabit/s, and a supervisory control and data acquisition software. The PLC controlled almost everything—variable-speed drives, temperature, mixing times, packaging lines, etc.—and also provided alarming, monitoring, and data collection capabilities. Energy- and utility-related equipment such as boilers, air compressors, and chilled water systems can also be monitored and controlled by PLCs. Distribution control systems in the older food plants are more common, as important processes in the plants can be controlled.

B. Operators

Still, most food plants are run more by operator knowledge than by scientific knowledge [10]. Many procedures have traditionally been based on the use of operators' sense of sight, smell, experience, and touch to detect the product quality characteristics during processing. Many of the operators' tasks will be taken over by computerized control systems, resulting in modification of operators' roles and tasks. Automation decreases the number of operators in a plant and increases their workloads as well as the number and complexity of their tasks. Thus labor feels that their jobs are in jeopardy, and training problems are also felt. Careful planning is required to avoid operator dissatisfaction. The following should be considered in such cases for an easy transition from manual to automatic control [20]: (1) Only the minimum required information should be shown to operators at a given time (2) The information to the operator should be conveyed in an appropriate manner—visually, audibly, or by other means. (3) Suitable methods should also be used for the operator's interaction with the process controller. (4) Proper procedures are needed to appraise the operators' performance. (5) Do not overmodernize the processes. The design objective for automation should be increased efficiency and yield, and decreased waste [24]. All the parameters should be programmable and displayed to the operator in real-time basis. (6) Loss in production time should also be included in the estimated costs of automating a plant or a process. In some cases this can be a considerable part of the total cost. Training costs should also be included [25]. Any new technology must be easy to use. (7) The process data and information to the operators should be provided in a familiar form to avoid misjudgments.

Operator safety should be a major consideration when designing automated process machines. To diagnose process problems correctly and efficiently, system operators/programmers should know the effects of variations in product and process parameters outside the influence of the control system [15]. They should also be familiar with the practical implementation of the control hardware and software. The system operator must know the basic tuning adjustment parameters to tune a controller—gain, reset, and rate.

C. Flexibility

Flexibility means control can be adapted to multiple purposes and to changing control needs. In such situations, new technology can easily be integrated with minimum change and cost, thereby taking advantage of the latest hardware developments. Controllers should be scalable, flexible, and modular. For a scalable controller, only the hardware, software, and necessary services for a process are needed. Usually, computer-based systems provide more production and product flexibilities. With a minimum of changeover time between products, a large number of products can be manufactured on the same production line. Thus flexibility of these systems decreases the cost of future modification in process conditions or products.

V. THE FUTURE

Newer controls, sensors, and accessories are more accurate, reliable, flexible, compact, portable, cost effective, easy to maintain, fast, and modular. Smart, low-cost sensors and hardware will be available in the future. These will help provide product consistency, zero judgment error, and nonstop processing. Better process models will allow implementation of advanced control algorithms. In the future, the process controls should control the operating conditions at the desired levels at minimum operating cost. Future technological breakthroughs and reductions in cost of digital electronics, computers, and sensors will further affect process control in the food industry.

REFERENCES

1. R. H. Caro and W. E. Morgan, Trends in process control and instrumentation, *Food Technol. 45*(7):62 (1991).
2. Anon., Using computers in the meat industry, *Meat Process. 24*(2):44 (1985).
3. W. Dietrich and R. Kline, Digital control optimizes spray dryer efficiency, *Food Process. 45*(3):100 (1984).
4. L. D. Pedersen, W. W. Rose, and H. Redsun, Status and needs of sensors in the food processing industry, Proceedings of Food Process Automation, Am. Soc. Agric. Engg., St. Joseph, Michigan, 1990, p. 10.
5. F. LaBell, Low-cost interactive digital control provides math and sequence functions for small to medium applications, *Food Process. 45*(1):96 (1984).
6. R. J. Swientek, Computerized syrup operation cut batch time in half, *Food Process. 45*(3):96 (1984).
7. J. P. Lobe and R. J. Swientek, Microcomputer process controllers increase production 30%, *Food Process. 45*(1):90 (1984).
8. Anon., Computers help citrus plant handle 4300 tons of oranges daily, *Food Engg. 56*(4):135 (1984).
9. Anon., Avocado packer initiates operations with electronic packing system, *Food Engg. 56*(3):161 (1984).
10. C. Mey, K. Huffman, and R. J. Swientek, Net and component weight control program reduces averages from 11.4 to 3.2%, *Food Process. 45*(8):58 (1984).
11. D. S. Hammond and R. F. Ellis, Versatile robot palletizer stacks cases automatically, *Food Process. 45*(9):136 (1984).
12. Anon., Robots answer cost, health and labor concerns, *Prepared Foods 163*(9):77 (1994).
13. C. E. Morris, Global new plant design—32000 hogs per day, *Food Engg. 65*(8):61 (1993).
14. P. Demetrakakes, Graders appeal to fruit, vegetable market, *Food Process. 56*(2):49 (1995).
15. G. Brown, Process control microcomputers in the food industry, *Developments in Food Preservation—4* (S. Thorne, ed.), Elsevier Applied Sci, New York, 1987, p. 35.
16. J. R. Getchell, Computer process control of retorts, *Food Engg. News 5*:4 (1981).

17. K. J. Hannigan, Process control: A changing picture, *Food Engg.* 58(7):68 (1986).
18. C. E. Morris, Making the automation choice, *Food Engg.* 66(1):74 (1994).
19. G. Stephanopoulos, *Chemical Process Control*, Prentice Hall, Englewood Cliffs, New Jersey, 1993, p. 696.
20. M. C. Beaverstock, Process control, *Computer Aided Techniques in Food Technology* (I. Saguy, ed.), Marcel Dekker, New York, 1983, p. 361.
21. R. A. Wenstrup, Programmable logic controllers meet smart transmitters, *Food Process.* 55(4):72 (1994).
22. Anon., Programmable controllers boost cookie-line efficiency to 99%, *Food Engg.* 58(5):198 (1986).
23. B. Swientek, Placing a premium on real-time management, *Prepared Foods* 162(5):74 (1993).
24. S. Berne, Modernizing yesterday's plants to meet tomorrow's demands, *Prepared Foods* 164(2):78 (1995).
25. J. Wagner, What process controls really offer, *Food Engg.* 57(2):79 (1985).

2

Computer-Based Instrumentation: Sensors for In-Line Measurements

Gauri S. Mittal
University of Guelph, Guelph, Ontario, Canada

I. INTRODUCTION

A sensor, or instrument, is a device for measuring the magnitude of a physical variable. Sensors play an important role in the food industry. They are required to provide raw data to process monitor and control systems. Sensors can be contacting or noncontacting devices. The automation of food processing requires reliable sensors for the measurement of process variables such as pressure, temperature, flow, moisture, color, and viscosity. Sensors are also needed in food manufacturing for securing food quality and safety, and in meeting productivity and environmental/ energy concerns. In-line sensors help to produce more consistent food products while increasing productivity and decreasing production costs. Due to their diversity, food manufacturing facilities require various sensors for detection and measurements. Sensors are the eyes and ears of the manufacturing processes [1]. Thus these are the interface modules between the physical world and computers.

All sensing is using physical measurement of phenomena. Sensor developments have been greatly influenced by the progress in solid state electronics. Solid state electronics and chip technology have assisted in miniaturization of various sensors. For example, gas chromatography column is available on a small silicon wafer [2]. Human judgment is always the hardest thing to duplicate in any machine. However, advantages of in-line sensors include [3] real-time analysis and process control, improved product uniformity and quality, and conversion of batch operations to semicontinuous or fully continuous processes.

Sensors and their installation must confirm to the sanitary standards of part 110, 21 CFR Chapter 1, Code of Federal Regulation Current Good Manufacturing Practices in Manufacturing, Packing or Holding Food (FDA GMPs) [4]. The use and types of thermometers, pressure gauges, and temperature sensors must conform to the requirement of parts 113 and 114, 21 CFR Chapter 1, relative to the safe processing of low-acid and acidified foods. Sensors and their installation in the dairy industry must also conform to 3A standards of the Dairy Industry Committee, International Association of Milk, Food and Environmental Sanitarians, U.S. Public Health Service, U.S. Department of Agriculture (USDA), and Dairy and Food Industries Supply Association.

Vibration, dust, moisture, wide temperature cycles, and chemical sanitizing and cleaning affect sensor performance and useful life [4]. Many food processes involve harsh or difficult environmental conditions. Thus the selection and development of new sensors for the food industry should consider these situations. Sensors should be properly sealed to withstand harsh conditions and repeated washings. Sensors should be free from crevices from which food material cannot be removed easily during washing. Sensors should be immune to shock and vibration when installed on equipment.

Physical attributes—temperature, pressure, mass flow rate—and chemical attributes—composition (fat, protein, moisture, vitamins, minerals), etc.—must be monitored to control food processes. Techniques and sensors based on infrared, near infrared (NIR), refractometry, machine vision, X-rays, fiber optics, nuclear magnetic resonance spectroscopy, rheology, etc., are used in the food industry for process and product quality monitoring and control.

A. Terminology

1. *Span*: the range of measured variable that a sensor can measure.
2. *Least-count*: the smallest difference of measured variable that can be detected by a sensor.

3. *Readability*: the closeness with which the scale of a sensor can be read in analog output. A sensor with a 30-cm scale would have a higher readability than a sensor with a 15-cm scale and the same span. In digital output, readability will be the relative size of the letters.
4. *Sensitivity*: the change in output of the sensor with the unit change in input variable to be measured; e.g., if a 1-mV recorder has a 5-cm scale length, its sensitivity would be 5 cm/mV.
5. *Accuracy*: the deviation of the output of a sensor from a known measured input. Accuracy is usually expressed as a percentage of full scale reading; e.g., a 100-kPa pressure transducer having an accuracy of 1% would be accurate within ±1 kPa over the entire range.
6. *Precision*: the ability of a sensor to reproduce a certain output with a given accuracy.

The difference between precision and accuracy: Consider the measurement of a known temperature of 100°C with a certain transducer. Five observations are recorded, and the indicated values are 103, 105, 103, 105, and 103°C. These values show that the accuracy of the transducer is 5% (5°C), and its precision is ±1%, since the maximum deviation from the average reading 104°C is only 1°C. The transducer can be calibrated and then can measure temperature within ±1°C. Thus accuracy can be improved by calibration up to, but not beyond, the precision of the transducer.

7. *Threshold*: If the input of a sensor is very gradually increased from zero, there will be some minimum value below which no output can be detected. This minimum value is the threshold of the sensor.
8. *Resolution*: the input increment that gives some small but definite numerical change in the sensor output. Thus resolution is the smallest measurable change, while threshold is the smallest measurable input.
9. *Hysteresis*: A sensor exhibits hysteresis when there is a difference in readings depending on whether the values of the measured variable are approached from above or below.
10. *Linearity*: the maximum deviation of any calibration point from the linear relationship. This may be expressed as a percent of the actual reading or as a percent of the full scale reading.
11. *Error*: a deviation of the measured value from the true value. Errors are divided into three broad classes—gross, systemic,

and random [5]. The gross errors are mostly human errors such as incorrect reading, adjustment, and application of instruments. Loading effects, due to improper circuit impedance, also give this type of error. Systemic errors are related to the functioning of the instruments and their mechanical or electrical structure and calibration errors, and to the effect of environment on the equipment performance. These errors related to the instrument can be avoided or reduced by proper calibration of the instrument, and by selecting a suitable instrument. Environmental errors are reduced by using the instrument at the recommended conditions of temperature, relative humidity, pressure, etc. Random errors are due to unknown causes. They can be reduced by proper design of the instrument, and by taking more readings and using statistical methods to correct the observations.

1. Uncertainty Analysis

Uncertainty analysis should be carried out in measuring a variable because the magnitude of an error is always uncertain. Suppose that a result or sensor output (R) is a function of the independent variables $X_1, X_2, \ldots, X_n$. If U_R is the uncertainty in the output of a sensor and $U_1, U_2, \ldots, U_n$ are uncertainties with the same odds in the independent variables influencing the sensor output, then the U_R having these odds is

$$U_R = \sqrt{\left(\frac{\delta R}{\delta X_1} U_1\right)^2 + \left(\frac{\delta R}{\delta X_2} U_2\right)^2 + \cdots + \left(\frac{\delta R}{\delta X_n} U_n\right)^2} \tag{1}$$

Thus results can be improved by improving the instrumentation or measuring techniques connected with the relatively large uncertainties.

B. Transducers

Transducers are devices that transform values of physical variables into equivalent electrical signals—e.g., thermocouples and strain gauges. Two broad categories of transducers are passive and active. A passive transducer supplies output energy entirely or almost entirely by its input signal—e.g., thermocouples. An active transducer, on the other hand, requires an auxiliary source of power to supply a major part of the output signal, while the input signal supplies only an insignificant portion—e.g., strain gauges. Some of the basic transducers are as follows.

1. Variable-Resistance or Potentiometric Transducer

A variable-resistance transducer converts either linear or angular displacement into an electric signal. This may be in the form of a moving contact

on a slide wire or on a coil of wire, through either linear or angular movement (Fig. 1). Depending on the type of resistance wire wound, the change
may be linear, logarithmic, exponential, etc. Pressure can be converted
to a displacement through mechanical methods so that the device can be
used in force and pressure measurements. These provide high electric

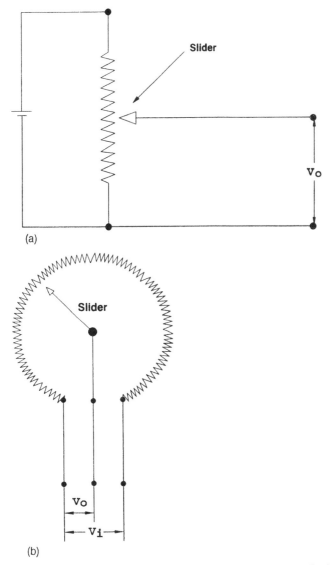

(a)

(b)

FIGURE 1 (a) A potentiometeric transducer for linear displacement measurements. (b) A potentiometric transducer for angular displacement measurements.
V_0 and V_i are output and input voltages.

efficiency; however, life is limited due to wear, and noise will develop due to wearout of the element.

2. Linear Variable Differential Transformer

A linear variable differential transformer (LVDT) is a special case of a differential transformer. In an LVDT, three coils are arranged linearly with a freely movable magnetic core inside the coils (Fig. 2). An ac input voltage is applied on the center coil, and the output voltage from the two end coils depends on the magnetic coupling between the core and the coils. This coupling, in turn, depends on the position of the core. Thus the output voltage of the LVDT is an indication of the core displacement. The output is nearly linear when the core remains near the coil center. There is a 180° phase shift from one side of the null position to the other. This transducer provides continuous resolution and low hysteresis, but it is sensitive to vibration.

3. Capacitive Transducer

The capacitance between the two parallel plates is given by $C = 9.85 (10^{12})\epsilon A/d$, where C = capacitance (F), d = distance between the plates (m), A = overlapping area (m^2), and ϵ = dielectric constant of the material between the plates. Thus d, A and ϵ can be measured with a change in capacitance due to a change in d, a change in A due to a relative lateral movement of the plates, and a change in ϵ of the material between plates. The capacitance can be measured with a bridge circuit. The output impedance of the capacitor is $Z = 1/(2\pi f C)$, where Z = impedance (ohm), f = frequency (Hz), and C = capacitance (F).

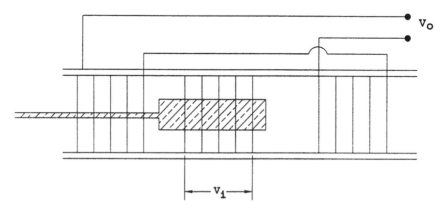

FIGURE 2 A typical linear variable differential transformer (LVDT). V_o and V_i are output and input voltages.

Capacitive transducers have excellent frequency response. Their limitations are sensitivity to temperature variations, and distorted signals due to long lead length.

4. Piezoelectric Transducer

When a force or pressure is applied to a crystal, a corresponding deformation occurs. With asymmetrical or piezoelectric crystals (such as quartz, Rochelle salt, and barium titanite) this deformation will produce a potential difference at the surface of the crystal. The induced charge on the crystal is $Q = kF$, where Q = electric charge (C), F = force (N), and k = piezoelectric constant (C/N).

The output voltage of the crystal is $E = shp$, where E = voltage (V), s = voltage sensitivity (V·m/N), h = crystal thickness (m), and p = applied pressure (Pa). The s for quartz is 0.05 V·m/N, and for PbZ, 0.001 V·m/N [6].

These crystals are widely used as pressure transducers for dynamic measurements. The transducers provide good high-frequency response, and the output is affected by temperature. They are not suitable for measuring at static conditions.

5. Photoelectric Transducer

A photoelectric transducer converts a light beam into an electric signal. The photoemissive cathode releases electrons when light strikes. These electrons are attracted toward the anode and produce an electric current. The cathode and anode are enclosed in a glass or quartz envelope, which is either evacuated or filled with an inert gas. The photoelectric sensitivity is given by $I = S\phi$, where I = photoelectric current (A), S = sensitivity (A/lumen), and ϕ = illumination of the cathode (lumen).

Photoelectric transducers are quite useful for light intensity measurement, and they can be used for counting purposes through periodic interruption of a light source, i.e., to detect light pulses of short duration.

6. Photoconductive Transducer

When light strikes a semiconductor material (such as cadmium sulfide) where voltage is applied, resistance is proportionately decreased, thereby increasing the current. Transducers based on photoconductivity are useful for measurement of radiation at all wavelengths and in light control.

7. Photovoltaic Cell

A voltage is generated when light strikes the barrier between a transparent metal layer and a semiconductor material. A solar cell is a type of photovoltaic cell that is used to convert solar energy into electric energy.

8. Inductive Transducer

In an inductive transducer, the change in inductance with a change in physical variable is observed. Hysteresis errors are limited to the mechanical components. Inductive transducers respond to both static and dynamic measurements, and they provide continuous resolution with high sensitivity. Frequency response is limited by the device converting force to displacement, and output is influenced by external magnetic fields [5].

9. Displacement Transducer

A displacement transducer converts an applied force into an equivalent displacement. For this purpose, diaphragms, bellows, and various shaped tubes are used. The displacement created by these devices is then converted into electrical parameters by using other transducers such as those of capacitive, inductive, or piezoelectric type. Diaphragms are constructed of stainless steel, bronze, or tantalum in any of four forms—flat, dished, capsule, and corrugated.

10. Velocity Transducer

A moving coil is suspended in a permanent magnetic field. The motion of the coil in the magnetic field generates voltage that is proportional to the coil velocity. The output is relatively more stable at various temperatures.

11. Others

A combination of light-transmitting optical fiber and miniature silicon sensors can be used to measure temperature, pressure, and refractive index [7]. These sensors are unaffected by electromagnetic and radiofrequency interferences, fiber bending, connector losses, and source and detector aging. Fiber optics transmits light through long, thin flexible fibers of glass, plastic, or other materials via total internal reflection.

Ultrasonic sensors are rugged, accurate for all types of objects regardless of reflectivity, color, or texture, and have high noise immunity.

II. SPECIFIC SENSORS

A. Temperature

In the food industry, temperature is the most common physical parameter that must be measured. Commonly used temperature sensors in computerized control systems are as follows.

1. Thermocouples

The seebeck thermal emf is the voltage produced between two junctions of dissimilar metals in a circuit when they are at different temperatures.

The generated voltage depends on the temperature difference and the junctions' materials. Thus, a known reference temperature is required to measure a temperature. Nine types of thermocouples using specific metals or alloys have been adopted and are represented by letters (B, E, J, K, N, R, S, T, and W-Re). Types J, T, E, and K are popular because of their relatively high seebeck coefficient and low cost. Type E is suited for detecting small temperature changes because of its highest seebeck coefficient and low thermal conductivity. The size of thermocouple wire is determined by the application and can range from AWG#10 wire in rugged environments to fine #30 for biological measurements. Electronic ice-points are available for thermocouples as reference junctions.

An exposed thermocouple junction is suited for noncorrosive environments where fast response is needed (Fig. 3). An underground junction is required for corrosive environments where the thermocouple is electrically isolated from and shielded by the sheath. The thermocouple is insulated from its sheath by MgO powder. The grounded junction is used for the corrosive environment and for high-pressure applications. The junction is welded to the sheath. The sources of errors in thermocouple measurements are as follows [8]: (1) Poor junction connection. To avoid this, the junction should be welded to measure higher temperatures. Commercial thermocouples are welded using a capacitive-discharge technique to ensure uniformity. (2) Thermal shunting. This occurs when small-diameter wire is joined to larger-diameter wire. (3) Noise. A thermocouple acts as an antenna, picking up noises from electromagnetic radiation in the radio, TV, and microwave bands. To avoid this, the following noise reduction techniques are recommended: The extension wires should be twisted and wrapped with a grounded metallic foil sheath; the junction is grounded at the measurement point.

Thermopiles are thermocouples connected in series (Fig. 4). Total output voltage from these thermocouples will be the algebraic sum of the individual voltages. Thermocouples connected in parallel (Fig. 5) will provide the average voltage of the junctions as long as the resistances of each circuit are the same.

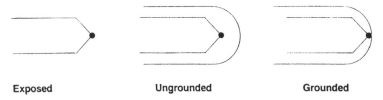

Exposed   Ungrounded   Grounded

FIGURE 3 Exposed, covered, and grounded thermocouples.

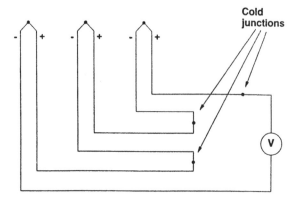

FIGURE 4 A series connection of thermocouples is a thermopile.

ASTM Standard E608—Metal-sheathed base-metal thermocouples:
This covers the requirements for sheathed thermocouples for high-
reliability applications [9]. It also covers the selection of thermo-
elements, insulation, and sheath material, measurement of junction configuration, and thermocouple assembly length. Other standards are as follows:

E220—Calibration of thermocouples by comparison techniques [10]:
This gives techniques of thermocouple calibration based on comparison of the thermocouple output with that of a standard ther-
mometer. Specifications are given for stirred liquid baths, uniformly heated metal blocks, tube furnaces, and dry fluidized baths.

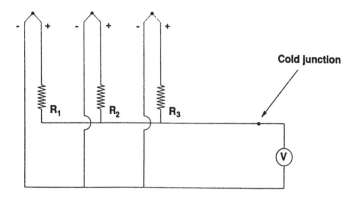

FIGURE 5 A parallel connection of thermocouples.

E230: Temperature–emf tables for thermocouples.

E344: Definitions of terms relating to temperature measurement.

E563: Recommended practice for preparation and use of freezing point reference baths.

E585: Specification for sheathed base-metal thermocouple material.

Overall, thermocouples are inexpensive and rugged, and they provide good stability and fast response.

2. Resistance Temperature Detector

The resistance temperature detector (RTD) and thermistor are based on the changes in the electric resistance of certain materials with temperature. An RTD is a platinum or nickel wire, having a positive resistivity coefficient. Platinum can withstand high temperatures and is very repeatable, stable, sensitive, and expensive. It has a temperature coefficient of about 0.4%/°C. Nickel is also very sensitive but less expensive; however, the stability is not so good. Platinum RTDs are also available in the form of thin films on ceramic substances, for smaller size, lower cost, and increased ruggedness. Typical response time of an RTD is 0.5 to 5 s. A bridge circuit is generally used to detect the changes in resistance of an RTD. The Callendar–Van Dusen equation is used to approximate the RTD calibration curve [8]:

$$R_T = R_0 + R_0 \alpha \left[T - \delta \left(\frac{T}{100} - 1 \right) \left(\frac{T}{100} \right) - \beta \left(\frac{T}{100} - 1 \right) \left(\frac{T^3}{100} \right) \right] \qquad (2)$$

where R_T = resistance at temperature T, T = temperature in °C, R_0 = resistance at 0°C, δ = 1.49 for platinum, α = temperature coefficient (for platinum, 0.00392), β = 0 at $T \geq 0$, and 0.11 for $T < 0$. A typical RTD will have a range of -100 to 650°C.

ASTM standard E644—Industrial resistance thermometers: This describes the principles, apparatus, and procedures for calibration and testing of resistance thermometers [11]. Various possible errors during calibration include (1) immersion error, caused by heat conduction and/or radiation between the element and external environment; (2) self-heating error, caused by the variations from the calibration conditions in the self-heating of the element at a particular current; (3) thermoelectric effect error, caused by a thermal emf in the measurement system; (4) connecting wire error, caused by connecting wire resistance; and (5) bath gradient error, caused by temperature differences in the work area of the bath.

3. Thermistor

Similar to an RTD, the thermistor is also a temperature-sensitive resistor. They are made up of semiconductor materials, and most have a negative temperature coefficient. Its output is highly nonlinear. The common thermistor forms include disks, beads, and rods of various sizes. These are manufactured from oxides of nickel, iron, copper, cobalt, manganese, magnesium, titanium, and other metals. Thermistors are small in size, sensitive, and fast acting, and they provide stable output. Thermistor output can be described by Steinhart–Hart equation [8]:

$$\frac{1}{T} = A + B \ln R + C(\ln R)^3 \tag{3}$$

where T = temperature in K, R = thermistor resistance in ohms, and A, B, and C are constants.

   The same precautions that apply to thermocouples also apply to RTDs and thermistors; i.e., use shielded wires, use proper sheathing, and avoid stress and steep gradients.

4. Optical Pyrometry

Optical pyrometers are noncontact sensors in which emitted energy from an object is proportional to T^4, where T is temperature of the object in K. Thus temperature is measured by measuring the radiated energy. These sensors measure the emitted energy, particularly the radiation in the infrared portion of the spectrum of the emitted radiation [8]. The electromagnetic spectrum band of 0.5 to 20 μm is used for infrared temperature measurement. Infrared thermometers can measure the temperature of most chilled and frozen foods to an accuracy of $\pm 2.5°C$ [12]. With certain combinations of foods and instruments, this accuracy is $\pm 1°C$. Better accuracy is obtained for chilled foods than frozen ones. An emissivity setting of 1 provided accurate readings. Infrared thermometers are unsuitable for measuring low temperatures of foils and other packaging materials of low emissivity.

5. Solid State Temperature Sensor

A solid state temperature sensor is an integrated circuit temperature transducer; it is available in both voltage- and current-output configurations. In both types, outputs are linearly proportional to absolute temperature. Typical values of the proportionality constant are 1 μA/°C and 10 mV/°C. When output is a current and its impedance is high, long leads may be used without errors from voltage drops or induced voltage noise.

Surface temperature is measured by sensors based on optical pyrometry or contact methods. Contact sensors should be placed on the surface without disturbing the surface or conducting heat. A constant emissivity is required when using optical pyrometry. A temperature sensor can lose or gain heat by conduction, convection, and radiation, thus influencing measurements.

A miniature microprocessor-based temperature sensing and recording system (33 cm³ volume) was developed by Ball Electronic Systems Division, Broomfield, Colorado [13]. It can store up to 1000 data points, is hermetically sealed, and is programmable for start time, sample interval, run identification, and security passwords. It eliminates connectors and wiring. It is ideally suited for conducting heat penetration tests. A portable console is used for programming and data retrieval.

For temperature sensing using refractive index, a layer of temperature-sensitive silicon can be inserted in the optical path, and a second rigid glass reflector used. The refractive index of silicon changes with temperature, thus changing the effective path length [7].

B. Strain Gauges

The strain gauge measures force indirectly by measuring the deflection (strain) it produces in a calibrated carrier. Advantages are good frequency response, fast output, good resolution and accuracy, and low impedance. Kinds of strain gauges include resistance type, semiconductor type, and piezoelectric type. The resistance strain gauge is a resistor whose resistance changes with applied strain caused by applied force. Unbonded gauges are of a wire stretched between two points. Widely used bonded gauges are made of a grid of very fine wire or foil bonded to the backing or carrier matrix. These are low in cost, can be made with a short length of wire or foil, are only moderately affected by temperature changes, have small physical size and low mass, and have fairly high sensitivity to strain. These foil-type gauges are produced by photo-etching techniques. They have a large ratio of surface to cross-sectional area, which permits the gauge to follow the specimen temperature and facilitates the dissipation of self-induced heat. They are more stable under high temperature and prolonged loading. The gauge resistances range from 30 to 3000 ohms, with 120 and 350 ohms being the most commonly used.

1. Semiconductor Strain Gauges

Typically, semiconductor strain gauges consist of a strain-sensitive crystal filament. The resistivity of this material ranges from 0.001 to 1.0 ohm·cm.

The lower the resistivity, the better the linearity. Nonlinearity increases with increasing strain. These gauges may have 30 times the sensitivity of the metal film gauges, but their temperature sensitivity is not easy to compensate for. Resistance values range from 60 to 10,000 ohms and gauge length from 0.25 to 6.35 mm. The strain limit for these is in the 1000 to 10,000 $\mu\epsilon$ range, with most tested to 3000 $\mu\epsilon$ in tension.

2. Gauge Factor

The gauge factor defined as the ratio of the fractional change in resistance to the fractional change in length (strain) along the axis of the gauge. The larger the gauge factor, the higher the resistance change and the resulting output, resolution, etc. Metal wire and foil gauges have gauge factors between 2 and 5, while semiconductor gauges have gauge factors between 45 and 200. Higher-gauge-factor materials are more temperature sensitive and less stable than low-gauge-factor materials.

3. Load Cell

A commonly used form of strain-gauge-based transducer converts an applied force or load into a bridge output voltage. In a load cell, the strain gauges are mounted on some form of mechanical sensing element (beam, column, etc.), and the gauges are usually connected into a bridge configuration. Compensations for temperature and nonlinearity are also provided. Load cells are used in weighing of tanks, reservoirs, and bins.

4. Measurement and Temperature Compensation

A wheatstone bridge is the most commonly used circuit. However, the choice of circuit configuration depends on the application, strain gauge type, and readout type used. Bridge circuits are made using one to four strain gauges, at least one of which is active. The resistance and gauge factor of strain gauges change with temperature. These changes are generally larger for semiconductor gauges than for metal gauges. A thermal strain is also experienced from the differential thermal expansion of gauges and the metals to which they are bonded. Bridge circuits are used to compensate for the temperature effects. The resistance changes of two gauges in adjacent arms of a bridge will subtract if of the same polarity, and add if opposite. Thus if two gauges are subjected to the same temperature, their apparent strain contributions will cancel.

Differences in ground potential, capacitive and magnetic coupling to long cables, and electrical leakage are a few sources of error. For gauge lengths from 3.81 to 6.35 mm, power dissipations from 20 to 50 mW will generally provide acceptable stability and prevent gauge damage.

C. Pressure

All pressure-measuring devices determine a change of differential pressure between that being measured and a reference. The reference can be absolute zero pressure or atmospheric pressure. Most pressure transducers require the transduction of pressure into displacement. Elastic elements such as diaphragms, bellows, or their combinations are used for this purpose. The movement or deformation of these elements is transformed into an electrical signal by one or more primary transducers—strain gauges, capacitive, piezoelectric, LVDT, and resistance. An elastic element—diaphragm—is suitable for dynamic pressure (force) measurement up to 500 kHz. Pressure up to 150 MPa can be measured.

1. When using a piezoelectric transducer, the pressure is applied through a diaphragm to a stack of quartz crystal disks. Piezoelectric transducers are widely used in transient pressure measurements because of the nature of piezoelectric crystal. They respond well to high shock levels and have relatively small errors from temperature changes. The output is generally lower and impedance is high; therefore, a high-impedance amplifier is needed for signal conditioning.

2. In a capacitive pressure transducer, capacitance linearly ($\pm 0.5\%$) changes with the movement of a diaphragm relative to a fixed electrode because of change in pressure. The range is 70 Pa to 70 MPa with a typical error of 0.25%. Advantages are small size, high frequency response, high temperature resistance, and good linearity and resolution. Disadvantages are temperature drift, sensitivity to vibration, and the need for complex electronic equipment.

3. Similarly, in an inductive pressure transducer, an LVDT is used to measure the displacement of bellows, capsule, or Bourdon tube due to pressure changes. Differential pressure of 0.025 mm of water can be measured by this device. Advantages are high output, response to both static and dynamic pressure, and continuous resolution. Limitations are limited frequency response (50 to 1000 Hz), need of ac excitation, transient errors from the nearby magnetic field, and wear and errors due to mechanical friction. Errors due to hysteresis and nonrepeatability may be $\leq 0.2\%$ [14].

4. A resistive pressure transducer uses strain gauge elements: (1) bonded type mounted on a flexing element, or (2) semiconductor type diffused onto a diaphragm. With resistive pressure trans-

ducers, stainless or silicon diaphragms are generally used. They
have good accuracy, higher output, lower cost, corrosion resistance, a wide range of output, and good temperature stability.
Limitations are high friction, finite resolution, high noise level
due to contact wear, sensitivity to vibration, low frequency response, and large size. Typical transducers have an error of
$\pm 1\%$, resolution 0.2%, linearity $\pm 0.4\%$, hysteresis 0.5%, and
temperature error $\pm 0.8\%$.

D. Flow

Flow-measuring devices/instruments/sensors are required in processing
operations where fluid transfer of any kind is involved. Many types of
flow meters are available for pipe flow. Generally, these meters are classified into differential pressure, positive displacement, velocity, and mass
meters. Flow sensors that can be used with the computerized control
systems are discussed in greater detail in this chapter. Commonly used
flow meters are rotameters and electronic mass flow meters. Readings of
rotameters are affected by temperature and pressure conditions; however,
these can be accurately calibrated. Electronic mass flow meters are more
accurate ($\pm 1\%$ of full scale) and are unaffected by temperature and
pressure.

1. Rotameters

A rotameter is a low-cost, simple, and precise device based on a variable-
area principle. It consists of (1) a uniformly tapered tube, (2) a spherical
float, and (3) a scale (Fig. 6). The tube is installed vertically with the
smaller-diameter end at the bottom for fluid inlet. The float is lifted to a
certain height based on the flow rate, and fluid flows between the float
and the tube wall. Thus the position of the float indicates the fluid's flow
rate, given by [15]

$$Q = CA \sqrt{2gh} \tag{4}$$

where Q is volumetric flow rate (m³/s), C is flow coefficient, A is annular
area at float position (m²), g is gravitational constant (m/s²), and h is
differential pressure across float (m). The value of C depends on the fluid
properties (density, viscosity), Reynolds number, pressure, temperature,
and float form. Typical curves are available with the rotameter. For better
accuracy, however, it should be calibrated with the fluid of interest, at
the required pressure and temperature. Rotameters are usually used for
relatively clear fluids. Vertical oscillations in float occur due to unstable
or pulsating flow.

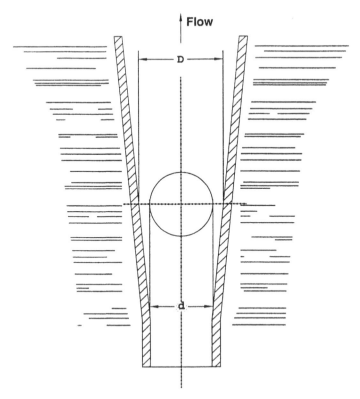

FIGURE 6 A typical construction of a rotameter. D and d are pipe diameter and float diameter, respectively.

The g (m/s^2) values vary with latitude (θ, in degrees) and height above sea level (h, in m) of the location. The g at sea level and 45° latitude is 9.80665 m/s^2, and can be calculated at other locations by [16]

$$g = 9.78049(1 + a\sin^2\theta - b\sin^2 2\theta) - (c + d\cos 2\theta)h + eh^2 \qquad (5)$$

where $a = 5.3024E - 3$, $b = 5.9E - 7$, $c = 3.0855E - 6$, $d = 2.2E - 9$, and $e = 7.2E - 13$.

2. Positive Displacement Type Flow Meters

Positive displacement type flow meters are highly accurate under steady flow conditions and provide relatively low pressure drops. They are not adaptable to metering slurries or fluids with large amounts of suspended particles. These meters include reciprocating piston, nutating disk, rotary piston, and rotary vane meters. Limitations are high cost and large maintenance requirement.

3. Vortex Flow Meters

Vortex flow meters employ the principle of vortex shedding. Eddies or vortices are generated when a fluid flows around a bluff object, and these vertices shed alternately downstream of the object [17]. The flow rate is directly proportional to the frequency of the vortex shedding. This sensor is of medium cost and maintenance, high performance, and wide range, is highly accurate, and has good reliability without any moving parts [18].

4. Hot Wire Anemometer

The decrease in temperature of a resistive element depends on the mass velocity of the fluid, the specific heat of the fluid, the heat transfer coefficient of the element, and the temperature and pressure of the fluid. Thus fluid velocity can be measured by detecting the resistance change with temperature. The resistive element is heated by passing a current through it to increase the temperature above that of the fluid. The temperature of the resistance will decrease due to heat loss, and the corresponding change in resistance is used to calculate fluid velocity. These anemometers can measure gas velocities from 0.1 to 500 m/s at temperatures up to 750°C. Liquid velocities of 0.01 to 5 m/s by a wire probe, and up to 25 m/s with a film probe, can be measured [16]. The electronic package includes the flow analyzer, a temperature compensator, and a signal conditioner that gives a linear output proportional to mass flow. Advantages are fast frequency response (up to 1 MHz), high accuracy, good signal resolution, wide range of flow over a wide range of fluid properties, and miniaturized sensor (typical size: 5 μm diameter and 2 mm length). Limitations are in situ measurement, fragile sensor, contamination affects calibration, and sensitivity to large fluid temperature changes.

Hot film sensors can be used to measure the flow of liquids and gases with particle contamination. A conducting film is deposited on an electrical insulator. This allows flexibility in its shape and robustness. A large diameter of film sensors is advantageous in many respects. The conducting film is of platinum or nickel, and the insulating materials used are quartz or pyrex. Accuracies better than $\pm 1\%$ are possible.

5. Turbine Meter

The turbine meter consists of a multibladed metal rotor in a stainless steel pipe. This is mounted perpendicular to the liquid flow. The fluid passing through the blades will rotate the rotor, and the rotational speed is proportional to the flow rate. The speed can be measured by photoelectric cell, magnetic pickup, or other electrical device. A magnetic pickup, containing a coil and magnet, converts rotor passage into a frequency signal proportional to fluid flow rate. The frequency signal is converted to flow rate and displayed on the indicator. This speed is monitored and transmitted to the LCD by a low-voltage Hall effect switch [19]. This is widely used in the food industry. Turbine flow meters have a wide range of models, high linearity and repeatability ($\pm 0.2\%$), fast response (2–3 ms), low cost, and high accuracy ($\pm 0.25\%$ of flow) for flow measurement of liquid foods such as milk, whey, syrup, and oils. The USDA accepts them for use in federally inspected meat and poultry processing plants. They are also approved for measuring milk and liquid milk products under 3-A Sanitary Standards No. 28-00 by appropriate committees of the International Association of Milk Sanitarians, U.S. Public Health Service, and the Dairy Industry Committee [19]. Limitations: relatively expensive, unsuitable for high-viscosity foods, calibration needed, and subject to wear.

6. Magnetic Flow Meter

A magnetic flow meter is an obstructionless volumetric measuring device. It works on Faraday's law of electromagnetic induction: a voltage will be induced when a conductor moves through a magnetic field. In this meter, liquid serves as the conductor. The generated voltage (E) is proportional to liquid velocity (V), magnetic field strength (B), and distance between the electrodes (D), i.e., $E \propto VBD$. The meter consists of (Fig. 7) (1) an insulated lined metal or unlined fiberglass-reinforced plastic tube, (2) a pair of cylindrical electrodes mounted diametrically opposed in the tube wall, and (3) electromagnetic coils outside the tube to create magnetic field perpendicular to the fluid flow. A voltage is generated across the electrodes when a conductive fluid (minimum conductivity of 10^{-8} mho/cm^3) passes through the magnetic field. The generated voltage is propor-

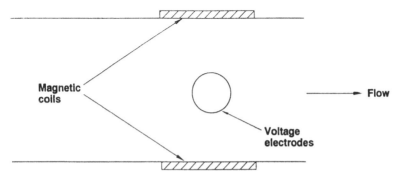

FIGURE 7   Schematic of a magnetic flow meter.

tional to the volumetric flow rate. The flow tube sizes are from 15 to 80 mm in diameter, and it measures flow rates from 0.04 to 30 L/s. The materials of construction are consistent with U.S. FDA regulations for food applications and comply with 3-A Sanitary Standards of design for dairy equipment. The readings are not affected by liquid density and viscosity. Thus it is suitable for most liquid and semiliquid foods, such as catsup and corn syrup, and viscous and sticky liquids. Advantages: flexibility, simplicity, low maintenance with no moving parts, high accuracy, linear output, insensitivity to density, pressure, temperature, and viscosity, minimum exposure of meter components to the fluid; limitations: sensitive to large changes in fluid conductivity, expensive, power supply required, and calibration needed.

7. Ultrasonic Type

These meters measure the time difference (Δt, in s) between transmitting an ultrasonic beam upstream and downstream across a fluid flow in a pipe (Fig. 8a). The Δt is given by

$$\Delta t = \frac{2D \, (\cos \alpha)V}{c^2} \tag{6}$$

where c = sound velocity in the fluid (m/s), D = pipe diameter (m), V = fluid velocity (m/s), and α = angle of ultrasonic beam with pipe wall. Keeping other parameters constant, the velocity will be proportional to Δt. These are also known as "time-of-travel" meters. The transducers are mounted on each side of the pipe in such a way that the sound waves are at a 45° angle to the direction of liquid flow.

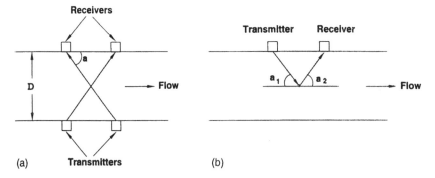

FIGURE 8   Schematics of flow meters based on ultrasound—transmission time (a) or frequency shift (Doppler) (b).

8. Doppler

A narrow beam of radio, light, or ultrasound is directed at a fluid flow. The beam will be scattered by the particles in the fluid and cause a frequency shift due to the flow of fluid causing the reflection (Fig. 8b). Thus this flow meter uses Doppler frequency shift of mostly ultrasonic signals from the discontinuities (suspended solids, bubbles, eddies) in the flowing liquid. An ultrasonic beam from a piezoelectric crystal is transmitted through the pipe into the fluid at an angle to the flow [19]. Another piezoelectric crystal receives reflected signals. An electronic circuit or microprocessor compares these transmitted and reflected signals. The frequency shift is proportional to the flow rate. The output is not affected by suspended solids. The meter temperature should be below 83°C, and the liquid should contain at least 2% suspended solids or bubbles of 30 μm diameter or larger.

The characteristics of the fluid to be measured, the fluid flow parameters, and the working environment are the factors in the selection of a flow meter. Table 1 provides guidance and data for the selection of flow meters [18].

E. Rheological Properties

Controlled shear stress and controlled shear rate methods are used to measure rheological properties. Various types of rheometers are available for the food industry. Their design is based on cone and plate, *concentric cylinders*, and *parallel plates*. There are limitations and advantages for these procedures. Concentric cylinders are most commonly used design in many rheometers for low- and medium-viscosity foods.

Cone and plate systems are also used to measure rheological properties. The sample is placed in the space created by upper cone and lower plate of the spindle and the inner wall of the cylinder. This avoids the generation of eddy currents by the spindle inserted in the sample. This eddy current may place an additional drag on the probe and thus may indicate an overly high viscosity value [20]. The advantages of cone and plate rheometers are ease of cleaning, small sample size requirement, and constant shear rate throughout the gap with small angles. The limitations are uniform sample requirement and lack of suitability for coarse dispersions and nonhomogeneous samples. In parallel plate rheometers, it is easy to extend the shear rate range suitable for coarse dispersions [21].

In a *rolling ball viscometer*, a small steel ball is inserted in one end of a capillary tube containing the sample. The tube is tilted by a fixed angle, and the time required for the ball to move through a fixed distance in the sample is observed. This time is an indirect indication of the viscos-

TABLE 1 Flow Meter Characteristics and Selection Guide

Type	Used for	Pressure loss	Accuracy, %	Viscosity effect	Relative cost
Positive displacement	Clean, viscous liquids	High	±0.25 of flow rate	High	Medium
Turbine	Clean, viscous liquids	High	±0.25 of flow rate	High	High
Vortex	Clean, dirty liquid	Medium	±0.5 of flow rate	Medium	High
Electromagnetic	Clean, dirty, viscous conductive liquids and slurries	None	±0.5 of flow rate	None	High
Ultrasonic (Doppler)	Dirty, viscous liquids and slurries	None	±5 of full scale	None	High
Ultrasonic (travel time)	Clean, viscous liquids	None	±1 to 5 of full scale	None	High
Mass (thermal)	Viscous liquids and slurries	Low	±1 of full scale	None	High

Source: Modified from Ref. 18.

ity [22]. Similarly, a falling needle rheometer also measures the rheological properties of Newtonian and non-Newtonian fluids.

Modified steady-shear *rotational rheometers* can also measure viscoelastic properties. One such modification is an oscillatory system that provides sinusoidal oscillation to the rotational part. Controlled shear–stress rheometers are also used for viscoelasticity and rheological parameters. Creep measurements can also be undertaken using modern rheometers.

In-line rheometers assist in quality control in food processing. With continuous food processing in place of batch processing, foods are processed in pipelines where in-line rheometers are necessary to monitor the product's rheological parameters continuously [23]. For this purpose, *coaxial cylinder* types are most accurate, but they are unsuitable for foods with suspended particles and high viscosity, and they can clog due to narrow gap between rotor and cup. In coaxial cylinder rheometers food flows through slits in the cylinder and encounters the rotor. On the other hand, pressure drop types provide moderate accuracy and reproducibility, but they are highly sensitive to process noise. Two to three pressure transducers are used to measure the pressure drops in a pipeline of at least two diameters. *Vibratory rod* types are not affected by process noise and are maintenance-free; however, they measure the product of apparent viscosity and density. Changes in these parameters are indicated by changes in the rod amplitude.

Concentric cylinder rheometers take long times for measuring parameters and are difficult to clean when used to measure dough properties [24]. In-line slit rheometers are more suitable for this purpose and for extruded foods. In this, two channels in the adapter plate deliver the dough to the rheometer entrance. The two variable-speed gear pumps force the dough along the slit die at controlled rates. Six sensor ports are placed along the slit surfaces. Four of them are for observing pressure drops and controlling flow rates, and the other two are for thermocouples.

F. Level

Level transmitters are needed to detect and transmit information on reservoir contents to controllers. Level controllers monitor and regulate the level of conductive liquids, and they are used in beverage bottling applications where a distinction between foam and liquid is to be made.

1. Diaphragm Level Probe

In a diaphragm level probe, a diaphragm detects the pressure exerted by the material in a container. A large diaphragm is needed for lower material density.

2. Conductivity Level Probe

A conductivity level probe has the following advantages: low cost, simplicity, and no moving parts. Its limitations: the resistivity of the foods should be $<10^8$ ohm/cm, and problems of corrosion and electrolysis.

3. Float Type

This is a common level transducer. The float can control a potentiometer or a rheostat, which provides an electric output. By replacing the potentiometer by a set of switch contacts, discrete and continuous outputs are obtained. If the conductance between two rods, immersed in a liquid, is sensed, then the liquid acts as a rheostat.

4. Capacitance Level Sensor

The probe and the container (tank) wall form the two plates of a capacitor, with the food acting as a dielectric. For a nonconductive container wall, either dual probes or an external conductive strip is needed. Appropriate electronics converts the capacitance changes to a direct current, which then can be measured. A capacitance level sensor can be used to measure levels of liquids, slurries, powders, and granular materials in containers.

The capacitance values change when the food material rises between two plates, exchanging air with a material of different dielectric constant. For material level monitoring, capacitive barrel style sensors are mounted in tanks, hoppers, vats, or pipes to sense levels of liquids, powders, grains, or other materials [1]. Advantages: simplicity, no moving parts, and proximity design. Limitations: relatively expensive, and accuracy is affected by changes in the dielectric properties of foods.

5. Ultrasonic Level Sensor

Ultrasonic level sensors consist of sensor, analog signal processor, microprocessor, and an output driver circuit. A series of transmit pulses and a transmit gate signal are generated by the microprocessor, and these are routed through the analog signal processor to the sensor. The sensor transmits an ultrasonic beam to the surface level, and the returned echo from the surface is detected by the sensor and routed to the microprocessor, which converts the signal into digital form giving the distance between the sensor and the surface level. Thus the ultrasonic level sensor operates on the principles of acoustical pressure wave reflection. The sensor can be used for level measurement and control of liquid, slurry, and solid levels. Advantages: no contact with the material, and unaffected by the physical and chemical properties of most materials. Such sensors can be used to measure the level of solid as well as liquid foods. Accuracy of these units is $\pm 0.1\%$ with temperature compensation. These are flexible and easy to install.

G. Relative Humidity and Water Activity

Electrolytic and dielectric type sensors have been used for a long time, but their response is slow and their life is short. They have good stability but are sensitive to contaminants. Solid state devices are more suitable for in-line relative humidity (RH) measurements. A change in RH will change the resistance or dielectric constant of a hygroscopic salt such as lithium chloride. Other materials for RH sensors are carbon strips, aluminum oxide, electrostatic conductive elements (phosphorus pentoxide), ceramic elements, silicone polymers, polyvinyl chloride elements, crystals, etc. For example, a thin layer of aluminum oxide is formed when surface of an aluminum strip is anodized. Outside of this oxide layer, a thin layer of aluminum or gold is deposited to act as one electrode. The other electrode is the aluminum strip. The change in the impedance of oxide layer due to the change in RH is measured to calculate RH.

Oscillation frequency of hygroscopic crystals changes with RH. The RH sensor based on this principle may compare the changes in frequency of two hygroscopically coated quartz-crystal oscillators. This arrangement changes the circuit frequency when water vapor is absorbed by the crystal. Many RH sensors based on solid state electronics are now available. These are costly, but more accurate and reliable, and they have long life.

Electric hygrometers based on capacitance or conductance are used to measure water activity of foods. These sensors measure the relative humidity of the sealed chamber when food is in equilibrium with the chamber environment.

H. pH and Other Ions

Chemical sensors are classified into ion selective electrodes, ion sensitive field effect transistors, and metal oxide gas sensors. Ion selective electrodes (e.g., pH probes) measure the concentration of ionic species. Ion sensitive field effect transistors are semiconductor chips upon which an ion selective membrane has been attached. These are more flexible than ion selective electrodes [25] because of their small size, ruggedness, and less complex calibration. Semiconductor gas sensors are based on the surface properties of oxides of tin or zinc.

The problem with conventional pH sensors is the weak signal resulting from high impedance. This signal is susceptible to noise. To avoid this problem, a miniaturized amplifier has been incorporated inside the electrode [26]. This provided a low impedance signal with a very high immunity to noise. The sensitive bulb was protected by the epoxy body housing.

Some pH sensors combine measurement electrode, reference electrode, and temperature compensator in one compact unit [27]. Recently [28], a sterilizable pH electrode without glass in its construction was introduced by Leeds and Northrup (North Wales, Pennsylvania) using ion sensitive field effect transistor (ISFET) technology. Sentron [29] has also developed a pH sensor based on ISFET. It has fast response (3 s) and eliminates the hassles associated with traditional glass electrodes by replacing glass electrodes with silicon ISFET microchips. The probe also includes a reference electrode and a temperature compensator. The current varies with the changes in H^+ concentration (Fig. 9). Portable handheld, pocket size, waterproof, and desktop models are available.

I. Color

Product color is very important for profitability, yield, and customer satisfaction. Color is a good indicator of the degree of baking, frying, or roasting. The color system of the Commission International de I'Eclariage (CIE) uses a standard source of illumination and a standard observer (L^*,

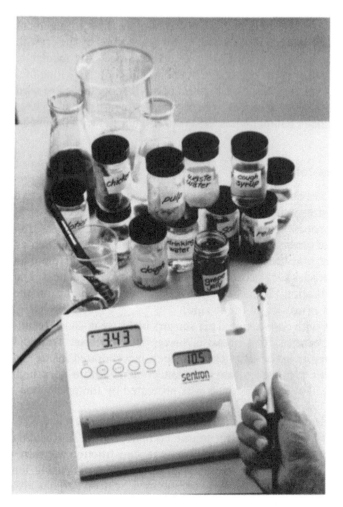

FIGURE 9 pH sensor based on ion sensitive field effect transistor (ISFET) technology. (Courtesy of Sentron, Federal Way, Washington.)

a^*, b^*). A similar color scale is Hunter L, a, and b. Both measure the degree of lightness (L^*, L), the degree of redness or greenness ($\pm a^*$, $\pm a$), and the degree of yellowness or blueness ($\pm b^*$, $\pm b$). Moreover, the magnitudes of a^* (a) and b^* (b) denote color intensity. Other important terms used are hue = $\tan^{-1}(b/a)$, chroma = $(a^2 + b^2)^{0.5}$, and total color difference $\Delta E = (\Delta L^2 + \Delta a^2 + \Delta b^2)^{0.5}$. The CIE color parameters can also be used to calculate these terms. Color-measuring instruments are based on the principles of a colorimeter or spectrophotometer. A spectrophotometer measures a ratio of the light reflected or transmitted from a product to that from a reference standard [30]. Fiber optic color recognition systems can sense and classify color by reflectance value as well as hue and chroma or relative brightness. Most color sensors emit a light beam from a probe, collect reflected light from the product, and break it into a spectrum. Some of these sensors are described here [31]:

1. Spectrophotometers with diffuse reflectance sphere accessories for color analysis are available from Varian Instruments, Toronto, Ontario, Canada.

2. An in-line color monitoring system, consisting of a vision unit, control unit, and programming unit, is marketed by Peerless Control Systems, Scarborough, Ontario, Canada. It is suitable for color measurement in baking, toasting, roasting, and frying processes.

3. A narrow-area viewer for the color evaluation of homogeneous and high-reflectance products (flour, powder, paste, peanut butter) is supplied by Filper Magnuson, Reno, Nevada. It measures reflectance at specific narrow-band wavelengths by using monochromatic light and narrowband optic filters.

4. An electrooptical color sorter is available from ESM, Houston, Texas, to separate fruits and vegetables based on color. It uses 22 photosensors for fast operation.

5. Tintometer Co., Williamsburg, Virginia, has designed a colorimeter to measure the color of edible oils and many other products in the yellow, amber, or brown range. It includes a tungsten-halogen lamp with an integral reflector focusing the light on to a 7.5-cm-diameter pinhole and passing it through a collimating lens to give a parallel beam through the sample.

6. A surface color meter for food products is supplied by Minolta Corp., Ramsey, New Jersey. A single lens reflex optical system and three highly sensitive silicon photocells combine to eliminate interference from outside light.

7. A full-scanning spectrophotometer (Spectrogard) for fast food color measurement is available from Pacific Scientific Instrument Division, Silver Spring, Maryland. It utilizes double-beam technology to measure reflectance and transmittance.

8. Reflected color of foods can also be detected by a Spectro-colorimeter marketed by Hunter Assoc. Lab., Inc., Reston, Virginia. The sensor uses 0° illumination and 45° circumferential viewing via fiber optics, and it senses the visible spectrum, from 400 to 700 nm. A constant-current tungsten-halogen lamp is filtered to simulate daylight.

Typical product standards such as color variation, product dimensions, and defects are established by showing the product to a scanner [32] and defining and adjusting classification parameters. The scanner can resolve 16.7 million colors and has a dimensional accuracy of 0.305 nm. Similarly, a scanner was developed [33] that includes a 1024-pixel, three-color (24-bit resolution), linear charge-coupled device (CCD) camera mounted over a 61 cm wide conveyor belt; a fiber optic transmitter sends the images to the computer at high speeds. Machine vision systems use spectral reflectance to monitor the color of food products.

Presently, no standard is available to select optical parameters when measuring the color of various foods. Moreover, parameters such as viewing geometry, illumination type, aperture, illumination area, and sample temperature are selected based on the recommendations of the manufacturers of the instruments. The calibration procedure is also not standardized.

J. Composition

In-line compositional analysis can be monitored by measuring refractive index, electrical impedance, infrared adsorption, etc. In-line fiber optic refractive index sensing has been used to optimize hydrogenation of edible oils and sugar, and chocolate blending for improved product uniformity and energy savings [34]. It can be used to measure the concentration of salt, sugar, and flavoring.

The refraction index of a medium is the ratio between the speed of light in a vacuum and the speed through the medium. Thus it will be always >1, as light travels fastest in a vacuum. The refractive index of a liquid is a function of both its concentration and temperature. For refractive index measurement, liquid food enters the cavity through capillary action, and it changes spectral modulation due to change in effective path length. Refractive index sensing assists in controlling processes such as extraction and evaporation, as refractive index is a function of concentration. Similarly, real-time monitoring of oil hydrogenation is also feasible [7]. An in-line refractometer has been used to measure the cholesterol concentration in egg products, and Brix levels in orange juice [3].

A critical angle, microprocessor-based, digital refractometer has been developed for liquid concentration analysis by Maselli Measurements, Stockton, California. [35]. The sensor is unaffected by optical system aging, and particulate or entrained air in the product. Digital technology greatly improved the shadow-line resolution and temperature compensation. The sensor has wide concentration (0 to 85 °Brix; °Brix is the percent by mass of sucrose dissolved in water) and temperature compensation (-5 to 121°C) ranges. Optimum accuracy is 0.05 °Brix, and repeatability is 5% of the accuracy. The sensor has applications in concentrating, diluting, reconstituting, blending, and alarm/divert/data collection. Model UR-15 mounts directly on the process line, and it continuously measures the concentration, displays the temperature-compensated concentration value, transmits the value to both digital and analog remote receivers, and activates alarm relay outputs (Fig. 10). Model IB-01 continuously measures in-line concentration and carbonation in beverages.

FIGURE 10 A digital refractometer for measuring continuously concentration of liquid foods or carbonation in beverages. (Courtesy of Maselli Measurements, Inc., Stockton, California.)

Near-infrared (NIR) spectroscopy analysis can monitor in-line the composition of food. Thus quantitative determination of moisture, fat, oil, sugar, and protein in many foods is feasible. These analyses can be performed accurately and rapidly without touching the food. The NIR instruments scan in the range of 700 to 2500 nm and are based on the principle that molecular vibrations excited in the mid-infrared region provide information on molecular structure [25]. The NIR instruments based on transmission techniques scan in the 900 to 1025 nm region. The major limitation of NIR measurement is the limited penetration of the radiation into the product. Thus composition near the product surface is measured. This is not a problem for homogeneous samples and products with fine particle size.

Katrina, Inc., Hagerstown, Maryland, [36] has developed a near-infrared sensor to measure moisture, solids, fats, sugar, oil, protein, acid, and starch in a variety of products (Fig. 11). This sensor is available in various configurations: (1) flow cells are suitable for liquids, pastes, or slurries in a pipeline; (2) immersion probes are used for liquids, pastes, or slurries within batch, holding, or mixing tanks; and (3) power cells are used for solid powders and granular materials. Based on this sensor, systems are available for specific processes, e.g., (1) in-line process control of cheese by monitoring solids and moisture; (2) control of fat and/ or solid level in milk, and (3) in-line process control of confectionary and chocolate by controlling solids and moisture.

Guided microwave spectrometry (GMS) measures the composition of food products including moisture, salt, sugar, acid, and fat contents. In this method, the microwave amplitude spectrum is measured when the energy passes through a food mixture in a waveguide [37]. More than 1700 discrete frequencies are measured over a 3000-MHz band in the microwave region. The conductivity or salinity level decreases the microwave amplitude. Other parameters measured are dielectric constant, dielectric loss factor, and molecular relaxation time. The large size of the microwaves decreases scattering, which is a problem in NIR devices.

K. Moisture

The moisture content can be measured by measuring the dielectric constant (or capacitance) and dielectric loss factor of the food materials. For measuring moisture in baked goods, a capacitance-based sensor has been used [28]. The microwave-based drying system measured moisture contents of many products such as beef, tomato paste, gravies, salad dressings, and dairy products.

FIGURE 11   A near-infrared sensor for measuring composition of foods in-line. (Courtesy of Katrina, Inc., Hagerstown, Maryland.)

The MDA-1000 microprocessor-controlled microwave dielectric analyzer (KDC Technology Corp., Acton, Massachusetts) measures moisture in foods, oils, and grains noninvasively (Fig. 12). The sensor determines true dielectric content differences to detect bulk moisture content and density at the same time; this provides density compensation. The sensor can be flush mounted in pipes, conveyors, shakers, hoppers, and

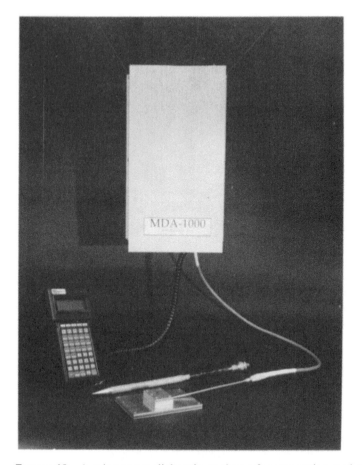

FIGURE 12 A microwave dielectric analyzer for measuring moisture in foods. (Courtesy of KDC Technology Corp., Acton, Massachusetts.)

containers for continuous in-line or batch control. When the sensor is in contact with a food material, its resonant frequency and reflection coefficient at the resonant frequency are influenced by the food's dielectric constant and loss factor. The analyzer provides the sensor with a low-level microwave signal. The signal is stepped through a narrow band of frequencies covering the sensor's resonant frequency [38]. The sensor's resonant frequency and return loss are determined by monitoring the reflected wave from the sensor. From these, dielectric components, moisture content, and density are determined. Thermocouples are also provided in the sensor for temperature compensation, as dielectric properties are temperature-dependent. Typical moisture content accuracies are between 0.01 and 0.5% with sensitivity >0.001%. Accuracy and resolution improve with decreasing bulk dielectric properties. This sensor can be used for continuous in-line measurement of moisture content and density of nonconducting fluids, particulates, and solids [39]. Each measurement cycle takes about 0.5 s.

Similarly, Dantec Systems Corp., Waterloo, Ontario, Canada, has developed a moisture sensor that uses radio waves to measure the product's dielectric properties [40].

L. Biosensors

Bioelectronics is the merging of molecular biology and electronics to produce new sensors and equipment for a wide range of applications. Biosensors are bioelectronic devices with real-time detection and measurement. They use biomaterial as their detection component and electronic components to transduce and condition the detection signal.

Biosensors are used in food analysis, diagnostics, and as feedback elements for automated food processes [41]. Food analysis includes the detection and quantification of food ingredients, additives, and contaminants. Current costly, time-consuming methods for food analysis use analytical chemistry including spectroscopy and chromatography. Biosensors based on ion-specific electrodes (Fig. 13) or enzyme thermistors provide more rapid food analysis.

Biosensors are available for the analysis of sugars, amino acids, fats, proteins, nucleic acids, alcohols, drugs, pesticides, and other chemicals. Recent advances in transducer technology have assisted further developments in biosensors. One limitation is the mass production of the bioactive component. Five approaches for biomaterial immobilization are classified as [41] (1) membrane entrapment, (2) binding or entrapment into films, (3) lipid bilayer entrapment (4) covalent binding onto a transducer, and (5) covalent binding to a membrane/film covering a transducer. Table 2 summarizes a few biosensors developed for food industry.

An amperometric ethanol sensor was constructed [42] by co-immobilization of alcohol dehydrogenase (ADH) and bovine serum albumin (BSA) in carbon paste containing entrapped NAD^+. The sensor provided fast response and high sensitivity as incorporation of both ADH and NAD^+ resulted in better relationships between the enzymatic reaction and the electrochemical process. A filtration probe from Advanced Biotechnology Corp., Buchheim, Germany, equipped with a hydrophilized polypropylene filter was used [43] to monitor ethanol, glycerol, and acetal-

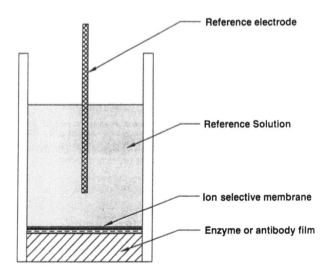

Reference electrode

Reference Solution

Ion selective membrane

Enzyme or antibody film

FIGURE 13 A biosensor based on an ion selective membrane and an enzyme or antibody film.

dehyde during industrial fermentation. Immobilized aldehyde dehydrogenase was used to sense acetaldehyde and glycerokinase for glycerol.

Enzyme-based biosensors are being applied to food freshness monitoring such as fish and meat (adenosine triphosphate), fruit (alcohol), fats (aldehydes), and milk (lactose) [41]. Enzyme biosensors are available for in-line determination of lactose, lactate, and glucose in milk [4]. Nucleotide degradation is a reliable indicator of the raw fish freshness [44]. After the death of the fish, adenosine triphosphate (ATP) starts degrading to uric acid.

Bioaffinity sensors, including those which are antibody and receptor based, have great promise for rapid food analysis. They are used to detect toxins and microbes using either specific antibody- or class-detecting receptor-based biosensors. Biosensors can also mimic human sensory capabilities such as taste and smell after immobilization of biological receptors. Thus, in the future, biosensors will be critical detection and control devices in food processing. In-line measurement of important biological metabolites in a broth is still not possible [43].

Biosensors offer great selectivity and can detect small amounts of a substance. Generally biosensors cannot be sterilized, as biomaterial is inactivated by strong acids, bases, and other harsh process conditions. These have limited life also. Further details are given elsewhere [45].

TABLE 2   Typical Biosensors Developed for Food Industry

Parameter measured	Sensor system
Amino acids	Immobilized blue crab antennule on a selective electrode
Ascorbic acid	Ascorbate oxidase on an enzyme thermistor
Aspartame	Carboxypeptidase A and L-aspartase on an ammonia electrode
Biotin	Displacement of avidin-catalase from biotin analog membranes on an oxygen electrode
Candida albicans	Piezoelectric crystals coated with *C. albicans* antibody
Fish freshness (nucleotide degradation)	Xanthine oxidase + nucleotide phosphorylase + 5'-nucleotidase on an O_2 electrode
Galactose	Galactose oxidase on a platinum electrode or an enzyme thermistor
Glucose	Fluorescence detection of dextran-FITC displaced from concanavalin A–coated optical fibers
Glutamate	Glutamate oxidase on an O_2 electrode or *E. coli* on a CO_2 electrode or yellow squash mesocarp on a CO_2 electrode
Lactose + glucose	β-Galactosidase + glucose oxidase on a platinum electrode
L-amino acids	L-amino acid oxidase on an O_2 electrode
L-lysine	Lysine decarboxylase on a CO_2 electrode
Meat freshness (tyramine content)	Monoamine oxidase on a CO_2 electrode
Organosphosphorus and cabamate pesticides	Acetylcholine receptor immobilized on an interdigitated electrode capacitor
Riboflavin	Displacement of aporiboflavin binding protein from electrodes coated with flavin membranes
Sucrose	Invertase + mutarotase + glucose oxidase on a platinum electrode or Invertase on an enzyme thermistor
Thiamine	Apopyruvate decarboxylase on a CO_2 electrode or *Lactobacillus fermenti* on an electrochemical sensor
Trypsin	Aprotinin-coated titanium wire electrode
Yeast mannan	Concanavalin A coated platinum electrode

Source: Ref. 41, taken with permission from the American Society of Agricultural Engineers, St. Joseph, Michigan.

Immunoassays are more sensitive, faster, and less expensive than other analytical methods such as chromatography. In food industry, they can be used to detect toxicants, pesticides residues, and microbial contaminants [30]. They use antibody–antigen reactions to identify specific chemicals. Isotopic immunoassays measure antigens, radioimmunoassays use antibodies bound to radioactive-labeled antigens, fluoroimmunoassays use fluorescein and rhodamine as labeling molecules, and enzyme immunoassays use enzymes such as alkaline phosphatase or glucose oxidase [30]. Immunosensors can monitor the variations in physical properties—such as acoustic, electrical, or optical—induced by an antigen binding event. Thus these are the combinations of biosensors and immunoassays.

M. Special Sensors

Special sensors are needed for food quality evaluation and for automated packaging. These sensors include electrical and photoelectric sensors for counting, fill height detectors, optical and proximity sensors, and magnetic inspection devices for metallic contaminants [4]. Sensors based on photoelectric, proximity, and limit switches are used to detect the presence of a product/material. Capacitive proximity sensors are used where inductive or photoelectric sensors cannot work.

Electrooptical and electromechanical sensors can detect the presence of a material piece and signal its arrival at a particular location. Mechanical probes can check the completion of a required task. Air-gauging equipment can measure the dimensions of a product without touching any part of it. Imaging sensors consist of line and matrix cameras, ultraviolet, infrared and NIR, X-ray, structured light, etc. [28]. Vision-based systems are used to determine fill levels, to determine positioning of labels on packages, to detect bone particles in meat, and to grade fruits and vegetables. X-ray-based systems are used to detect stones and debris in raw material, glass and metal pieces in packaged foods, and bone fragments in meat products. The presence of holes in potatoes can be detected using special ultrasonic equipment. This method is less expensive than X-ray machines [46]. By reducing the frequency and by using means to reduce the area of insonification, the hollow hearts in potatoes can be detected.

Digital image processing is used for in-line sorting of good and defective products. Applications of neural networks with computer vision systems are gaining popularity, and a commercial system is available for apple sorting [47].

Mark detectors detect marks in every color of the spectrum, operate on a physical-optical principle, and are suitable in product packaging. A transmitter provides a transmitter wavelength of between 470 nm (blue) and 630 nm (red). Monochromatic light beams are reflected or absorbed in varying levels by colored surfaces. The receiver of the register mark scanner measures only the gray light value of the colored mark and light it reflects [1].

III. HARDWARE AND SOFTWARE

Proper care is required to avoid impedance mismatching when connections are made between electrical devices. The general principles of impedance matching are that the impedance external to an electronic device or sensor should match its internal impedance for maximum energy transmission. Also, the external impedance should be large compared with the sensor's internal impedance when the internal voltage of the sensor is measured.

In data acquisition, peripherals are required to get information and reports from the data loggers through computers, and to direct the data logger what to do and how to do it. Printers are used to log alarms and printout various reports. CRTs and keyboards are used for display, and to allow the operator to interface directly with the process. Many times, operators just touch the part of the screen where the system is displayed to obtain information or enter new instructions. This touch screen option permits fast and direct call-up of operating data and displays. TV monitors with mouse give an overall view of various processes.

Interfacing a microprocessor system to external devices involves both hardware and software. The commonly used bus standards are RS232C, IEEE488, and CAMAC. RS232C is the classical binary serial interface, and transmission speed up to 9600 baud is used. IEEE488 is also known as "General Purpose Interface Bus (GPIB)" or ASCII bus. CAMAC is IEEE's Standard 583, and it specifies a parallel bus with specific physical modules. Analog signals from sensors must be converted to digital signals for further processing by the microprocessor. Digital output of the microprocessor is converted to analog form to actuate the valves and switches at the process equipment. Thus the process interface brings signals into and out of the control system. In addition to analog signals from sensors, these signals include contact inputs and outputs such as valve position, pump on/off status, etc. Further signal-conditioning modules are needed to provide bridge excitation, zero balance, gain, span adjustment, and low pass filtering. For interfacing, plug-in boards are available that provide programmable gain, analog to digital, and digital to analog conversions. Solid state relays are used for applications such as contact closures, interfacing to low power state circuit, maximum speed and control of contact closure, etc. Complex control strategies can be implemented simply by changes in software, once the inputs and outputs are connected to the microprocessor.

IV. FUTURE TRENDS

Few sensors have met their published specifications consistently and repeatably [48]. Improvements are needed in the long-term reliability of sensors. Progress in sensor technology is in the direction of low fabrication cost, miniaturization, use of fiber optics, replacement of analog sensors with digital sensors, more accurate and precise sensors, and improved service life and repeatability. Miniaturization permits lighter packaging and sensor placement much closer to the parameter measured. Smart devices are made intelligent by incorporating microprocessors. Fiber optics enables remote instrumentation. Thus modern instruments are becoming smarter and easier to use.

Further development in sensor technology is needed to measure very high viscosities, moisture content, color of uneven surfaces, and thickness of material in motion [49]. In-line, nondestructive, and real-time measurement of enzyme activity, microorganism activity, flavor/aroma, and shelf life will enhance and ensure food quality, consistency, integrity, and safety [48].

Variables can be monitored indirectly, where appropriate sensors are not available, by combining computer calculation ability, process knowledge, and other sensors [17]. A neural network–based intelligent sensor to emulate the human nose has been described [50]. It is based on the interaction of volatile chemicals with many sensors, providing aroma in both digital and graphic format. In the future, ultrasound-based sensors may be available to measure the droplet size in oil-in-water emulsions, and crystallization characteristics in fats.

REFERENCES

1. P. MacInnis, Acquiring an intelligent sense of direction, *Manufacturing & Process Automation 5*(3):5 (1995).
2. G. A. Kranzler, Impact of electronics on materials handling and processing, *Agricultural Electronics—1983 and Beyond*, Amer. Soc. Agric. Eng., St. Joseph, Michigan, 1984, p. 614.
3. Anon., On-line sensors open up new window into the process, *Prepared Foods 163*(2):86 (1994).

4. L. D. Pedersen, W. W. Rose, and H. Redsun, Status and needs of sensors in the food processing industry, Proc. Food Processing Automation Conf., Am. Soc. Agric. Eng., St. Joseph, Michigan, 1990, p. 10.

5. A. D. Helfrick, and W. D. Cooper, *Modern Electronic Instrumentation and Measurement Techniques*, Prentice Hall, Englewood Cliffs, New Jersey, 1990, p. 446.

6. M. J. Usher, *Sensors and Transducers*, MacMillan, London, 1985, p. 163.

7. G. Yazbak, Fiberoptic sensors solve measurement problems, *Food Technol.* *45*(7):76 (1991).

8. Omega, Practical temperature measurements, *The Temperature Handbook*, Vol. 28, Omega Engg., Inc., Stamford, Connecticut, 1992, p. Z-11.

9. ASTM, *Standard Specification for Metal-Sheathed Base-Metal Thermocouples*, Standard E608, Am. Soc. Testing Materials, Philadelphia, 1979.

10. ASTM, *Standard Method for Calibration of Thermocouples by Comparison Techniques*, Standard E220, Am. Soc. Testing Materials, Philadelphia, 1972.

11. ASTM, *Standard Methods for Testing, Industrial Resistance Thermometers*, Standard E644, Am. Soc. Testing Materials, Philadelphia, 1978.

12. J. A. Evans, S. L. Russell, and S. J. James, An evaluation of infra-red non-contact thermometers for food use, *Automatic Control of Food and Biological Processes* (J. J. Bimbenet, E. Dumoulin, and G. Trystram, eds.), Elsevier, New York, 1994, p. 43.

13. W. R. Cross and D. R. Lesley, Self-contained microcircuitry probe acquires and records food-process temperature data, *Food Technol.* *39*(12):36 (1985).

14. M. F. Hordeski, *Transducers for Automation*, Van Nostrand Reinhold, New York, 1987, p. 301.

15. J. Okladek, An overview of flow metering devices, *Amer. Lab.* *20*(1):84 (1988).

16. G. C. Barney, *Intelligent Instrumentation*, Prentice Hall, Englewood Cliffs, New Jersey, 1985, p. 532.

17. Omega, Pressure measurement, *The Pressure, Strain and Force Handbook*, Vol. 28, Omega Engg., Inc., Stamford, Connecticut, 1992, p. Z-7.

18. M. C. Beaverstock, Process control, *Computer Aided Techniques in Food Technology* (I. Saguy, ed.), Marcel Dekker, New York, 1983, p. 361.

19. N. P. Cheremisinoff and P. N. Cheremisinoff, *Flow Measurement for Engineers and Scientists*, Marcel Dekker, New York, 1988, p. 392.

20. S. D. Christian, New instrumentation for viscosity measurements, *Amer. Lab.* *21*(19):10 (1989).

21. C. F. Shoemaker, J. I. Lewis, and M. S. Tamura, Instrumentation for rheological measurements of food, *Food Technol.* *41*(3):80 (1987).

22. S. M. Block and A. Spudich, The rolling ball viscometer, *Amer. Biotechnol. Lab.* *5*(4/5):38 (1987).

23. M. Jackman, In-line viscometers help achieve perfect products, *Food Technol.* *45*(7):90 (1991).

24. S. Bhakuni and C. G. Gogos, Development and use of an on-line viscosity sensor for continuous food processing equipment, Proc. Food Processing

Automation Conf., Am. Soc. Agric. Eng., St. Joseph, Michigan, 1990, p. 206.

25. J. Giese, On-line sensors for food processing, *Food Technol. 47*(5):88 (1993).

26. B. Moshiri, The amplifier electrode: An advance in pH technology, *Amer. Lab. 20*(19):16 (1988).

27. Anon., Sensors resist harsh process, *Food Engg. 64*(5):70 (1992).

28. Anon., On-line sensors open up new windows into the process, *Prepared Foods* 163(2):86 (1994).

29. H. C. G. Lightenberg and F. Scheper, ISFET pH measurements in dairy products, *Sentron Application Notes*, Sentron, 33320 1st Way South, Federal Way, Washington, 1995.

30. J. Giese, Rapid techniques for quality assurance, *Food Technol. 47*(10):52 (1993).

31. Anon., Instrumentation for color measurements, *Food Technol. 41*(4):91 (1987).

32. M. Scher, Process line QC monitor sees, shows, and tells, *Food Process. 54*(1):132 (1993).

33. H. R. Ritchie, Color scanner, advanced imaging software improve product quality, *Food Process. 55*(4):15 (1994).

34. Anon., Sensors that see and perceive, *Prepared Foods 160*(6):73 (1991).

35. R. Ver Mulm, Advantages of using digital refractometer technology for liquid concentration analysis, Maselli Measurements, Inc., Stockton, California, 1995, p. 25.

36. Anon., On-line process monitors for the food, dairy, chemical and pharmaceutical industries, Katrina, Inc., P.O. Box 418, Hagerstown, Maryland, 1995.

37. Anon., Guided microwave spectrometry provides breakthrough analytical technique, *Food Engg. 66*(8):21 (1994).

38. The MDA-1000, microwave dielectric analyzer for industrial process monitoring and control, KDC Technology Corp., Acton, Massachusetts, 1995, p. 2.

39. R. J. King, Continuous moisture and density measurement of foods using microwaves, Proc. Food Processing Automation Conf., Am. Soc. Agric. Eng., St. Joseph, Michigan, 1994, p. 10.

40. Anon., Colonial cookies controls product moisture and color on-line in real time, *Food Engg. 66*(8):37 (1994).

41. R. F. Taylor, Applications of biosensors in the food processing industry, Proc. Food Processing Automation Conf., Am. Soc. Agric. Eng., St. Joseph, Michigan, 1990, p. 156.

42. M. Boujtita and N. ElMurr, Biosensors for analysis of ethanol in foods, *J. Food Sci. 60*:201 (1995).

43. M. Rank, J. Gram, K. Stern Nielsen, and B. Danielsson, On-line monitoring of ethanol, acetaldehyde and glycerol during industrial fermentations with *Saccharomyces cerevisiae*, *Appl. Microbiol. Biotechnol. 42*:813 (1995).

44. J. H. T. Luong, K. B. Male, and A. L. Nguyen, Development of a fish freshness sensor, *Amer. Biotechnol. Lab. 6*(12):38 (1988).

45. G. S. Mittal, *Food Biotechnology*, Technomic Pub., Lancaster, Pennsylvania, 1992, p. 380.

46. K. C. Watts, L. T. Russell, and G. C. Misener, Ultrasonic detection of hollow heart in potatoes, Paper No. 86-414, Can. Soc. Agric. Engg., Saskatoon, Saskatchewan, Canada.

47. S. Gunasekaran and K. Ding, Using computer vision for food evaluation, *Food Technol. 48*(6):151 (1994).

48. J. W. Newton, Identifying sensor technologies for the food industry—a Campbell Soup Company perspective, Proc. Food Processing Automation Conf., Am. Soc. Agric. Eng., St. Joseph, Michigan, 1990, p. 86.

49. R. H. Caro and W. E. Morgan, Trends in process control and instrumentation, *Food Technol. 45*(7):62 (1991).

50. Anon., Going the nose one better, *Food Engg. 66*(11):25 (1994).

3

Food Process Modeling and Simulation

Gilles Trystram and Francis Courtois
High School of Food Science and Technology, National
Institute for Agronomical Research, Massy, France

I. INTRODUCTION AND DEFINITIONS

Food process engineering has three inherent objectives: to understand the phenomena that occur during processing, to design unit operations, and to control them. Control is a difficult job, in which the main goal is to achieve some objectives without human intervention. In the case of food processes, the objectives are generally complex and numerous. Due to the difficulty of measuring product properties during or at the end of a process, the control is often performed directly by human operators based on their own subjective evaluation. Nevertheless, control tasks, including operator decision support, are difficult. It is often necessary to predict the future behavior of the process to make the best decision for piloting a single operation or the whole plant. This task is generally well performed by the operator, but with a high time constant. Improvements could be obtained, if tools are available, for reliable prediction of the process behavior and coupling with optimization tools.

On the other hand, one of the main jobs on the line is the evaluation of the present state of desired properties and of process safety. A goal for control is to do it on line, without the need of people, in real time if possible, to make the best decision at each process step. The prediction of new variables from measured ones becomes important here too.

Many processes are considered as steady state processes. During the life of the process, in fact, most of the operations are performed during transient states. They are generally not visible for human observation, and the use of sensors permits a new understanding of those processes. Such transient behaviors are encountered for startup and shutdown procedures, for batch processes, during product changes, cleaning, etc. Even if a process is considered a stable one, small transient behaviors are realized. It is well recognized today that process optimization should take into account those phenomena. But without tools and simulation support, it is difficult to analyze transient behavior. A model able to represent the dynamics of these phenomena is called for. The same kind of model is necessary for the design and the tuning of the controller that should replace the human decision to maintain transient behavior in steady state ones. The design of such a controller requires the calculation and the representation of the process and of the influence of important variables on the objectives. Prediction of the process evolution seems to be necessary.

How can one predict process evolution and phenomena? Many ways have been developed: simple measurements, use of short-time tendencies of process variables, simulation of process, diagnosis from specific variables, etc. In all cases, in fact, models are used, even if often they are not written as classical food-engineering models.

A good, and concise, definition of *modeling is the building of tools able to predict the future of the process with good accuracy.* On the basis of this definition it becomes crucial for control purposes to design good models of a process in order to make the best decision for controlling tasks.

Models are well known in process engineering. For a long time, they were only used for design purposes, or for the interpretation of phenomena carried out during processing. Numerous calculations are performed for the choice of exchange surface, nature of materials, kind of unit operations, cleaning frequency, and accurate design computations. Specific complex models are developed. Are these models usable for control purposes? Not necessarily. Because the goals of control are different from the goal of design [1], the tools developed for control must be different. Figure 1 proposes several uses of models for process control purposes. As can be observed, different needs are encountered that imply, certainly, several kinds of models are appropriate for each need. The purpose of this chapter is to deal with modeling tasks, to present, describe, and illustrate methods for dynamic modeling of food processes, and to introduce the use

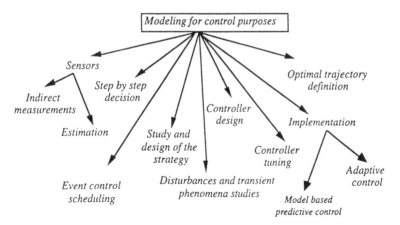

FIGURE 1   Objectives of modeling for process control applications.

of these models for prediction of process evolution in simulation packages, eventually performed on line.

A. Modeling: A Two-Step Job

Before dealing with modeling and appropriate methods, we must understand the scope of modeling. Our definition is a very inclusive one. All tools able to predict the behavior of several variables can be called models. Modern computer technology improves the development of numerical models, but a quick, small process able to present the same behavior as the real one is a physical model. Today it is better to build modeling approaches only from the use of computers. But if the power of computers is increasing, we still must introduce, as much as possible, process knowledge and understanding during the modeling step.

During modeling it is important to *develop a two-step approach*. The first step concerns the analysis of the variables of the process, which includes analysis of the kinds of relationships among variables; it is the qualitative part of modeling that concerns the choice of the model structure. The second step is the quantitative one in which the parameters that characterize the concerned unit operation are calculated and the model is then validated. This is the identification step. As indicated in Fig. 2, these two steps are always encountered during modeling. It is simple to understand that many approaches are available both for the qualitative and quantitative parts of modeling. The choice of a structure implies gener-

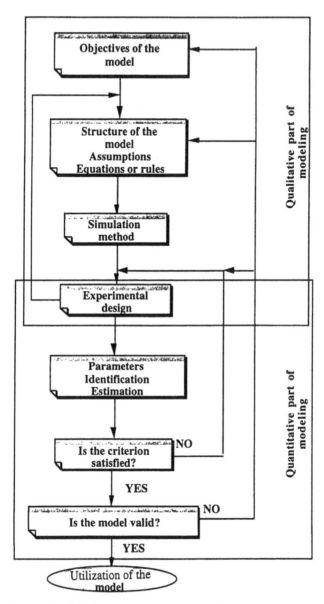

FIGURE 2   Principle and steps of modeling.

ally some constraints for the choice of an identification methods, even if general rules are still available.

A systems approach permits the control engineer to identify the key variables in a process, from a control point of view, without detailed understanding of the process. Figure 3 presents four types of variables. Input variables are those influencing the future, or present, behavior of the process. Two kinds of input variables are encountered. First, decision variables are those on which we can act. Generally actuators are used for controlling them. Another kind of input variable is the disturbances. These are variables that modify the process. It is impossible to act on them (sometimes because the cost of actuators is too high, or because their evolution depends on uncontrolled situations). The consequence of evolution of both decision variables and disturbances is that some important process variables are modified. These are state variables. The evolution is sometime helpful, sometimes not. State variables are outputs, and some of them are objectives for the process; these must be controlled. Those specific variables are generally called output variables (Fig. 3).

The classification of variables is illustrated in Fig. 4 using the example of a heat exchanger. Input temperatures for product and hot water are not controlled. They depend on upstream operations, and so they are considered as disturbances. For decision variables, two examples are illustrated in Fig. 4. First, because no controller is implemented for the product flow rate, the valve positioner appears to be the right decision variable. The product flow rate (Fi) becomes a consequence of the valve behavior; it is then a state variable. Second, on the other hand for the water flow rate, a controller is used to control the valve with respect to a given flow rate setpoint independently of all disturbances that could occur. The right decision variable in this case is the flow rate setpoint, and the valve positioner then follows an evolution that is the consequence of the controller decision.

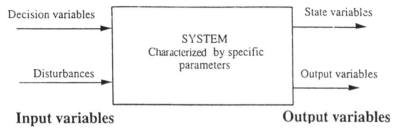

Input variables   **Output variables**

FIGURE 3 Systems characterization of a process—position and definition of variables.

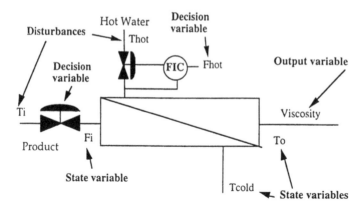

Figure 4 Variable analysis and classification for a systemic approach of a heat exchanger.

Such a description is always available, even if no information about the kind of relationship between input and output variables can be developed. If we consider the case of a heat exchanger, the same description is available for different unit operations and different plants. Parameters of the system must also be considered. They are mainly design parameters such as exchange surface, sensors location, and kind of materials used. These parameters affect the behavior of the process [1]. They are specific to a given unit operation. The systems approach is an essential one to reduce the complexity of a modeling problem. After the analysis of the role of each variable, more advanced work is needed for the choice of the kind of model to be developed.

B. Different Kinds of Models

If several model structures are available, it is important to know these structures and to choose the best one (best in regard to a given objective of modeling). Table 1 presents some process charactertics and the consequences for modeling. Many combinations are possible. The best choice should be made on the process basis, in relation to the objectives of modeling. The characteristics of many food processes imply the need for nonlinear, multivariable models, even if most of the applications of process control are based on a linear monovariable representation of processes. The use of stochastic or deterministic models depends on the objectives of modeling, and mainly deterministic models are used. For control purposes, the principle of control is to obtain a given trajectory for output variables independently of evolution of disturbances. The evolution is

TABLE 1   Definitions and Consequences for the Modeling Job of Some
Process Characteristics

Process characteristics and definitions	Consequences for modeling
The influence of input variables is linear.	A linear model is sufficient.
The influence of input variables is nonlinear on the outputs.	The model should represent nonlinear relations or should be able to adapt its characteristics during the evolution of the process.
Strong interactions are present between several input variables and state or objective variables.	A simple single input, single output (SISO) model is not sufficient. The model should represent the interactions; a multiple input, multiple output model (MIMO) must be developed.
Interactions between variables are neglected.	An SISO model could be developed.
Random behaviors are encountered, for example, for raw materials or the process itself.	It becomes necessary to include the random evolution, if the goal of modeling requires it. Stochastic models are then necessary.
Only deterministic behaviors are recognized (or random phenomena are neglected).	Deterministic models are sufficient.

performed during time, and sometimes in space; models should then be able to predict in the next future interval the trajectories of the state variables of the process to choose the best behavior of the decision variables for actuator tuning. Dynamic models are the only ones capable of doing this, even if for some cases large time constants of processes permit use of steady state or quasi steady state models. The main question of modeling then becomes that of finding the best dynamic, multivariable, and probably nonlinear structure of relationship between input and output variables and the right identification procedure for the design of a good tool capable of accurately predicting the behavior of process variables.

C. Modeling the Dynamic Behavior of Food Processes

The purpose of this chapter is to survey the main ways available for modeling of food processes. Before dealing with the specific characteristics of food processes, we must understand how the problem of modeling is taken into account in other process engineering domains.

For a long time, chemical engineers have been developing large-scale simulation packages. There are many unit operation models and simulation tools for using them. But now numerous unit operations are considered. Initially such tools were developed only for the oil and chemical industries. The main advantage of such tools is that models are already written and the user only has to define parameters or build the network of unit operations that represent the process. After the development of steady state simulators (Aspen, Process, for example), dynamic simulators are proposed. Probably the main one is Speed up that permits simulation of the dynamic behavior of numerous operations and also permits new models developed for specific applications. For batch processes, a specific tool exists, the BATCHES software. The main approach of such tools is to use existing models. For the food industries, there are some limitations. Thermodynamic data banks are necessary, and for complex mixing and formulation of food products these are not available. The same is true for heat- and mass-transfer coefficients and other important parameters (water activities at high temperature, for example). Another limitation is the lack of good tools for reactor modeling which need to be developed and implemented specifically. We emphasize, too, the limitations of all existing software for hydrodynamic modeling, particularly with non-Newtonian liquids.

The use of such software is then difficult; nevertheless, applications for dynamic simulation are performed, for example, for unit operations in the sugar industry or freezing processes [2–4]. The main applications concern the study of batch processes for optimization of the operating conditions, the study and tests of control strategies, some studies on safety, and the training of operators.

This kind of approach is based on the use of a library of models. Many research centers are developing libraries, but their use is not easy and many parameters are often identified. These large-scale simulators are designed for a wide range of applications, and they are not necessarily the best approach for a given problem. On the other hand, for control purposes, it is important to have in mind that the on-line implementation of models could be a good technique for controller design (method of model-based control, for example; see Fig. 1). If the user of the model is not the designer, the implementation on line is difficult and time-consuming. Then, another way for modeling the dynamics of food processes is to consider a *library of methods* and to choose the best method for a given objective.

Approaches for the choice of dynamic models for the food unit operations and plants are presented in Fig. 5 [5]. The available methods are presented as three classes.

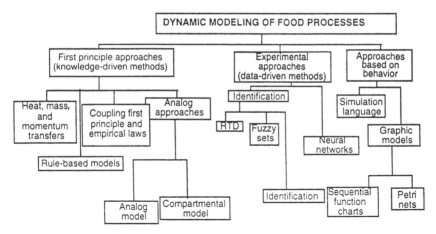

FIGURE 5 Principal approaches for dynamic modeling of food processes.

First, *knowledge-based methods* are considered. These are based on the classical approach in food engineering in which first-principle laws are written on the basis of understanding of phenomena. This is the so-called descending approach (or knowledge-driven models approach); the model is developed from assumptions about phenomena and then is validated with experimental data. Problems are encountered with the determination of physical and thermophysical parameters in such models. It is sometimes better to talk about an ignorance coefficient, and a large part of the problem in such cases is reported in the identification part of work. As examples will show later, some alternative approaches are easy to use in these kinds of methods. These methods include the rule-based approaches (Fig. 5).

Second, as shown in Fig. 5, many approaches are developed as ascending methods (or *data-driven models*): only experimental data are considered, without any assumptions about the underlying phenomena. This is often known as the black box modeling approach. It is easily used for steady state modeling (response surface methodology); applications are also numerous for dynamic models. The main advantage of this approach is that generally the chosen structure of the model is easy to simulate. This allows such models to be used on line, or to perform real-time identification in adaptive control methods. The many approaches available at this level are illustrated in Fig. 5, in which both linear (classical in a control science point of view) and nonlinear (necessary in a process control point of view) models are possible. Notice that such methods are closely related to the identification problem.

Third, Fig. 5 indicates approaches dedicated to the modeling of the process and control together. This is a classical way for the design of a sequential controller, and some explanations of such approaches will be given later. It is important to consider these approaches because probably in the future they will be required to study setpoint control and process design problems together.

Each of these ways needs to be described and illustrated. The next parts of this chapter illustrate which approaches are suitable for several classes of simulation and control problems.

II. KNOWLEDGE-BASED MODELING AND SIMULATION FOR CONTROL PURPOSES

Numerous works in the field of food process engineering are concerned with modeling, and the methods are well described in many books dealing with first principles of food engineering laws. A knowledge-driven model is first built on the basis of assumptions and the selection of some phenomena able to represent the process. Heat- and mass-transport phenomena are written as equations with respect to the classical principle:

Accumulation = input − output

A simple example is developed in the section concerning data-driven models. A systematic development of the equations is easy if the phenomena are well recognized. This, though is difficult for food processes. For dynamic modeling a point must be highlighted. Because the time evolution must be taken into account, time differential equations are used (or sometimes partial differential equations for lumped-parameter systems), and the choice of the appropriate spatial coordinates is important and initial conditions should be well known. Generally the model is much more complicated than a steady state one.

Nevertheless, such models are not difficult to write. Often the greater problems concern the phenomena and related assumptions that must be chosen. Progress in food engineering is occurring, but many food material transformations during processing are not understood at present. The most difficult are the determination of unknown physical and thermophysical parameters. An advantage of knowledge-driven models is the ability to perform experiments specifically for parameter estimation, independently of the process. But because numerous phenomena are carried out during the process, the validity of the equation is not necessarily established. The design of the simulation of the model could be a difficult task, too, because many constraints are encountered from model stability considerations.

Nevertheless, this is the most powerful approach because the range of applications is the greatest. It is specifically the case for the study of transient behaviors like startup, products changes, or shutdown steps.

Figure 6 illustrates an example, for the production of yeast from whey, at industrial scale. A model was developed both for physiological behavior of the yeast and gas transfer inside the fermenter [6]. The model is available even at the beginning of the process, where the production is a batch one characterized by an important range of variations in yeast content (1 to 5). The model permits analysis of several strategies for the startup procedure, and it improves significantly the yield of production. The accuracy of the knowledge-based model is 7% in this case.

As indicated in Fig. 5, other ways are available for dynamic modeling on a knowledge basis. Another example is the use of compartmental modeling. As an illustration, the dewatering and impregnation soaking process, useful for facilitating the achievement and the preservation of the quality of product, is studied for dynamic analysis in a modeling approach. When pieces of fruit are immersed in a concentrated solution of salt or sugar, simultaneous mass transfer occurs. Water is removed from the product, and sugar or salt enters the product. These phenomena are used before drying, for dewatering or for impregnation, to formulate specific products.

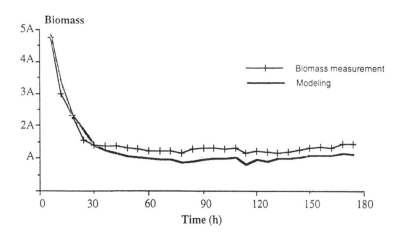

FIGURE 6 Example of the application of a knowledge-based model for the study of the startup procedure: case of continuous fermentation for yeast production from whey. (From [6].)

Classical modeling is based on the diffusion model, but the simulation is long, not necessarily easy, and diffusion mechanisms do not correctly represent all the situations. To obtain a reliable, dynamic model, a simple representation of the mass transfer is possible as indicated in Fig. 7a. The model is derived from previous study [7]; it is a compartmental model (two compartments). Four parameters are identified to describe internal and external mass transfers. Heat transfer is neglected. The main advantages of this approach are first that the model is simple, easy to

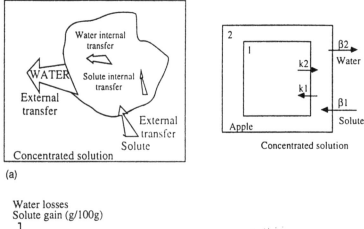

(a)

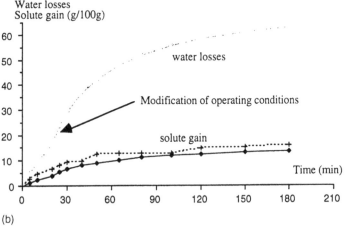

(b)

FIGURE 7   (a) Principle of the dehydration and impregnation soaking process and simple dynamic modeling of water and solute movements, compartmental approach. (b) An example of validation of the osmotic dehydration process compartmental model, for a variable-condition experiment: comparison of modeling and experimental results in the case of apple pieces in sugar-based solutions (From [8]).

implement on line, and second that only four parameters are identified. The dynamic performance is illustrated in Fig. 7b, where it is established that transient behaviors are predicted well by the simulation. This is a gray box model: an explanation of water movement inside and outside the product is used for model building.

The last way for dynamic modeling on a knowledge basis is the design of rule-based models. As a result of progress in artificial intelligence and computer science, new tools are available for knowledge representation. Expert systems and fuzzy inference methods are possible for modeling food processes, even for design or control purposes. More illustrations are presented in Chapters 5 and 6, dealing with fuzzy control, but one should understand that those ways for modeling are important and that probably in the near future significant progress will be made on this basis.

III. DATA-DRIVEN MODELS AND SIMULATION FOR CONTROL PURPOSES

The development of knowledge-based models is a time-consuming task. Due to the objectives of control, simple approaches are suitable and save substantial time. From the beginning of control science, many results have been available for linear controller design, tuning, and implementation. The use of computer-controlled systems permits implementation of the model and controller together to improve the performance, or use of computerized methods for the controller design. All of these methods are based on dynamic models, mainly linear ones.

A. Model Development

To explain and justify the concept of dynamic modeling classically used for control algorithm development, let us consider the case of a bioreactor as represented in Fig. 8. For control applications, a dynamic model should be derived.

Classically, the relation between level (H) and output temperature (T_s), the heat flux (V) and the input flow rate of substrate (Q_e) is derived on a knowledge basis, using heat and mass balances:

$$\frac{d}{dt}(AH\rho) = Q_e - Q_s \tag{1}$$

$$A\rho\frac{dH}{dt} = Q_e - Q_s \tag{2}$$

where ρ is density, and the addition of sodium hydroxide is neglected.

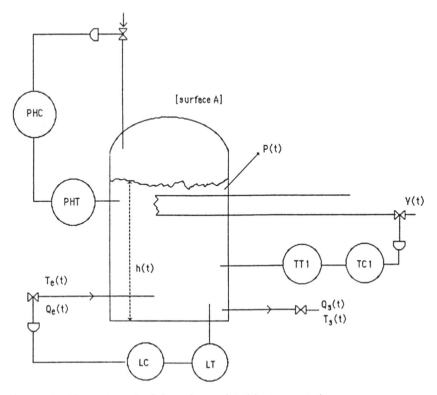

Figure 8   Bioreactor principle and associated instrumentation.

For the heat balance,

$$\frac{d}{dt} [\rho \, Cp \, AH(T_s - T_0)]$$

$$= Q_e \, Cp \, (T_e - T_0) - Q_s \, Cp \, (T_s - T_0) + W(t) - Pd \quad (3)$$

T_0 is chosen as the reference temperature ($T_0 = 0°C$), and Pd is energy losses. We have

$$W(t) = \frac{V^2}{R} \quad (4)$$

if we assume that the heater is an electric one, characterized with a value of electric resistance R. We further have

$$\rho \, Cp \, A \frac{d}{dt} (HT_s) = Q_e \, Cp \, T_e - Q_s \, Cp \, T_s + \frac{V^2}{R} - Pd \quad (5)$$

$$\rho \, Cp \, A T_s \frac{d}{dt} H + \rho \, Cp \, A H \frac{dT_s}{dt} = Q_e \, Cp \, T_e - Q_s \, Cp \, T_s + \frac{V^2}{R} - Pd \qquad (6)$$

$$A\rho \, Cp \, T_s \frac{dH}{dt} + A\rho \, Cp \, H \frac{dT_s}{dt} = Cp \, T_s(Q_e - Q_s) + A\rho \, Cp \, H \frac{dT_s}{dt} \qquad (7)$$

$$A\rho \, Cp \, H \frac{dT_s}{dt} = Q_e \, Cp \, T_e - Q_e \, Cp \, T_s + \frac{V^2}{R} - Pd \qquad (8)$$

This implies

$$\frac{A\rho H}{Q_e} \frac{dT_s}{dt} + T_s = T_e + \frac{V^2}{R \, Cp \, Q_e} - \frac{Pd}{Cp \, Q_e} \qquad (9)$$

The model of the bioreactor is

$$\frac{A\rho H}{Q_e} \frac{dT_s}{dt} + T_s = T_e + \frac{V^2}{R \, Cp \, Q_e} - \frac{Pd}{Cp \, Q_e} \qquad (10a)$$

$$A\rho \frac{dH}{dt} = Q_e - Q_s \qquad (10b)$$

which is a nonlinear model. As explained previously, this kind of model is not easy to implement for on-line and real-time use in control applications, even if this example is a quite simple one.

The equations could be written in another representation using the state vector $\mathbf{x}$. Let $\mathbf{x}$ be the vector $\begin{pmatrix} T_s \\ H \end{pmatrix}$, let $\mathbf{v} = \begin{pmatrix} Q_e \\ v \end{pmatrix}$ be the control vector, and let $\mathbf{r}$ be the vector of disturbances. The previous equation could be written as $f \, (\mathbf{x}, d\mathbf{x}/dt, \mathbf{v}, \mathbf{r}) = 0$.

The order of the model (state model) is the dimension of the state vector $\mathbf{x}$. The order depends on the required level of description. As an example, it could be emphasized that the controlled temperature is the temperature of the heat resistance T_R, which follows an equation such as

$$\frac{d}{dt} (M_R \, Cp_R \, T_R) = \frac{V^2}{R} - a_R A_R(T_R - T_s) \qquad (11)$$

where a_R is the heat transfer coefficient, A_R is surface area of the resistance, and M_R is the mass of the resistance. Then we have

$$\frac{M_R \, Cp_R}{A_R a_R} \frac{dT_R}{dt} = \frac{V^2}{a_R A_R R} - T_R + T_s \qquad (12)$$

A new state vector is then introduced as $\mathbf{x} = \begin{pmatrix} T_s \\ H \\ T_R \end{pmatrix}$.

A similar approach is possible if we consider the measured temperature, which is related to the kind of sensor used. Even if the simulation of such a model is possible, a linear description is easier and often sufficient.

In most applications, setpoint control is performed, and the behavior of the process is observed only in the neighborhood of the chosen setpoint. In the case of the bioreactor, we can consider the setpoint defined with steady state values H^*, T_s^*, T_e^*, Q_s^*, Q_e^*, V^*, and Pd^*. At the steady state setpoint, it becomes

$$\frac{dH^*}{dt} = 0, \qquad Q_e^* = Q_s^*, \qquad \text{and} \qquad \frac{dT_s^*}{dt} = 0 \tag{13}$$

Let us consider the error variable defined as the difference between measured and setpoint values:

$$\Omega_s = T_s - T_s^*; \quad \Omega_e = T_e - T_e^*; \quad q_e = Q_e - Q_e^*; \quad q_s = Q_s - Q_s^*;$$
$$p = Pd - Pd^*; \quad h = H - H^*; \quad v = V - V^* \tag{14}$$

If those new variables are introduced in Eqs. (10a, b), they become

$$A\rho \frac{d(H^* + h)}{dt} = q_e - q_s + Q_e^* - Q_s^* \tag{15}$$

$$A\rho \frac{dh}{dt} = q_e - q_s \tag{16}$$

$$A\rho(h + H^*) \frac{d(T_s^* + \Omega_s)}{dt} + (q_e + Q_e^*)(T_s^* + \Omega_s)$$

$$= \frac{(v + V^*)^2}{R\,Cp} + (q_e + Q_e^*)(T_e + \Omega_e) - \frac{Pd^* + p}{Cp} \tag{17}$$

$$A\rho \frac{d(T_s^* + \Omega_s)}{dt} + \frac{(q_e + Q_e^*)(T_s^* + \Omega_s)}{H^* + h}$$

$$= \frac{(q_e + Q_e^*)(T_e^* + \Omega_e)}{H^* + h} + \frac{(v + V^*)^2}{R\,Cp\,(h + H^*)} - \frac{Pd^* + p}{Cp\,(h + H^*)} \tag{18}$$

From a control point of view, variations around the setpoints are limited. If it is assumed that they are negligible, the following relation is applicable because of a Taylor development at first order (h/H^* is very small compared with 1):

$$\frac{1}{H^* + h} = \frac{1}{H^*} \frac{1}{1 + \dfrac{h}{H^*}} = \frac{1}{H^*}\left(1 - \frac{h}{H^*}\right) \tag{19}$$

Including the setpoint relations (13) and (14), this becomes

$$a_1 \frac{d\Omega_s}{dt} + \Omega_s = \Omega_e + a_2 q_e + a_3 v + a_4 Pd \tag{20}$$

with

$$a_1 = \frac{T_e^* - T_s^*}{Q_e^*} \qquad a_2 = \frac{2V^*}{R\,Cp\,Q_e^*}; \qquad a_3 = -\frac{1}{Cp\,Q_e^*}; \qquad a_4 = \frac{A\rho H^*}{Q_e^*}$$

The new linearized model of the reactor is obtained:

$$A\rho \frac{dh}{dt} = q_e - q_s \tag{21a}$$

$$a_1 \frac{d\Omega_s}{dt} + \Omega_s = \Omega_e + a_2 q_e + a_3 v + a_4 Pd \tag{21b}$$

The first result of this example is that it is possible to obtain a linear representation of a given process if only small variations around the setpoints are considered. A generalization is used for most control applications. Two ways are used for the representation of such a linear model. Using the state vector $\mathbf{x}$ and control vector $\mathbf{v}$:

(I) $\quad \dfrac{d\mathbf{x}}{dt} = A\mathbf{x} + B\mathbf{v} + C\mathbf{p} \qquad \mathbf{p}$ is the disturbances vector $\hfill (22a)$

$$(\text{II}) \quad \frac{d}{dt}\begin{pmatrix} \Omega_s \\ h \end{pmatrix} = \begin{pmatrix} \dfrac{-1}{a_1} & 0 \\ 0 & 0 \end{pmatrix}\begin{pmatrix} \Omega_s \\ h \end{pmatrix} + \begin{pmatrix} \dfrac{a_2}{a_1} & \dfrac{a_3}{a_1} \\ 1 & 0 \end{pmatrix}\begin{pmatrix} q_e \\ v \end{pmatrix}$$

$$+ \begin{pmatrix} 0 & \dfrac{1}{a_1} & \dfrac{a_4}{a_1} \\ -1 & 0 & 0 \end{pmatrix}\begin{pmatrix} q_s \\ \Omega_e \\ Pd \end{pmatrix} \tag{22b}$$

The state representation is obtained here with parameters derived from the knowledge modeling. In a more general case, matrices are characterized with unknown parameters:

$$\dot{x} = A\mathbf{x} + B\mathbf{v} \tag{23a}$$

$$Y = C\mathbf{y} \tag{23b}$$

where A, B, C are unknown matrices, and $\mathbf{y}$ is the output vector. In this example if $\mathbf{y} = \mathbf{x}$, then $C = I$. This is equivalent to

$$\sum_{i=1}^{n} a_i \frac{d^{(i)}y}{dt^i} = \sum_{k=1}^{m} - \left[\sum_{j=0}^{n} \left(b_{jk} \frac{d^{(j)}u_k}{dt^i} \right) \right] \tag{24}$$

where u_k is the kth input of the model; for example, in the case of the bioreactor, $u_1 = \Omega_e$, $u_2 = v$, $u_3 = p$, $u_4 = q_e$.

The second way for modeling a linear system is obtained using the discrete transfer function. Assuming the disturbances are negligible, $p = 0$, and $\Omega_s = 0$, the simplified model is

$$a_1 \frac{d^2 \, Tm}{dt^2} + a_2 \frac{d \, Tm}{dt} + a_3 \, Tm = b_1 q_e + b_2 v \tag{25a}$$

$$a_4 \frac{dh}{dt} = q_e - 0 \tag{25b}$$

Let consider the sampling time D, $t = t_0 + kD$, $k = 1, 2, 3, \ldots$. With $t_0 = 0$, $Tm(t) = Tm(kD)$ and $t \in [kD, (kM)D]$. Taking into account the classical finite difference scheme,

$$\frac{dh}{dt} = \frac{h(kD) - h[(k-1)D]}{D}$$

$$\frac{d^2 \, Tm}{dt^2} = \frac{Tm \, (kD) - 2 \, Tm \, [(k-1)D] + Tm \, [(k-2)D]}{D^2}$$

the following equations are derived:

$$A_1 \, Tm \, (kD) + A_2 \, Tm \, [(k-1)D] + A_3 \, Tm \, [(k-2)D]$$
$$= b_1 \, q_e(kD) + b_2 \, v(kD) \tag{26a}$$

$$A_4 h(kD) - A_4 h \, [(k-1)D] = q_e(kD) \tag{26b}$$

The delay operator q^{-1} is introduced in such a way that

$$Tm \, [(k-1)D] = q^{-1} \, Tm \, (kD) \tag{27}$$

The equations become

$$(A_4 - A_4 q^{-1})h(kD) = q_e(kD) \tag{28a}$$

$$(A_1 + A_2 q^{-1} + A_3 q^{-2}) \, Tm \, (kD) = b_1 q_e(kD) + b_2 v(kD) \tag{28b}$$

The discrete transfer function is defined as the ratio between output Tm (kD) or $h(kD)$ and considered inputs (control or disturbances), $q_e(kD)$, $v(kD)$:

$$Tm \, (kD) = \frac{b_1}{A_1 + A_2 q^{-1} + A_3 q^{-2}} \, q_e(kD) \tag{29a}$$

$$Tm \, (kD) = \frac{b_2}{A_1 + A_2 q^{-1} + A_3 q^{-2}} \, v(kD) \tag{29b}$$

and

$$h(kD) = \frac{1/A_4}{1 - q^{-1}} q_e(kD) \qquad (29c)$$

The general discrete transfer function of a dynamic model, represented in Fig. 9, is $A(q^{-1})Y(kD) = q^{-d}B(q^{-1})V(kD)$.

The simple relation presented in Fig. 9 is used classically as the dynamic model for a large number of studies and implementations of dynamic control for all kinds of processes, even in the food industry. An illustration of such a modeling approach is presented later, in Fig. 15, for the dynamic modeling of biscuit color versus air temperature variations. The knowledge model would be too complicated, and a discrete transfer function approach is sufficient. Because of the recurrent form of the model, the implementation into computer memory is simple.

B. Data-Driven, Nonlinear Dynamic Modeling

There are numerous methods for modeling the dynamics of processes. In the case of food processes, due to variability and nonlinear behavior of natural products, nonlinear methods are suitable. Neural networks are today recognized as good tools for dynamic modeling, and they have been extensively studied since the perceptron identification method was introduced [9]. The uses of such models include modeling without any assumptions about the nature of underlying mechanisms, the ability to learn even when the amount of available experimental data is low (particularly for dynamic models), and of course the ability to take into account nonlinearities and interactions between decision variables [10]. Recent results estab-

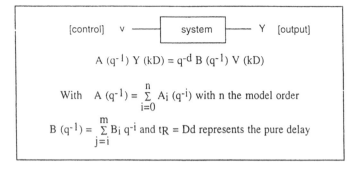

FIGURE 9 Discrete transfer function of a dynamic system.

lish that it is always possible to identify a neural model based on the perceptron model, with only one hidden layer, either for steady state models or dynamic ones (with recurrent models) [11]. For food processes, the applications of neural computations keep growing. Applications include fermentation, extrusion processes, filtration, drying, and for more complex problems, neural computations were used for the modeling of the sensory description of a food product based on composition [12].

1. The Principle

An artificial neural network is an association of elementary cells or "neurons" grouped into distinct layers and interconnected according to a given architecture. Each neuron computes the weighted sum of its inputs and transforms this sum using a transfer or "activation" function to give the output. The strength of the connection from neuron j to neuron i is given by the coefficient A_{ij}, called the weight of the connection. To represent phenomena with threshold or offset, additional dummy neurons with in-

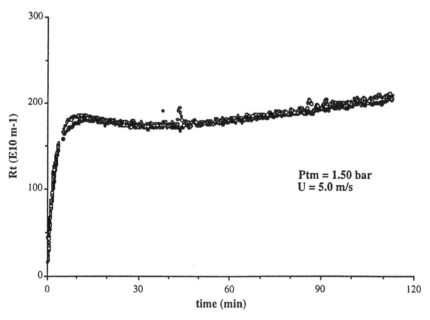

Figure 10 Dynamic modeling of the microfiltration of sugar products using neural networks. Hydraulic resistance is plotted as a function of time and operating conditions. The neural network model can predict the full trajectory of the flux during the batch process of microfiltration. (From [14].)

puts and a constant output of 1 are used. The strength of the connection from a dummy neuron to neuron i is given by the coefficient B_i, called the bias of neuron i.

The most popular architecture is the multilayer perceptron introduced by Rosenblatt [13]. Theoretical results [11] showed that most multi-input, multi-output functions encountered in practice can be well approximated by a feed-forward neural network with at least one hidden layer. Chapter 5 presents the design, identification, and procedures for neural dynamic modeling. The applicability of these approaches is illustrated in Fig. 10, which represents the dynamic modeling of the microfiltration process for a sugar product.

IV. BEHAVIOR-BASED MODELING AND SIMULATION FOR CONTROL PURPOSES

The most common controllers in food industries are programmable logic controllers (PLCs). For a long time, control engineers have developed methods for realization of logic functions able to control sequential and on–off processes. New methods were developed 20 years ago to implement software into PLC memories. These methods model both the process and the controller together. The principle is simply to establish the sequence of states characterizing the process and the events that permit evolution between states. Graphical methods are proposed. The graphic is in fact the dynamic model of the controlled process. The main tool is the sequential function chart, the principle of which is explained here.

Figure 11 presents the description of a mixing process for food ingredients. The principle is the weighing of product A and B in the vessel D, the weighing of B and C in the vessel E, and the final mixing, after transferring the contents of D and E into tank F. The control, and the related dynamic process, consists of the description of valves and agitator behaviors (the actuators) versus time. It is easy to define a state as an invariant behavior between two events. For example, during the transfer and weighing of product A, valve V1 is open. This valve remains open until the required weight is reached. For each invariant state, a graphical symbol is used: the square. The way between two states is represented using a line, and the attempted event is indicated as a small line associated with a logic condition. The next state is reached when both previous state and logic condition are realized. Figure 12 explains the principle of the model structure and parameters. The sequential chart model of the process is represented in Fig. 11.

Dynamic models of a sequential and event-based process, for control purposes are simply performed using sequential function charts. More

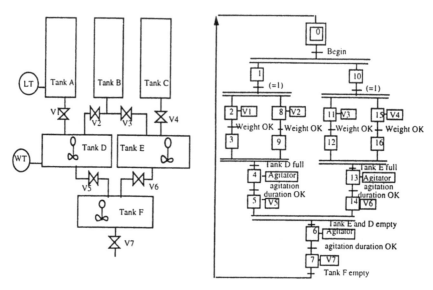

FIGURE 11 Example of a sequential chart model of a simple mixing station. The model here is nonoptimal and simplified. It describes all states that can be reached during the process and the conditions of evolution between states and time due to several events.

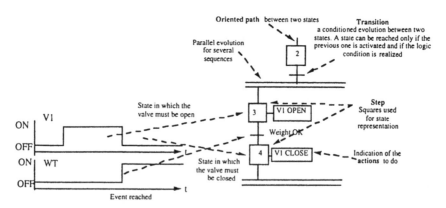

FIGURE 12 Principle of the sequential chart model for the example of valve V1 in the mixing process of Fig. 11.

advanced tools are available in this field of methods. The power of the approach is that the model can be easily implemented into PLC because PLC languages include specific instructions for sequential chart descriptions. The simulation of such a model is simple, and process behavior is studied easily.

Other tools, such as petri nets, are available but are more complicated and not so frequently used. If they are to permit optimal description, the simulation and implementation are quite difficult, and so they are not used in the food industry. Manufacturing process engineers develop other approaches called simulation languages based on several mathematical tools: Markov chains, waiting files, expert systems, etc. Mainly used for simulation purposes, the scope of applications of such tools is production management, scheduling, etc.

V. IDENTIFICATION IN DYNAMIC MODELS

As discussed in the introduction of the chapter, modeling is a two-step job in which both parts have the same importance. After the choice and the design of the model structure, one must identify the values of unknown parameters. This is the identification procedure. Of course several methods have been developed, and most of them are related to the model structures that are chosen. Identification is a general problem for modeling, but the dynamic models have specific requirements.

Two main methods are encountered for identification. In the first case, the unknown set of parameters is estimated only from measurements on the process. Estimation procedures are then used; the least squares method is the most popular tool even if the estimator introduces a bias. On the other hand, if it is possible to simulate the model, the model-based method can be used. The principle is presented in Fig. 13, a general and classical overview of identification in a model. Some differences could be encountered, but the main steps are represented. The same inputs (mainly decision variables) are introduced in the process through experiments and in the model for simulation. A criterion is evaluated as a function of the comparison between reality and model simulation. If a difference is observed, the model parameters are adapted, until the criterion is satisfied. This problem is exactly the same as all optimization problems. The adaptation loop can be performed off-line or on-line depending on the use of the model or the specific problems that could be encountered. For example, in the case of adaptive control, implemented with a model, or in the case of diagnosis applications, on-line identification is realized. In most cases, off-line identification is chosen. The main advantage of the model-based

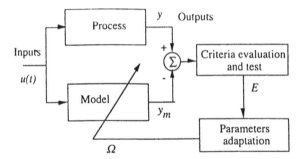

FIGURE 13   Principle of the identification of the parameters of a dynamic model.

methods is the large number of choices for adaptation algorithms. Both linear and nonlinear algorithms are available.

A.   The Choice of a Criterion

Before presenting the classical approaches for adaptation algorithms, one must choose a criterion to be minimized during the optimization procedure. Often, the criterion is defined as a simple quadratic function of the error between modeling and process outputs, or states. Using the notation presented in Fig. 13, the criterion E is, for example,

$$E = \int_0^{t_{max}} (y_m - y)^2 \, dt \qquad (30)$$

The search using a dynamic model implies the necessity of using a time integral of the error, not only a simple value obtained at a given time. Because the algorithm is implemented on a computer, a discrete form is preferred:

$$E = \sum_{i=1}^{N} [y_m(i) - y(i)]^2 \qquad (31)$$

where N is the number of points that are considered for identification (i.e., the time horizon), and i represents the sampling time.

B.   The Adaptation Algorithm

The least squares method is the most frequently used for the estimation of the set of parameters. This method is available when the model is a linear one. Write the model as $y_m(i) = \theta^T x(i)$, where θ^T is the vector of the parameter and $\mathbf{X}$ the vector of observation:

$$\mathbf{X}^T(i) = (y(i - 1), y(i - 2), \ldots, y(i - n), v(i - 1), \ldots ; v(i - m)) \quad (32)$$

The dynamic model presented in Fig. 9 is written as $y_m(i) = a_1 y(i - 1) + a_2 y(i - 2) + \ldots + a_n y(i - n) + b_1 v(i - 1) + \ldots + b_m v(i - m)$, where $y_m(i)$ is only a function of the measured outputs and inputs. The problem is then the estimation of the set of parameters θ^T. The criteria E could be rewritten as $E = \sum_{i=1}^{N} [\theta^T \mathbf{x}(i) - y(i)]^2$. The best choice of the set of parameters is then characterized by $dE/d\theta = 0$. That means $\sum_{i=1}^{N} [\theta^T \mathbf{x}(i) - y(i)] \cdot \mathbf{x}(i) = 0$. The solution is

$$\theta = \left[\sum_{i=1}^{N} \mathbf{x}(i)\mathbf{x}(i)^T \right]^{-1} \left[\sum_{i=1}^{N} y(i) \cdot \mathbf{x}(i) \right] \quad (33)$$

This quantity is easy to calculate, and provides a good, though biased, estimate of the parameters θ^T. Often this estimation is used for initialization of other algorithms. Many other methods are available, the most popular after the least squares method being the instrumental variables method [15]. In the case of more complicated models, including for example, the modeling of noise, of measured disturbances, the generalized least squares method becomes available. The principle is the same as least squares method, but the vectors are more complicated.

A real-time implementation of an estimation algorithm is sometimes necessary. Some solutions are proposed. The general form of these algorithms is, for example, the recursive least squares formulation, which is the most popular for adaptive implementation of a controller. The principle is the step-by-step modification of the vector θ^T, using a recursive formula, only on the basis of measured values:

$$\theta^T(i) = \theta^T(i - 1) + \frac{P(i - 1)[y(i) - \theta^T(i - 1) \cdot \mathbf{x}(i)] \cdot \mathbf{x}(i)] \cdot \mathbf{x}(i)}{\lambda + \mathbf{x}(i)^T \cdot P(i - 1) \cdot \mathbf{x}(i)} \quad (34)$$

$$P(i) = P(i - 1) - \frac{P(i - 1) \cdot \mathbf{x}(i) \cdot \mathbf{x}(i)^T \cdot P(i - 1)}{\lambda + \mathbf{x}(i)^T \cdot P(i - 1) \cdot \mathbf{x}(i)} \quad (35)$$

If λ is chosen to be less than 1, the values in the past may be forgotten.

The principle of the *model-based method* is in fact the same, but the estimation of the present value y_m of the output is made on the basis of the predicted values of y_m and not from the measured values y. This method can be shown as a more powerful one. It is possible to perform both linear or nonlinear identification using appropriate algorithms. Classical approaches use gradient (first or second order), heuristic (such as the simplex method), or other methods [16].

Numerous software packages are available for performing such calculations. The most popular, and most powerful, is the Mathlab library.

C. The Nature of Experiments

If a steady state model is suitable, a database is easily obtained, performing simple and classical experiments. Well-known tools are helpful for the design of experiments (experimental design theory). When a dynamic model is the object of the work, it is more difficult to design the best experiment because there is a close relation between the experiments that are carried out and the performance of the model. If the frequency domain is considered, the process should be excited over the full range of frequencies that are supposed to be encountered during the control. Generally, this range is not known exactly. Nevertheless, methods are proposed to find the best excitation to the variable for performing an exploration of all the frequencies probably important for the control problem. Simple step response tests are not sufficient, and the amplitude of the excitation is often too high for industrial processes (the induced disturbance is not suitable for the operating conditions and constraints). The accuracy of such an approach is recognized to be poor, one, and model validation can be difficult.

Pseudo-random binary sequences (PRBS) are the most commonly used technique. It consists in the design of a sequence of process excitation inputs characterized by varying time periods. Most often the excitation is a step-periodic one, as presented in Fig. 14. In fact an iterative procedure is probably the most efficient. Some preliminary experiments permit definition of the range of frequencies that are important for the process. In a second step, an optimal sequence of excitations is performed

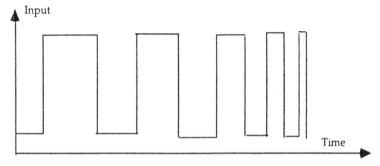

Figure 14   Example of an excitation signal for identification in a dynamic model.

and a first identification is realized. Then, the analysis of the model permits the proposal of a more accurate experimental protocol to obtain the best set of parameters, or eventually to design a better model structure. The amplitude of the excitation signal is not necessarily high. It depends on the signal-to-noise ratio. The control engineer should take a critical view of the choice of the identification procedure to optimize the amount of testing and the cost of the test for obtaining the best dynamic model.

D. Three Specific Problems in Dynamic Models Identification

Because a dynamic model is a representation versus time of the process behavior, some specific problems should be examined. The first problem we must deal with is the choice of the sampling time. If many computers are used, the power of the computer permits the use of very small sampling times. It has been demonstrated (Shannon theorem, [15,16]) that a bad choice of sampling period comprises two problems. First, if the sampling period is too large, information can be lost. A classical example is the case of a sinusoidal periodic signal for which a sampling period equal to the signal period is chosen. The observation of the recorded signal is a constant value. A lower sampling rate implies a loss of information. Second, it is established that a shorter sampling period could introduce disturbed phenomena and create noise, which is not representative of the original signal. The Shannon theorem allows the choice of the best sampling period [15]. The sampling frequency must be greater than twice the system's bandpass.

A practical way of choice is proposed by Landau [17], who recommends taking four time samples during a time constant of the process (between two and nine samples during a rising time after a step excitation of the input variable). Such a choice gives, for example: 1 second for flow rate control, 20 to 45 seconds for dryers, 10 to 180 seconds for distillation columns.

The second difficult problem, in the case of black box model identification, is the determination of the pure delay value. Due to the use of computers, and the work with sampled models, the pure delay is often chosen as a constant multiple of the sampling time. A test and errors procedure is the best way for choosing the unknown constant value, except if a simple step response permits estimation of the pure delay of the process.

The third problem is related to the difficulty of interaction between noise and signal. If the identification is performed on a noisy signal, the results are poor. The use of filtering is necessary, but a strong filtering

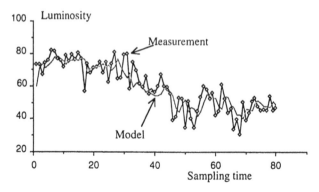

FIGURE 15   An example: the identification of the luminosity of biscuits measured using an on-line color sensor and modeled as a function of air temperature inside the continuous oven.

operation could introduce a loss of information. The simplest filter is a moving average of the sensor signal. Often this is enough. In most complicated cases, specific filters must be studied.

E. An Example

An illustration of the identification of a dynamic model of a food process is presented in Fig. 15. It is the case of the evolution of the luminosity of a biscuit measured at the end of a continuous baking oven and modeled as a function of air temperature in a given zone of the oven. The original signal was filtered; the sampling period is 30 s. A second-order transfer function, with no pure delay, is used as the model. Identification is performed using the PIM software [17]. The adaptation algorithm is a least squares method, performed off line.

VI. CONCLUSIONS

Modeling is an important task for control engineers. Numerous types of works and of approaches are available for the several objectives. The ability to design a dynamic model of the process is an important condition in deciding the performance achievable through automatic control. This chapter illustrates several applications in which models are developed and used. Two steps are highlighted. The first concerns the use of knowledge for the design of the model structure. The second is the realization of experiments and the calibration of the model's unknown parameters. Both steps are important. If the work is not well prepared for identification, it is not possible to obtain a reliable quantitative model.

The choice of a particular method is not easy. Figure 16 summarizes the main ways proposed here and some classes of applications. For design and tuning of controllers, for classical control, the main approach is data-driven modeling, mainly using linear models (transfer function or state representation). It is the form in which the greatest number of algorithms are available. Many software packages are available for parameter identification, even for the best design of experiments. A trial-and-error based method is generally used for the design of the model structure.

If more advanced control functions are appropriate, it is better to design a knowledge-driven model. The range of applicability of the model is greater, and it permits, for example, studies of transient phenomena like startup or shutdown of the process. In the case of batch processes, these approaches seem to be better, even when applications using adaptive linear black box modeling are available.

New tools, coming from artificial intelligence, are becoming available for quantitative nonlinear modeling. The range of applications is the same as in all the kinds of data-driven models. Probably, in the near future, the ability to be combined with such dynamic models and real-time optimization methods, will allow better-performing controllers capable of taking into account more constraints about product quality.

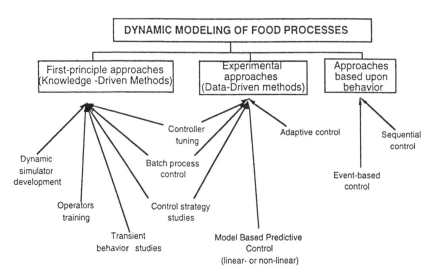

FIGURE 16   Examples of applications of dynamic modeling and simulation for food processes. Classification of the recommended ways for some types of applications.

The third way for the dynamic modeling of food processes is based on the analysis of the controlled system. Such approaches are implemented and used. The correspondence between model and implementation into PLC is an important part in the choice of such methods, specifically for event-based control applications.

Modeling is the main part of controller design and tuning. It is a job that requires cooperation between process engineer, control engineer, and people working on the line for experiments and implementation of models. Much progress is realized during process control studies because of the variety of information brought forth during dynamic modeling and simulation. It is thus important to give attention to the modeling part of the work in automatic process control.

NOMENCLATURE

All units are SI.

A	Surface
A_i, a_i	*Parameters*
Cp	*Thermal capacity*
H, h	Level
k	Sampling discrete time
Pd	Energy losses
Q, q	Flow rate
R	Electric resistor (Ohm)
T	Temperature
Tm	Measured value of temperature
V	Tension
ρ	Density
q^{-1}	Delay operator
Ω	Error
x	State vector
v	Command vector
y	Output vector
r	Disturbance vector

Subscripts

e	Input
s	Output
R	Related to heat resistor

REFERENCES

1. G. Trystram and F. Courtois, Food processing control: Reality and problem, *Food Research International 27*:173 (1994).

2. S. Banoune and D. Depeyre, Dynamic simulation and real time expert system for food freezing process control, *Automatic Control of Food and Biological Processes* (J. J. Bimbenet, E. Dumoulin, and G. Trystram, eds.), Elsevier Applied Science, Amsterdam, 1994, p. 159.

3. M. Bouaramni, A. Isambert, and D. Depeyre, Dynamic simulation for sugar crystallisation plant control. A comparative study of specific and generalised approaches, *Automatic Control of Food and Biological Processes* (J. J. Bimbenet, E. Dumoulin, and G. Trystram, eds.), Elsevier Applied Science, Amsterdam, 1994, p. 167.

4. Z. Bubnik and P. Kaldek, Modelling and dynamic simulation of cooling crystallisation of sucrose, *Automatic Control of Food and Biological Processes* (J. J. Bimbenet, E. Dumoulin, and G. Trystram, eds.), Elsevier Applied Science, Amsterdam, 1994, p. 175.

5. J. J. Bimbenet, G. Trystram, A. Duquenoy, F. Courtois, M. Decloux, M. L. Lameloise, F. Giroux, and A. Lebert, Dynamic modelling and simulation of food processes, *Developments in Food Engineering* (T. Yano, R. Matsuno and K. Nakamura, eds.), Blackie Academic and Professional, London, 1994, p. 981.

6. G. Trystram and S. Pigache, Modelling and simulation of a large scale air lift fermenter, *Computer Chemical Engineering*: S171 (1993).

7. A. L. Raoult-wack, F. Petitdemange, F. Giroux, S. Guilbert, G. Rios, and A. Lebert, *Drying Technol.* 9(3):613 (1992).

8. F. Giroux, S. Guilbert, and G. Trystram, Dynamic modelling of fruit dehydration, *Developments in Food Engineering* (T. Yano, R. Matsuno, and K. Nakamura, eds.), Blackie Academic and Professional, London, 1994, p. 421.

9. D. Rumelhart and D. Zipner, Feature discovering by competitive learning, *Cognitive Sci.* 9:75 (1985).

10. C. M. Bishop, Neural networks and their applications, *Rev. Sci. Instrum.* 65(6):1803 (1994).

11. K. Hornik, M. Stinchcombe, and H. White, Multilayer feedforward networks are universal approximators, *Neural Networks* 2:359 (1989).

12. I. Bardot, N. Martin, G. Trystram, J. Hossenlopp, M. Rogeaux, and L. Bochereau, A new approach for the formulation of beverages, Part II; Interactive automatic method, *Leben Wiss. u. Technol.* 27:513 (1994).

13. F. Rosenblatt, The perceptron: A probabilistic model for information storage and organisation in the brain, *Psychol. Rev.* 65:386 (1958).

14. M. Dornier, M. Decloux, G. Trystram, and A. Lebert, Dynamic modeling of crossflow microfiltration using neural networks, *J. Membrane Science* 98: 263 (1995).

15. K. Astrom and B. Wittenmark, *Computer Controlled Systems: Theory and Design*, Prentice Hall, New York, 1984.

16. P. Eykoff, *System Identification*, Wiley, New York, 1974.

17. I. D. Landau, *Identification et commande des systèmes*, Hermes, Paris, 1994.

4

Computerized Control Systems: Basics

Gordon L. Hayward
University of Guelph, Guelph, Ontario, Canada

I. INTRODUCTION

Before a process is designed, the goals to be achieved by the process are specified. The design parameters, however, are chosen to allow for disturbances and unforeseen circumstances. The process is, therefore, overdesigned by some factor of safety to ensure operation under worst-case operating conditions. Under normal conditions, this could result in an overprocessed product, both increasing the production costs and reducing the value of the product.

The goals to be achieved are typically specified in terms of *process variables* related to the state of the process or to product quality attributes. Process control systems included in the design adjust the process to keep these process variables at the desired levels, or *setpoints*, by changing one or more of the process inputs. These inputs, or *manipulated variables*, may be varied *manually*, where an operator adjusts the process, or *automatically*, where some form of computer does the adjustment. While an operator is capable of very complex reasoning, automatic controllers can react much faster to process upsets and can give more consistent results. The choice must be based on the requirements of the process.

Example 1: A grain dryer is designed to handle the largest quantity of the wettest grain expected. A safety factor is also included. Under normal conditions, the dryer must be operated at less than its maximum capacity or else the grain will be overdried, cracked, or perhaps scorched. The operating conditions are decided by checking the moisture content (process variable) of the dried grain, comparing it with the required moisture content (setpoint), and adjusting the fuel flow (manipulated variable) to the burner. The fuel flow will almost always be less than its maximum value.

Deciding how to set the manipulated variable can be done in one of two ways. In *open loop* or *feed-forward* control, the disturbances are sensed and the process set to an appropriate operating condition before the process variable changes. This can provide effective control; however, unmeasured disturbances will not be corrected. *Closed loop* or *feedback* control operates by measuring the process variable and taking corrective action based on the departure from its setpoint (*process error*). Since this corrective action also affects the process variable, a loop is formed, as shown in Fig. 1. All of the disturbances that can affect the process variable are included in the loop. Indeed, minor design shortcomings are also corrected by the feedback action. Feedback control is, therefore, the most used type of control system.

The characteristics of a feedback control system must match those of the process to achieve good performance. If the control system is too

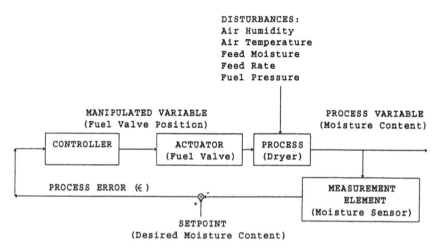

FIGURE 1 Moisture control loop.

aggressive—that is, takes large control action quickly for a small disturbance—instability may result. On the other hand, if it is too sluggish, the disturbance will not be corrected quickly. This chapter presents some of the classical approaches to control system analysis and design and extends these to computerized controllers.

II. PERFORMANCE ANALYSIS

One of the main tools used in the analysis and design of control systems is the block diagram. Figure 1 is a schematic diagram of a control system. More information may be conveyed by placing mathematical models of the process elements in the appropriate blocks. These operate on the input to the block to produce its output; the contents of the block are thus a *transfer function* relating its output to its input.

The mathematical models may be based on a theoretical analysis of the process element within the block or on an experimental measurement of the dynamics of the process element. In both cases it is common to transform the data into the Laplace domain. The definition of the *Laplace transform* is

$$F(s) = \int_0^\infty f(t)e^{-st}\,dt \tag{1}$$

where $F(s)$ is the function in the Laplace domain and $f(t)$ is the function in the time domain. The integration operation removes the time terms from the function, leaving only s, the analogous Laplace parameter. Laplace transforms of various functions are widely published; a few of the more important ones are given in Table 1. It can be seen from the table that the derivative and integral operators are transformed to simple multiplication and division by s, respectively. Transfer functions in the Laplace domain become simple algebraic operations:

$$\frac{\text{Output}(s)}{\text{Input}(s)} = G(s) \tag{2}$$

where $G(s)$ is the transfer function. The Z transform is analogous to the Laplace transform, but it is for sampled data (digital) systems rather than continuous (analog) systems.

Purists will note that the initial condition terms of the Laplace transforms in Table 1 have been omitted. This is deliberate, because control systems are usually analyzed in terms of *perturbation variables*. These represent the departure from a chosen steady state, usually taken as the setpoint, rather than the absolute value of the variable. By assuming the system to be at the steady state before any change occurs (at times less

TABLE 1 Selected Transforms

	Time domain	Laplace domain	Z domain
Derivative	$\dfrac{df(t)}{dt}$	$sF(s)$	
Integral	$\int f(t)\,dt$	$\dfrac{1}{s}F(s)$	
Step function	0 for $t < 0$ 1 for $t > 0$	$\dfrac{1}{s}$	$\dfrac{1}{1 - z^{-1}}$
Sine function	$\sin(at)$	$\dfrac{a}{s^2 + a^2}$	$\dfrac{z^{-1}\sin(aT)}{1 - 2z^{-1}\cos(aT) + z^{-2}}$
Exponential	e^{-at}	$\dfrac{1}{s + a}$	$\dfrac{1}{1 - e^{aT}z^{-1}}$
First order	$\dfrac{1}{a}(1 - e^{-at})$	$\dfrac{1}{s(s + a)}$	$\dfrac{(1 - e^{-aT})z^{-1}}{a(1 - z^{-1})(1 - e^{-aT}z^{-1})}$
Dead time	$f(t - D)$	$e^{-sD}F(s)$	

than 0), all of the perturbation variables have initial values of 0, and thus the initial condition terms are omitted. The steps involved in setting up a model of a process are illustrated in the following example:

Example 2: Consider a steam heated stirred tank with constant level and liquid flow where the inlet varies in temperature and the steam flow is used to control the tank temperature. This system is shown in Fig. 2a. The theoretical model may be developed by following these steps:

1. Write the heat balance equation:

$$F\rho C_P T_{IN} \quad - \quad F\rho C_P T_{OUT} \quad + \quad \begin{array}{c} F_{STEAM} \\ \Delta H_{VAP} \end{array} \quad = \quad \rho V C_P \frac{dT_{OUT}}{dt}$$

Inlet sensible heat	Outlet sensible heat	Condensation heat from steam	Heat accumulation

where F is the volume flow of liquid through the tank, C_P is the heat capacity of the liquid, ρ is the density of the liquid, ΔH_{VAP} is the heat of vaporization (condensation) of the steam, and V is the volume of liquid in the tank. T_{IN} is the inlet temperature (disturbance), T_{OUT} is the outlet temperature (process variable), and F_{STEAM} is the mass flow of steam (manipulated variable).

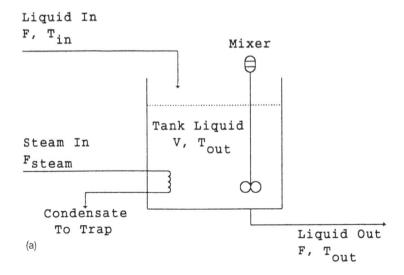

(a)

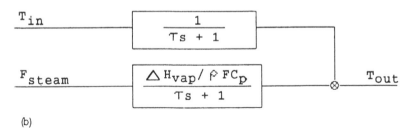

(b)

FIGURE 2 (a) Steam heated stirred tank. (b) Block diagram.

2. Subtract the steady state equation to get perturbation variables:

    $F\rho C_P(T_{IN} - T_{IN,S}) - F\rho C_P(T_{OUT} - T_{OUT,S})$

    $$+ (F_{STEAM} - F_{STEAM,S}) \Delta H_{VAP} = \rho V C_P \frac{d(T_{OUT} - T_{OUT,S})}{dt}$$

    where the ,S subscript means at steady state where the time derivatives are 0. (The steady state term appears inside the derivative since $d(T_{OUT,s})/dt = 0$.)

3. Take the Laplace transform:

$$F\rho C_P[T_{IN}(s) - T_{OUT}(s)] + F_{STEAM}(s) \Delta H_{VAP} = s\rho V C_P T_{OUT}(s)$$

where each variable is the Laplace transform of the corresponding perturbation variable.

4. Rearrange so that the terms can be put into a block diagram:

$$T_{OUT}(s) = \frac{1}{\tau s + 1} T_{IN}(s) + \frac{\Delta H_{VAP}/F\rho C_P}{\tau s + 1} F_{STEAM}(s)$$

where τ is the time constant of the process, which is equal to the liquid residence time, V/F.

5. Set up the block diagram as in Fig. 2b. Note how this diagram matches the process block in Fig. 1.

When the differential equation from step 1 contains a nonlinear term, its Laplace transform cannot be found. In this case, the term may be expanded as a Taylor series about the same steady state used in step 2. The result is a constant term that will be eliminated in step 2, a perturbation variable term, and higher-order terms. The higher-order terms may be truncated if the departure from the steady state is small. Fortunately, the operation of the control system keeps the process error small, justifying this assumption.

Example 3: Consider the flow of liquid out of a tank given by $kh^{1/2}$, where h is the liquid head in the tank and k is the orifice constant of the outlet. Linearizing this term by a Taylor series gives $kh_0^{1/2} + 0.5kh_0^{-1/2}(h - h_0) + \cdots$. The first term is the steady state part, which will be eliminated in step 2. Because h is close to h_0, the higher-order terms will be very small and may be neglected.

Experimental measurements to determine the transfer function can be carried out by sending a known input signal to a process element and measuring the output. Because the analysis is based on perturbation variables, this input may be a small departure from normal operating conditions. The measurement may thus be carried out on-line with an operating process. The transfer function is determined from the input and output signals; note, however, that the simple ratio in Eq. (2) is only valid in the Laplace domain. Determining the transfer function from time domain data is more complicated.

One of the most used methods is also the simplest. A step disturbance is applied to the input and the output fitted to a predefined function by nonlinear regression techniques. The function used is the time domain solution of the output for a particular transfer function. This could require many trial functions; however, the determination of the exact transfer function is, fortunately, typically unnecessary.

A. Step-Response Performance Criteria

The inversion of a Laplace domain transfer function into the time domain may be done in several ways; however, the partial fraction expansion is the most generally useful. The Laplace function is expanded as a series of terms:

$$F(s) = \frac{N(s)}{D(s)} = \frac{A_1}{s - R_1} + \frac{A_2}{s - R_2} + \cdots \tag{3}$$

where $F(s)$ is the response function in the Laplace domain and $N(s)$ and $D(s)$ are its numerator and denominator expressed as polynomials. The A terms are constants, and the R terms are the roots of the characteristic equation $D(s) = 0$. Since the Laplace transform is linear, each term may be inverted into the time domain individually and the results added to give the overall result.

The roots may be real or complex. A real root gives an exponential solution, while complex roots occur as conjugate pairs that give sinusoidal solutions. The real part of the root determines the decay (or increase). Negative real parts give a decaying solution, while positive real parts give an exponential increase in the response. Information regarding the time domain response may be obtained from the roots without having to do the inversion into the time domain.

Most systems give a response resembling that of a first- or a second-order system. More complex systems usually have terms that decay quickly, leaving a dominant response that can be used to characterize the system. This type of approximation is very useful in describing control systems.

B. First-Order Systems

A first-order system is one that contains one storage element or capacitor. Example 2 presents such a system. The general first-order transfer function is

$$G(s) = \frac{k}{\tau s + 1} \tag{4}$$

where $G(s)$ is a transfer function, k is the gain, and τ is the time constant. Figure 3 shows the response of a first-order system to a step input. The *time constant* is a measure of how fast the system responds, and the gain

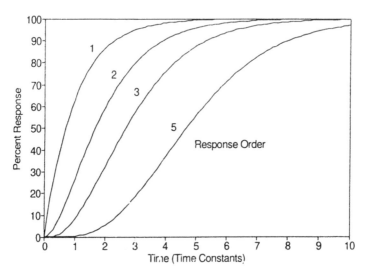

FIGURE 3   Real root response.

a measure of how much it responds. As can be seen, most of the response of a first-order system is complete after 5 time constants.

Figure 3 also shows some higher-order systems made up of first-order systems in series. These are the most common form of response found in physical systems. The initial "tail" is the result of the higher order, but rather than fitting high-order models, a *first-order with dead time* approximation is often used. Here, the tail is modeled as a delay after the step input occurs and the response is modeled as a first-order system with a fitted time constant.

C. Second-Order Systems

The higher-order system responses shown in Fig. 3 include second-order systems with real roots; a more interesting case, however, is the second-order system with complex roots. Physically, this corresponds to two storage elements such as capacitance and inductance in electrical systems, or to spring tension and inertia in mechanical systems. Energy is transferred back and forth between these, giving oscillatory behavior. This is also the typical response of closed loop control systems, where the system and controller act as two storage elements.

The general second order transfer function is given by

$$G(s) = \frac{k}{\tau^2 s^2 + 2\tau\zeta s + 1} \tag{5}$$

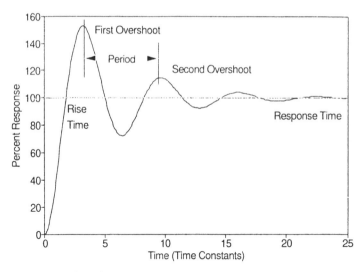

FIGURE 4 Complex root response.

where ζ is a *damping coefficient*. When ζ is greater than 1, the system is overdamped and behaves as two first-order systems in series, as above. When ζ is between 0 and 1, the system is underdamped, and overshoot behavior is observed.

A typical second-order step response is shown in Fig. 4. Several of the features of the response are also listed in Table 2[1]. The first *overshoot* is the amount the response goes over the final response value. This can be a critical specification in the design of a controller where the process variable cannot be allowed above a set maximum. A good example is a heating process where exceeding a maximum temperature will cause protein coagulation. The *decay ratio* is the ratio of the second overshoot peak to the first. This gives a measure of how fast a disturbance will die out. From the relations in Table 2, a specification of ζ and τ can be translated into the overshoot amount, or an overshoot can be translated into values of ζ and τ.

TABLE 2 Underdamped Response Parameters

First overshoot	$\exp(-\pi\zeta/\sqrt{1 - \zeta^2})$
Decay ratio	$\exp(-2\pi\zeta/\sqrt{1 - \zeta^2})$
Oscillation period	$\sqrt{1 - \zeta^2}/\tau$

TABLE 3   Integral Performance Measures

ISE	Integral Squared Error	$\int \epsilon^2 \, dt$
IAE	Integral Absolute Error	$\int |\epsilon| \, dt$
ITAE	Integral Time · Absolute Error	$\int |\epsilon| \, t \, dt$

The *rise time*, wherein the response first reaches the final value, is also a critical specification. This is the main parameter that determines the speed of a control system. To decrease the rise time, stronger initial control action is required. This will, however, increase the overshoot. The *response time* required for the oscillations to become less than some percentage of the total response, say 5%, will also increase with more aggressive control action. Control design is, therefore, a trade-off between fast response and good overshoot behavior.

D. Integral Performance Measures

It is sometimes necessary to characterize the entire response by a single number. This is particularly useful in adjusting controllers to optimize performance. To generate this number, the area between the response

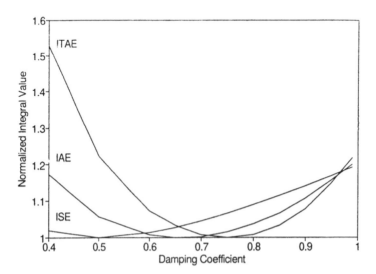

FIGURE 5   Integral performance measures.

curve and the 100% response (final state) shown in Fig. 4 is obtained by integration. Since the area has positive and negative components, some form of absolute value must be taken. Three integral methods are summarized in Table 3, but this list may be extended to include other appropriate functions. Each of these provides a different weighting of the response. ISE is most affected by large overshoots, while ITAE is most affected by a slow response or by sustained oscillations. Figure 5 is a plot of these measures for the second-order system shown in Fig. 4 with different damping coefficients. Each gives a different optimum damping coefficient, reflecting the trade-off between response speed and overshoot.

III. FEEDBACK CONTROL SYSTEMS

Figure 6 presents a simplified block diagram of a generic control system. The task in designing a control system is to determine a controller transfer function Gc(s) that will give the required performance in compensating for load disturbances (*regulatory* control) or setpoint changes (*servo* control). The most common controller is the *PID* or proportional–integral–derivative system, which can be tuned to match many different processes.

A. Proportional Control

The simplest feedback controller type takes the error signal, multiplies it by a constant gain, and uses the result as the manipulated variable. This works well in many applications; however, it has a serious drawback: offset. *Offset* is the difference between the final steady value of the process variable and the setpoint.

Example 4: Consider the process in Fig. 6 to be the steam heater discussed in Example 2 with a proportional controller with a gain of Kc. If the tank is at its setpoint temperature, the error is 0 and so the manipulated variable, or steam flow, is also 0. This cannot maintain the setpoint; therefore, some error is required to maintain some steam flow. This can also be shown mathematically using the Laplace final value theorem for a step input in the setpoint:

$$\lim_{s \to 0} s \quad \frac{1}{s} \quad \frac{1}{s(\tau s + 1 + Kc)} = \frac{1}{1 + Kc}$$

| Final value theorem | Step input | Closed loop transfer function | Steady state process variable |

The change in setpoint was 1, while the result is less than 1. This is offset.

As the gain of the controller is increased, the same control action will be produced for smaller errors, so that the offset is reduced. Unfortunately, increasing the gain has two rather nasty consequences. The first is an increase in the sensitivity to noise. All measurements have *noise*, comprising a random component generated by thermal effects in the measuring system or by such phenomena as fluid turbulence, and a manmade component, which is more properly termed *interference*. As the controller gain is increased, even small amounts of noise can cause large control actions.

The second consequence is *instability*. The closed loop transfer function of Fig. 6 with a proportional controller, $Gc(s) = Kc$, is

$$\text{Goverall} = \frac{Kc\ Gp(s)}{Kc\ Gp(s) + 1} \tag{6}$$

As above, the form of the response depends on the roots of the characteristic equation

$$Kc\ Gp(s) + 1 = 0 \tag{7}$$

The real parts of the roots determine the exponential decay terms; however, if they are positive, the decay will become an exponential increase. This can, of course, have catastrophic effects on a process as the process variable increases without bound. The Routh test [1] can be used to determine the maximum gain that can be used while maintaining stability. As the gain is increased toward this limit, the offset decreases, but the overshoot increases.

B. Integral Control

To eliminate offset, *integral* action may be added to the controller. By integrating the error, sustained errors such as offset are reduced as the integral moves the manipulated value to achieve the setpoint. Here, the error is 0, and so the integral no longer changes. In Example 4, the required steam flow at the setpoint is the value of the integrated error term.

The integral time, τ_I, is a constant that determines how fast the integration occurs. The shorter the time, the faster the offset is removed, but the response becomes more oscillatory and overshoots more. This effect is somewhat alleviated by decreasing the proportional gain, which is possible since the offset has been removed by the integral action. *PI* controllers are perhaps the most popular form of continuous controller.

C. Derivative Control

To reduce overshoot, *derivative* action may be added to the controller. The derivative of the error provides a prediction of the size of an overshoot; that is, large peaks have large initial slopes. By opposing the change, derivative action reduces the overshoot. The amount of opposition is set by the derivative time, τ_D. In position control systems, the proportional term is the distance away from the target and the derivative term is the velocity of approach. By slowing the approach, overshoot is reduced.

IV. PID CONTROLLER TUNING

The PID controller transfer function is given by [2]

$$Gc(s) = Kc \left(1 + \frac{1}{\tau_1 s} + \tau_D s \right) \tag{8}$$

By combining the three modes, a good compromise among the features of each can be reached. Proportional control reduces the error, integral control eliminates the offset, and derivative control reduces the overshoot. To match the controller to a particular process it must be tuned, that is to say, Kc, τ_1, and τ_D must be chosen.

Several classical methods have been published [3–5], but these are based on approximations of the process involved. In a real situation, these may be taken as starting points and optimized manually when the physical process is operating.

A. Ziegler–Nichols Tuning

Ziegler and Nichols [3] proposed one of the first general tuning methods. Using a closed loop test where the controller is set to a proportional-only mode, they proposed that the controller gain be increased to the point where oscillations are just sustained. This is the controller gain at the limit of stability, or Ku. For a proportional controller they suggested that the operating gain be half of Ku, where they postulated that the decay ratio would be 0.25. The exact nature of the process that they were considering was not specified; however, this information is imbedded in the experimental determination of Ku. To determine the integral and derivative times, the period of the oscillations, Pu, at the point of sustained oscillations is also required. Their recommended settings are given in Table 4. These settings are consistent with the foregoing description of the operation of the three control modes. The gain of the PI controller is reduced from that of the P controller because the integral mode increases the overshoot. The gain of the PID controller is increased above that of the P controller because the derivative action reduces the overshoot.

Ziegler and Nichols also presented a set of tuning parameters based on a first-order with dead time process model determined from the re-

TABLE 4 Ziegler-Nichols Controller Settings
Based on Closed Loop Tests

Control mode	Kc	τ_I	τ_D
P	0.5Ku		
PI	0.45Ku	Pu/1.2	
PID	0.6Ku	Pu/2	Pu/8

sponse to a step input or process *reaction curve*. The model equation is given by

$$G(s) = \frac{Kp\ e^{-Ds}}{\tau s + 1} \tag{9}$$

where Kp is the process gain, τ is the process time constant, and D is the dead time. These parameters were obtained from a graphical method wherein a line is drawn tangent to the inflection point of the response curve. This is often less than satisfactory [5]. A better quick approach is to use a trial-and-error method using a spreadsheet software package. A rigorous nonlinear regression is, of course, the best method, but it is also more difficult and time consuming. These tuning parameters are presented in Table 5.

B. Cohen–Coon Tuning

Cohen and Coon [4] used a first-order with dead time process model to develop a set of initial controller settings. The controller settings were derived by putting Eqs. (8) and (9) into the controller and process blocks of Fig. 6. This gives a response equation in the three unknowns, Kc, τ_1,

TABLE 5 Ziegler–Nichols Controller Settings Based on Process Reaction Curves

Control mode	Kc	τ_I	τ_D
P	τ/Kp D		
PI	0.9τ/Kp D	3D	
PID	(1.2 to 2)τ/Kp D	2D	0.5D

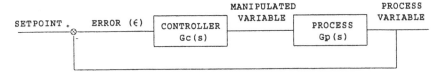

FIGURE 6 General control loop.

and τ_D. To provide two other equations, they designed the controller to give a decay ratio of 0.25 and to have a maximum stable gain. The result is given in Table 6. There is some similarity between the entries in this Table 6 and those in Table 5.

C.   Integral Performance Tuning

Miller et al. [5] have generalized the Ziegler–Nichols formulae (Table 5) into three equations:

$$Kc = \frac{1}{Kp} U \left(\frac{D}{\tau}\right)^{-V} \tag{10}$$

$$\tau_I = \tau W \left(\frac{D}{\tau}\right)^{X} \tag{11}$$

$$\tau_D = \tau Y \left(\frac{D}{\tau}\right)^{Z} \tag{12}$$

The coefficients are given in Table 7 for the three integral performance measures discussed above. The optimum values were obtained by com-

TABLE 6 Cohen–Coon Controller Settings

Control mode	Kc	τ_I	τ_D
P	$\dfrac{\tau/D + 0.33}{Kp}$		
PI	$\dfrac{0.9\ \tau/D + 0.082}{Kp}$	$\dfrac{\tau[3.33D/\tau + 0.3(D/\tau)^2]}{1 + 2.2D/\tau}$	
PID	$\dfrac{1.35\tau/D + 0.27}{Kp}$	$\dfrac{\tau[2.5D/\tau + 0.5(D/\tau)^2]}{1 + 0.6D/\tau}$	$\dfrac{\tau[0.37D/\tau]}{1 + 0.2D/\tau}$

TABLE 7   Integral Performance Optimization Coefficients

Control mode	Integral method	Coefficient value					
		U	V	W	X	Y	Z
P	IAE	0.902	0.985				
	ISE	1.411	0.917				
	ITAE	0.490	1.084				
PI	IAE	0.984	0.986	1.644	0.707		
	ISE	1.305	0.959	2.033	0.739		
	ITAE	0.859	0.977	1.484	0.680		
PID	IAE	1.435	0.921	1.139	0.749	0.482	1.137
	ISE	1.495	0.945	0.917	0.771	0.560	1.006
	ITAE	1.357	0.947	1.176	0.738	0.381	0.995

puter simulation using a first-order with dead time process model. Of these methods, the ITAE minimization was shown to be the best, giving better performance than Ziegler–Nichols, Cohen–Coon, or the other integral methods.

V.   DIGITAL CONTROL IMPLEMENTATION

Control systems implemented using digital computers have many advantages over the older analog units. Digital computers can use more complex control schemes to achieve better performance, can include alarms or fail safe procedures, and can log process data for report generation.

The components of a *direct digital control* (*DDC*) loop are shown in Fig. 7. The heart of the system is the control computer. In the past, these were general-purpose computers configured for control operation. Over the past two decades, however, the main features required, such as a processor, memory, communications, and analog to digital converters, have been combined in single integrated circuits. These *microcontrollers* are available for single unit prices in the tens of dollars. Computer control is, therefore, very cost effective.

A.   Conversion Between Analog and Digital Data

The main difference between analog and digital control is the format of the data used. *Analog* systems use data represented by voltage levels (or by pressures, as in the case of pneumatic instruments), which are *continuous* in both value and time. *Digital* data are represented by binary numbers, which are *discrete* in value and time.

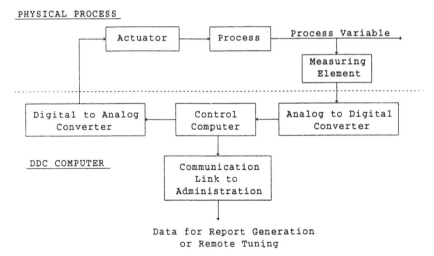

FIGURE 7   Direct digital control loop.

A binary number consists of a series of *bits* (0 or 1 values), which represent a number. Example 5 shows an 8-bit number, also called a *byte*, that represents the decimal number 181.

Example 5:   To convert the 8-bit binary number 10110101 to a decimal number, the following calculation is used:

First digit (right hand)	$1 \times 2^0 =$	1
Second digit	$0 \times 2^1 =$	0
Third digit	$1 \times 2^2 =$	4
Fourth digit	$0 \times 2^3 =$	0
Fifth digit	$1 \times 2^4 =$	16
Sixth digit	$1 \times 2^5 =$	32
Seventh digit	$0 \times 2^6 =$	0
Eighth digit (left hand)	$1 \times 2^7 =$	128
	Sum $=$	181

The binary number system works just like the decimal system, except that each digit can only be a 1 or a 0. Since each number has a fixed number of digits, it has a fixed *resolution*. In the case of an 8-bit number, the resolution is 1 in $2^8 - 1$, or 0.39%.

Example 6:   The liquid level in a tank 3 meters deep is measured as an 8-bit binary number. The resolution is $[1/(2^8 - 1)] \times 3$ meters = 0.012 meters, or 1.2 cm. A level change of less than 1.2 cm may not cause the binary level value to change. If level changes, smaller than 1.2 cm, are to be resolved, then the number of bits must be increased: If a 10-bit number is used, the resolution becomes $[1/(2^{10} - 1)] \times 3$ meters $= 0.003$ meters, or 3 mm.

This "granularity" of the measurement may reduce the precision of the measurement, but in control applications, the negative feedback will cause the system to cycle by ± 1 bit, averaging the effect of the resolution. In Example 6, this means that in the case of 10-bit numbers, the level can be expected to cycle by ± 3 mm.

The conversion from analog measurement to digital data is done by an *analog to digital (A/D)* converter. These sample the input and convert it to a number by one of several methods. The most common of these is *successive approximation*, wherein trial numbers are converted to an analog signal, which is compared with the input. This method is simple and fast. Since each conversion takes a finite amount of time, of the order of 50 μs, the result represents a snapshot of the input at the instant it is sampled. In typical control systems, these samples are taken at regularly spaced time intervals (the *sampling time*).

The opposite conversion, from digital numbers to an analog signal, which may be amplified to drive the *actuator* (the device that manipulates the process), is done by a *digital to analog (D/A)* converter. The typical D/A converter consists of a series of semiconductor switches and a resistor ladder network. Because the processor outputs the number only for a brief time, on the order of microseconds, the output must be held until the value is updated. While the hold may be included in the digital output circuits, its effect must be considered in the mathematical analysis of the control loop.

B. Digital PID Control Implementation

The PID control equation in the time domain is

$$
MV = Kc \left[\epsilon + \frac{\int \epsilon \, dt}{\tau_I} + \tau_D \frac{d\epsilon}{dt} \right] \tag{13}
$$

where MV is the manipulated variable, ϵ is the process error, and Kc, τ_I, and τ_D are the tuning parameters discussed earlier. To convert this into discrete time steps, the integration may be carried out using the rectangular rule and the derivative replaced by a simple slope:

$$
MV_n = Kc \left[\epsilon_n + \frac{\sum \epsilon_i \Delta t}{\tau_I} + \frac{\tau_D (\epsilon_n - \epsilon_{n-1})}{\Delta t} \right] \tag{14}
$$

where the subscript n is the current value, the subscript $n - 1$ is the previous value, and Δt is the sampling time. If the sampling time is small enough, this equation is a good approximation of Eq. (13). A relatively fast microcontroller (compared with the speed of the process) can, therefore, give very good PID control performance using Eq. (14).

There are three problems associated with PID control that can be easily solved in the digital mode. These are transfer bumps, integrator windup, and derivative fighting. The first two can be solved by rewriting Eq. (14) in terms of a change in the manipulated variable rather than its absolute value:

$$MV_n - MV_{n-1} = Kc \left[(\epsilon_n - \epsilon_{n-1}) + \frac{\epsilon_n \Delta t}{\tau_I} + \frac{(\epsilon_n - 2\epsilon_{n-1} + \epsilon_{n-2})\tau_D}{\Delta t} \right]$$

(15)

Equation (15) is the *velocity form* of the PID controller. The manipulated variable is obtained by summing all the calculated changes:

$$MV_n = \sum \Delta MV$$

(16)

The advantage of this form is that the integral represents the manipulated variable rather than the errors, so that the initial value can be easily specified and appropriate limits put on its value.

Transfer bumps occur when the controller is switched to automatic after a period of manual operation. The value of the integral in Eq. (13) represents the steady state action required to overcome offset. Setting its initial value may be difficult in that the value depends on Kc and τ_I, and so a value of 0 is usually assumed. This causes a disturbance that eventually loads the integral to its appropriate value. By using the velocity form, the initial value required is the manual mode manipulated variable, which is known, so that the disturbance is avoided.

Integrator windup occurs when there is a large sustained error such as that caused by an actuator power failure. When the actuator is restored, an opposite error is required to bring the integral term back to its appropriate value. It is difficult to place limits on the integral in Eq. (13) and so the value can become extremely large during a sustained failure. This is easily done in the velocity form by limiting the manipulated variable to its full on and full off values.

Derivative fighting occurs after a large process upset. While the upset is developing, the derivative action opposes the upset; during the recovery phase, however, the derivative will oppose the recovery. While the former is desirable, the latter is not. To solve this problem, the controller can be changed to a PI unit during the recovery phase simply by ignoring the derivative, the last term in Eq. (14). During the recovery phase, the sign of $\epsilon_n * (\epsilon_n - \epsilon_{n-1})$ is negative.

The implementation of the velocity form PID controller is a computer program run on the microcontroller. Figure 8 shows the *pseudo-code* for this program. Pseudo-code is a description of the computer operations required in a readable form, without the complexity inherent in a formal computer language. One interesting feature of the program described in Fig. 8 is that it never ends. Control loops run all the time.

Digital PID controllers may be tuned using the same methods as the analog controllers discussed in Section IV; however, the sampling time and integration method complicate the procedures somewhat. The effect of the hold on the manipulated variable output has been shown [2] to act as a dead time when the sampling time is small. The value of this additional dead time is half of the sampling time; therefore, it must be added to the dead time of the process when calculating the tuning parameters.

Equation (14) was developed using the rectangular rule integration method. By writing the equation as a polynomial in ϵ_n, ϵ_{n-1}, and ϵ_{n-2}, it can be shown that changing the integration method to one using the

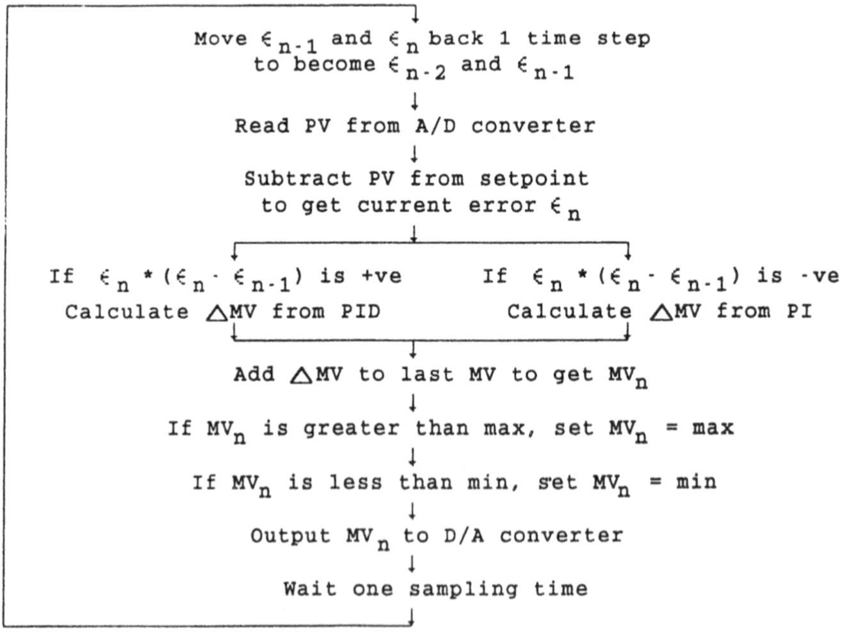

```
          Move ε_{n-1} and ε_n back 1 time step
               to become ε_{n-2} and ε_{n-1}
                          ↓
               Read PV from A/D converter
                          ↓
               Subtract PV from setpoint
               to get current error ε_n
                          ↓
  If ε_n * (ε_n - ε_{n-1}) is +ve      If ε_n * (ε_n - ε_{n-1}) is -ve
  Calculate ΔMV from PID               Calculate ΔMV from PI
                          ↓
             Add ΔMV to last MV to get MV_n
                          ↓
        If MV_n is greater than max, set MV_n = max
                          ↓
         If MV_n is less than min, set MV_n = min
                          ↓
              Output MV_n to D/A converter
                          ↓
              Wait one sampling time
```

Figure 8 Direct digital PID controller: pseudo-code for program.

trapezoidal rule gives the same result if the tuning parameters are changed. This means that the integration method changes the tuning values, but the performance does not change. The tuning methods, therefore, provide initial estimates, but the system performance must be optimized by on-line tuning.

C. *Z* Transform Control Design

A more rigorous method of controller design, which has the advantage of inherently compensating for long process dead times, involves the Z transform. Z transforms represent sampled data systems in a manner analogous to the Laplace representation of continuous systems. In fact, the Laplace and Z transforms are related by the mapping function

$$z = e^{\Delta Ts} \tag{17}$$

where s is the Laplace variable, z is the Z transform variable, and ΔT is the sampling time. Both s and z are complex variables.

A definition of the Z transform is given by

$$f(z) = f(0) + f(\Delta T)z^{-1} + f(2 \Delta T)z^{-2} + \cdots + f(n \Delta T)z^{-n} \tag{18}$$

where $f(n \Delta T)$ are the values of the time function at the sampling instants. Because n must be an integer, the Z transform contains no information relating to the system at times other than the sampling instants. The Z transforms of several useful functions are presented in Table 1. These may be derived from this definition of the Z transform by series contraction.

Z transforms may be used to develop block diagrams of digital systems, much like the Laplace domain block diagrams of continuous systems. There is a major difference, however, which is illustrated in Fig. 9. A Z transform is only valid for sampled data; therefore, a sampler (the switch symbol in Fig. 9) is required between each Z transform block to indicate that data is passed only at the sampling instants. Since the signal from the hold to the process is continuous, the blocks cannot be transformed separately. The output equation for the loop is

$$PV(z) = Z\{H(s)G(s)\} D(z) \{SP(z) - PV(z)\} \tag{19}$$

where $H(s)$ is the Laplace transform of the hold, $G(s)$ is the Laplace transform of the process, $D(z)$ is the Z transform controller, $PV(z)$ is the sampled process variable, and $SP(z)$ is the sampled setpoint. The Z operator is the Z transform from the Laplace domain.

The Laplace transform of the hold function is

$$H(s) = \frac{1 - e^{-\Delta Ts}}{s} \tag{20}$$

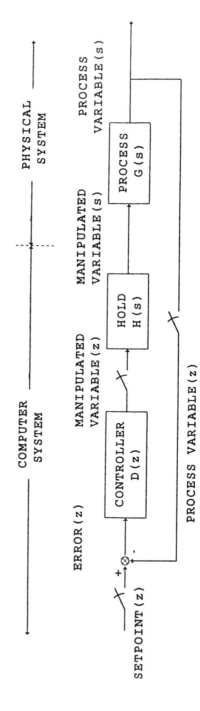

FIGURE 9   Z transform control loop.

which is equivalent to a return to zero from the previous sample and a step to the new sample at each sampling instant. Although this is a block in the continuous part of the loop, it is implemented as part of the micro-controller system. It is noteworthy that, while this is a "continuous" function in that data is held between sampling instants, there are corners in the values that violate the mathematical definition of a continuous function.

The block diagram equation can be rearranged to give the required controller transfer function as a function of the process model, $Z\{H(s)G(s)\}$, and a specified system servo performance, $PV(z)/SP(z)$:

$$D(z) = \frac{1}{Z\{H(s)G(s)\}} \frac{PV(z)/SP(z)}{1 - [PV(z)/SP(z)]} \tag{21}$$

The controller transform must be inverted into the time domain as a difference equation so that it may be implemented as part of a computer program such as the one shown in Fig. 8. This is done by writing the controller as

$$D(z) = \frac{MV(z)}{\epsilon(z)} \tag{22}$$

Using the equivalence between z^{-n} and a delay of n sampling times, the constants become terms of MV_n or ϵ_n, the terms of z^{-1} become MV_{n-1} or ϵ_{n-1}, and so on. This technique is the basis of the design of Z transform controllers.

Example 7:   Consider a first-order process with a gain of Kp and a time constant τ. The Z transform of the process with the hold is

$$Z \frac{1 - e^{-\Delta Ts}}{s} \frac{Kp}{\tau s + 1} = \frac{Kp\,(1 - e^{-\Delta T/\tau})z^{-1}}{1 - e^{-\Delta T/\tau}z^{-1}}$$

The response is specified to be equivalent to one sampling time of dead time (a deadbeat controller); therefore, $PV(z)/SP(z) = z^{-1}$. This gives a controller transfer function from Eq. (21) of

$$D(z) = \frac{1 - e^{-\Delta T/\tau}z^{-1}}{(1 - z^{-1})\,Kp\,(1 - e^{-\Delta T/\tau})}$$

Inverting this into the time domain as a difference equation, first multiply out the result and Eq. (22):

$$(1 - z^{-1})MV(z) = \frac{1 - e^{-\Delta T/\tau}z^{-1}}{(1 - z^{-1})\,Kp\,(1 - e^{-\Delta T/\tau})}\,\epsilon(z)$$

Then use the delay time definition to get

$$MV_n - MV_{n-1} = \frac{\epsilon_n - e^{-\Delta T/\tau}\epsilon_{n-1}}{(1 - z^{-1}) \, Kp \, (1 - e^{-\Delta T/\tau})}$$

It is interesting to note that this polynomial is the same in form as the PI controller part of Eq. (15). When the process has dead time, or is higher than first order, the controller becomes more complex. The implementation, however, remains the same. This difference equation is inserted into the program instead of the PID difference equation.

There are several types of Z transform controllers. The simplest of these is the *deadbeat* or minimal response controller [2] where the output is specified to follow a setpoint change after a delay of 1 sampling time. While this is often possible, the requirement of a one sampling period rise time may be too stringent. This is to say that real actuators may not have enough power to achieve this goal. Another problem arises when the process dead time is longer than the sampling time. In this case it will be impossible to meet the goal unless a prediction is made. Such controllers are *unrealizable*. The only way to design a controller in this situation is to relax the performance specification and allow a longer rise time.

Dahlin [6] proposed that the performance specification $PV(z)/SP(z)$ be that of a first-order system with a dead time equivalent to that of the process. This gives a realizable controller with a good chance of achieving the required rise time. The time constant of the first-order response was left as a tuning parameter. A long time constant gives sluggish control, while too fast a value causes the assumption of a first-order response to break down and overshoot occurs. Dahlin [6] suggests a tuning process where the time constant is gradually decreased until overshoot is just observed.

Both the deadbeat and Dahlin controllers show ringing in the manipulated variable. This can cause excessive wear in the actuator components. Dahlin suggested that replacing the root of the manipulated variable part of the controller equation responsible for the ringing with a constant would reduce this behavior. This works fairly well; however, Kalman [2] suggested a more elegant method where both the process and manipulated variable responses were specified, the latter without ringing. This approach also works well.

The major advantage of the Z transform controllers lies in their use of a model of the process. Process dead time compensation is inherent in their design; hence, their use is indicated where the process dead time is large. They also allow longer sampling times than can be used with approximations to continuous controllers. The main disadvantage of the Z transform controllers also stems from their use of a process model.

They are not general-purpose controllers but are specific to a particular process.

D. Adaptive Controllers

There are a number of processes in which a simple PID or Z transform controller will not perform well over a range of expected errors. The concept of changing a PID controller to a PI unit when large errors occur has been discussed above. This simple adaptive scheme avoids the derivative action opposing a return to the process setpoint. *Adaptive control* is the on-line adjustment of a controller to reflect changes in the process parameters.

pH control presents a good situation for an adaptive controller [7]. A typical plot of pH against the amount of added acid or base is shown in Fig. 10. Here the pH is the process variable, and the added acid or base is the manipulated variable. The gain of the process is the slope of the curve. Since the control loop performance depends on the overall loop again, high process gains require low controller gains and vice versa. This is evident in the PID tuning equations in Tables 5 and 6 as well as in the Z transform design, Eq. (21), where the controller gain is inversely proportional to the process gain. To maintain a constant loop gain, the

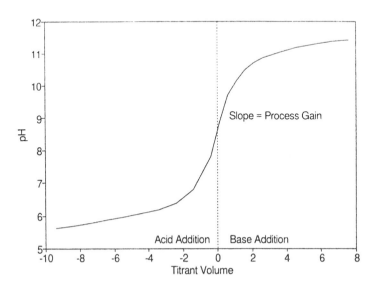

FIGURE 10 Neutralization curve.

controller gain is made a function of the process error rather than a constant value. This technique is *adaptive gain* tuning.

Controllers can also be designed to adapt to changing process dynamics. For an example, consider a changing level in the tank of Example 2. This changes the time constant, but not the gain. To adapt to this change, a *self-tuning* [8] controller may be used.

A self-tuning PID controller operates by minimizing an objective function, such as one of the integral criteria outlined in Table 3. The integral at a particular set of controller settings, $[\tau_I, \tau_D, Kc]$, is obtained by integrating a set of process errors using the rectangular rule approximation. A small change is made in the control settings to $[\tau_I + \Delta\tau_I, \tau_D + \Delta\tau_D, Kc + \Delta Kc]$, and a new measurement of the integral taken. Since there are three parameters, three such experiments must be made. From the result, the slopes of the integral values with changing control settings are obtained. Some single parameter curves are shown in Fig. 5. From the slopes a new set of control settings are calculated to move the operation toward the minimum integral point. This process repeats continuously. This scheme can optimize the controller performance; however, it is not particularly *robust*. There are several possible situations where small process upsets will give false slope values, which can lead to very poor controller settings.

VI. MULTIVARIABLE CONTROL

Multivariable systems that require the simultaneous control of two or more process variables are quite common. When each can be regulated by a manipulated variable that affects only the one process variable it is paired with, separate feedback loops can be designed and tuned independently. When the manipulated variables each affect more than one process variable, the loops *interact* and control becomes more difficult.

Example 8: Consider a shower where both the strength of the jet and the temperature are to be regulated. To increase the temperature, either the hot water flow must be increased or the cold water flow decreased. Both of these actions will change the strength of the jet. Similarly, to increase the strength of the jet, the hot or cold water flow must be increased, but either of these actions will change the temperature of the shower. To regulate this interacting system, both flows must be altered simultaneously when changing the temperature or jet strength.

Figure 11 shows a block diagram of a two-input, two-output process with interaction. The loops interact if the coupling gains K_{12} and K_{21} are both non-zero. Here, a disturbance traveling around either loop will be

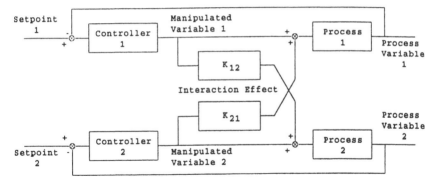

FIGURE 11 Block diagram of interacting system.

damped by the negative feedback; however, there is another possible path. A disturbance in the first loop will pass through K_{12}, travel around the second loop, and reenter the first loop through K_{21}. This combined loop contains two negative signs, which generate positive feedback. This is the origin of the control problem. The solution to this problem lies in feed-forward control.

A. Feed-Forward Control

Feed-forward controllers can be used to decouple interacting processes; they can also be very useful in single-loop systems where frequent large disturbances occur. Figure 12a presents the block diagram of a simple feed-forward controller. A disturbance entering the process will produce an error through the load transfer function. The feed-forward controller is designed to produce an equal but opposite effect through the servo transfer function so that the output remains unperturbed. This will be the case when the feed-forward element transfer function is

$$F(s) = \frac{-L(s)}{M(s)} \tag{23}$$

where $L(s)$ is the load transfer function and $M(s)$ is the servo transfer function of the process.

Example 2 considers a process with both of these transfer functions. In Fig. 2b, the output responds to a change in the inlet liquid temperature (the load) and to the flow of steam. Because the steam flow is the manipulated variable, the corresponding block is the servo transfer function.

Because $L(s)$ and $M(s)$ are models of the process, any approximations used in their derivation will produce error in the feed-forward com-

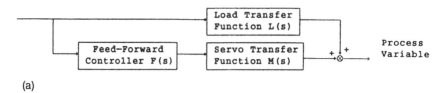

(a)

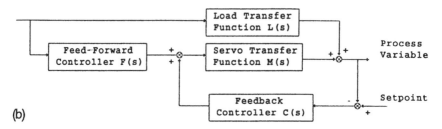

(b)

FIGURE 12 (a) Simple feed-forward control loop. (b) Combined feed-forward and feedback control.

pensation. Other disturbances that are not measured will not be compensated. To eliminate these effects, the feed-forward system can be coupled with a feedback controller as shown in Fig. 12b. If the feed-forward controller works perfectly, the feedback part remains idle. Unmeasured disturbances will be controlled by the feedback system. Because only the feedback loop transfer function determines the stability of the controlled process, it should be noted that a feed-forward element will neither improve nor reduce the stability of the controlled process. The feed-forward action will, however, greatly reduce the impact of frequent large load disturbances on the process performance.

B. Decouplers

The process shown in Fig. 11 can be *decoupled* by the addition of two feed-forward elements as shown in Fig. 13. Here $D_{12} = -K_{12}$ and $D_{21} = -K_{21}$. These prevent disturbances from crossing between loops, but the gain of each loop is changed by the $D_{12}K_{21}$ or $D_{21}K_{12}$ paths.

The analogy between a decoupler and feed-forward elements can be seen by considering K_{12} to be the load transfer function of the bottom process and D_{12} to be the feed-forward and servo blocks. The top part of the process is symmetrical with the bottom. The feed-forward analogy, while instructive, is not the best way to calculate the required decoupler

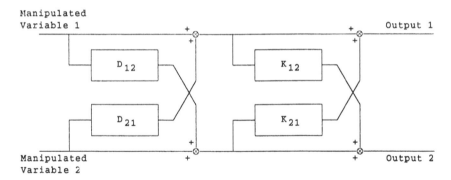

FIGURE 13 Decoupler.

functions. When the process contains more than two variables, the complexity increases greatly.

A better way of modeling the system is to write the transfer function as a matrix. For Fig. 11, the matrix is

$$[PV_1(s) \quad PV_2(s)] = \begin{bmatrix} G_{11}(s) & G_{21}(s) \\ G_{12}(s) & G_{22}(s) \end{bmatrix} \begin{bmatrix} MV_1(s) \\ MV_2(s) \end{bmatrix} \tag{24}$$

where $PV_1(s)$ and $PV_2(s)$ are the process variables, $MV_1(s)$ and $MV_2(s)$ are the manipulated variables, and the $G(s)$ terms are the process transfer functions. $G_{11}(s)$ is the relationship between MV_1 and PV_1, G_{21} is the relationship between MV_2 and PV_1, and so on. The decoupler for this process would be added as

$$PV(s) = G(s)D(s)MV(s) \tag{25}$$

where all of the functions in bold type are matrices like those of Eq. (24). So that there is no interaction, $G(s)D(s)$ must be a diagonal matrix. This is most easily found by making

$$D(s) = G(s)^{-1} \tag{26}$$

While making the terms of $G(s)$ and $D(s)$ the Laplace domain transfer functions is rigorously correct, the result is very difficult to compute. Great performance improvement can be achieved by using only the gain terms of the transfer functions in calculating the decoupler.

When only the gain terms of $G(s)$ are used, the matrix is referred to as the *steady state gains matrix*. To obtain the terms of this matrix experimentally, all of the control loops in the process must be placed on manual operation. A small step change is made in each manipulated variable successively, and the resulting process variable changes are measured. The individual gain terms are then calculated from

$$\mathbf{K}_{ij} = \left.\frac{\partial \, \mathrm{PV}_i}{\partial \, \mathrm{MV}_j}\right|_{\mathrm{MV}} \tag{27}$$

where $\mathbf{K}_{ij}$ is a term of the steady state gains matrix relating a change in process variable i to a change in manipulated variable j with all other manipulated variables held constant. The inverse of the steady state gains matrix thus obtained is the decoupler. Of course, the operations staff of an industrial operation will never allow such an experiment.

An experiment where only one control loop is opened at a time will probably be allowed. By making a small change in the manipulated variable of the opened loop, the response of that process variable will be seen, but all of the others will held constant by their controllers. The required information can be obtained by measuring the other manipulated variables, because these will change to hold the process variables constant. The data obtained will be terms of the matrix

$$\mathbf{H}_{ji} = \left.\frac{\partial \, \mathrm{MV}_j}{\partial \, \mathrm{PV}_i}\right|_{\mathrm{PV}} \tag{28}$$

which relates the change in the manipulated variable j needed to keep all but process variable i constant. Process variable i changes because of the experiment. The $\mathbf{H}$ matrix has been shown to be the inverse of the steady state gains matrix [7], and so it is the decoupler sought.

To implement this decoupling scheme, the process variables are read and the manipulated variables calculated in the manner shown in Fig. 8. Any of the controller types may be used. The resulting manipulated variables are arranged as a vector, which is multiplied by the decoupler to give a vector of corrected manipulated variables. These are sent via the D/A converter to the individual loop actuators.

VII. SUPERVISORY CONTROL

Setpoint supervisory control (*SSC*) is one level above the direct digital control (DDC) loops. Here, a computer with the task of optimizing a process passes setpoints to and receives data from the individual DDC computers. By separating these layers, the requirement that the supervisory calculations be performed within a sampling time is removed. The DDC computers operate in real time. The supervisor can take longer times and download the setpoints when necessary. Ginn [9] stresses that the supervisor should not try to track targets (setpoints), but should set the targets and leave the tracking to the DDC computers.

The recalculation of setpoints is necessary when any assumptions involving the input or environmental parameters used in operating a process change. An example would be a change in the size of a material to be dried. A size change could require a change in the drying temperature. This new temperature setpoint would be sent to the DDC computer, which adjusts the fuel flow. In this example, the supervisor computer responds to a change in the input material, but many other responses are possible. In addition to changing input parameters, the supervisor may respond to changes in material costs or product quality.

VIII. IMPLEMENTATION

The digital control concepts outlined in this chapter may be implemented in a number of ways [10]. For small systems, the measurement and actuator signals are commonly routed to a workstation computer located near the process. The computers are usually based on 80x86 processors because of their ready availability and low cost. Many manufacturers make industrial interfaces for these computers.

When greater distances must be covered, *SCADA* (supervisory control and data acquisition) systems are used. These are not much different from the workstation control systems; however, the control signals are collected at *remote terminal units* and transmitted to the control computer in digital form. To improve the reliability of remote control, *distributed control systems* may be used. Here, the remote terminals are replaced by local controllers. These transfer data to operator stations via a local area network, but they are capable of maintaining setpoints during a network failure.

Programmable logic control (*PLC*) systems are very common industrial controllers. Their operation is very similar to the distributed system. While they are capable of the continuous control described in this chapter, they are designed to handle large numbers of on/off signals. They are well suited to control batch formulation and other sequential operations.

IX. SUMMARY

Computerized control offers many benefits over conventional analog control. In simple implementations, their performance is similar but the cost of the digital systems is lower. Digital systems offer more advanced control types. Feed-forward systems compensate for frequent large disturbances. Decouplers improve the performance of multivariable processes.

Adaptive controllers adjust themselves to match changing process dynamics. The communications capability of digital systems allows report generation and remote operation. Setpoint control allows complex processes to be optimized by supervisory computer systems. These are a few of many possible control systems that can be realized using digital microcontrollers.

REFERENCES

1. D. R. Coughanowr, *Process Systems Analysis and Control*, McGraw-Hill, New York, 1991.
2. C. L. Smith, *Digital Computer Process Control*, Intext Educational Publishers, Scranton, Philadelphia, 1972.
3. J. G. Ziegler and N. B. Nichols, Optimum settings for automatic controllers, *Trans. ASME 64*:759 (1942).
4. G. H. Cohen and G. A. Coon, Theoretical consideration of retarded control, *Trans. ASME 75*:827 (1953).
5. J. A. Miller, A. M. Lopez, C. L. Smith, and P. W. Murrill, A comparison of controller tuning techniques, *Control Engineering*, December (1967).
6. E. B. Dahlin, Designing and tuning digital controllers, *Instruments and Control Systems 41*(6):77 (1968).
7. F. G. Shinskey, *Controlling Multivariable Processes*, Instrument Society of America, Research Triangle Park, North Carolina, 1981.
8. G. Stephanopoulos, *Chemical Process Control*, Prentice-Hall, Englewood Cliffs, New Jersey, 1984.
9. P. L. Ginn, *An Introduction to Process Control and Digital Minicomputers*, Gulf Publishing, Houston, 1982.
10. Anon., *Automation and Process Control Using Programmable Logic Controllers*, Instrument Data Communications, Richmond, British Columbia, Canada, 1995.

5

Neuro-Fuzzy Technology for Computerized Automation

Rakesh K. Singh, and Feng Ou-Yang
Purdue University, West Lafayette, Indiana

I. INTRODUCTION

The computerized automation of food processes is more challenging than that of chemical or pharmaceutical processes. Food processes largely rely on operator's rules of thumb and are not fully automated. Processes may be difficult to control or model by conventional methods, where simplifications or linearization are often made. Processes become too complicated because of tightly coupled control loops, nonlinear parameters around the operation point, or some parameters being subject to unpredictable noise. Some food processes are combinations of continuous and batch operations. Those processes, however, usually can be controlled by the experienced operators. It is not easy to transfer the knowledge from process experts to mathematical models when designing a control system for food processes.

The other problem in food process automation is that only a limited number of variables are measurable on line. Some of them, such as color, odor, taste, appearance, and texture, are subjective and usually evaluated qualitatively in linguistic terms or need a longer time to analyze in the laboratory. The properties of food material usually vary and depend on unpredictable factors such as seasons, location, and climate. Because of these reasons, automation of a food process may cost time and money. Fuzzy logic and neural network techniques, separately or combined, can be used to facilitate computerized automation.

The concepts of fuzzy logic were first proposed by L. A. Zadeh [1] as means of expressing the ambiguity and uncertainty in human thinking. Fuzzy logic can capture the approximate, inexact nature of the real world. In contrast to the conventional approach of process modeling or control, a fuzzy logic system is carried out by implementing linguistic rules that come from the experience of operators or the knowledge of experts. Therefore fuzzy modeling transforms the problems from building exact mathematical models to encoding a knowledge base containing inexact, commonsense information rules. It is similar to humans' strategies using imprecise models and decision rules to achieve fairly robust results. It increases the possibility of automating some complicated or ill-defined food processes.

An artificial neural network (ANN) is a computational model that mimics biological neural systems. The first artificial neuron model was presented by McCulloch and Pitts [2]. Its computational potential was recognized and became more widely used after the back propagation learning method for multilayer neural networks was proposed. Neural networks are famous for their learning or adapting ability, and unlike fuzzy systems they do not need much knowledge of underlying relationships between their input and output variables. The network learns from the input and output data of itself, repeatedly. It also can approximate any continuous or discontinuous, linear or nonlinear function. Therefore such networks are very useful for modeling some not-well-understood food processes, developing sensors for food qualities measurements, etc.

In this chapter, the basic principles of fuzzy logic and neural networks will be described in Sections II and III. The applications of fuzzy logic and neural networks to food process automation will then be introduced and summarized in Section IV.

II. INTRODUCTION TO FUZZY LOGIC

A. Fuzzy Sets and Fuzzy Set Operations

Some basic principles of fuzzy logic are summarized in this section. More complete theories can be found in [3–6].

1. Fuzzy Sets

A classical (crisp) set is normally defined as a collection of elements in a given universe of disclosure. The universe contains all the possible elements from which sets can be formed. Those elements are dichotomized into two groups, either members or nonmembers. The boundary for deciding whether a element belongs to a set is precise and without vagueness in the classical set.

Let U denote a given universe of disclosure, and u be an element, i.e., $u \in U$. A crisp set A is described by a characteristic function μ_A, which is defined by

$$\mu_A(u) = \begin{cases} 1, & \text{if and only if } u \in A \\ 0, & \text{if and only if } u \in A \end{cases} \tag{1}$$

The fuzzy set introduces the concept of vagueness by generalizing the boundary of a characteristic function that is either completely member (1) or nonmember (0) to that of a membership function that indicates the degree of membership of an element in a set. A *fuzzy set A* in the universe of disclosure U can be defined as a set of ordered pairs

$$A = \{(u, \mu_A(u)) \mid u \in A\} \tag{2}$$

$$\mu_A: U \rightarrow [0, 1] \tag{3}$$

where μ_A is called the membership function of fuzzy set A. μ_A represents the degree (or grade) of the membership of u in A, or the degree that u belongs to A. The membership function maps the universe U into the membership space $[0, 1]$.

Example 1: It is difficult to express some food quality factors using classical sets, because such factors are usually subjective, imprecise, and language-oriented. Consider expressing the property "very crunchy" of cookies at different shear stresses.

Let universe set $U = [0, 100] \text{ N/m}^2$. The degree of "very crunchy" of cookies is represented by assigning a value between $[0, 1]$ at each shear stress, as shown in Fig. 1a. In a classic set, we can only dichotomize U into two regions, as in Fig. 1b. If the shear stress is larger than 80 N/m^2, it is very crunchy. Otherwise, it is not.

The *support* of a fuzzy set A, Supp(A), is defined as the crisp set of the elements in A whose degrees of membership function are larger than zero:

$$\text{Supp}(A) = \{u \mid \mu_A(u) > 0, u \in U\} \tag{4}$$

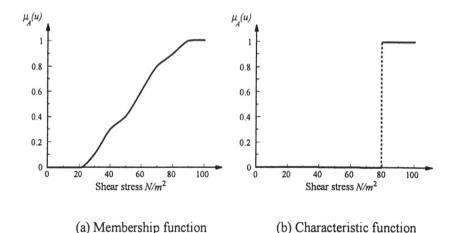

(a) Membership function   (b) Characteristic function

FIGURE 1   The fuzzy set (a) and nonfuzzy set (b) representing a food property term, "very crunchy."

A fuzzy set A can be represented by its support:

$$A = \frac{\mu_1}{u_1} + \frac{\mu_2}{u_2} + \cdots + \frac{\mu_n}{u_n} = \sum_{i=1}^{n} \frac{\mu_i}{u_i} \tag{5}$$

where $u_i \in \mathrm{Supp}(A)$, " $+$ " denotes the union of the elements, and $\mu_A(u_i)$ is the degree of the membership function of A at u_i. If X is a continuous, not a discrete, universe set, then A is represented by

$$A = \int_{u \in U} \frac{\mu_A(u)}{u} \tag{6}$$

where $\int$ represents the union of the elements in A.

   If a fuzzy set A whose support contains only one single element at $u = u_S$ and $\mu_A(u_S) = 1$, then the fuzzy set is called the *fuzzy singleton*. The fuzzy singleton is often used to fuzzify the crisp data into a fuzzy set in the applications of fuzzy logic control.

   Example 2: If the universe set is discrete in Example 1 and $U = \{0, 10, 20, \ldots, 90, 100\}$, then the membership function can be expressed by its support as in Eq. (5). The support of A is $\mathrm{Supp}(A) = \{30, 40, 50, 60, 70, 80, 90, 100\}$, and

$$A = \frac{0.1}{30} + \frac{0.3}{40} + \frac{0.4}{50} + \frac{0.6}{60} + \frac{0.8}{70} + \frac{0.9}{80} + \frac{1}{90} + \frac{1}{100} \tag{7}$$

The α-*cut* (α-*level set*) of fuzzy set A is a crisp set, denoted by A_α, the elements of which belong to the universal set U and have a degree of membership function greater or equal to the value α:

$$A_\alpha = \{u \mid \mu(u) \geq \alpha, u \in U\} \tag{8}$$

A fuzzy set A can be described in the form of its α-cuts,

$$A = \sum_{\alpha=0}^{1} \alpha A_\alpha, \quad \text{for a discrete universe} \tag{9}$$

or

$$A = \int_{\alpha \in (0,1]} \alpha A_\alpha, \quad \text{for a continuous universe} \tag{10}$$

Example 3: Consider a fuzzy set A defined in a discrete universe set U, $U = [0, 10, 20, 30, 40, 50, 60, 70, 80, 90, 100]$. The fuzzy set is represented by

$$A = \frac{0.3}{20} + \frac{0.4}{30} + \frac{0.6}{40} + \frac{0.8}{50} + \frac{1}{60} + \frac{0.8}{70} + \frac{0.6}{80} + \frac{0.4}{90} \tag{11}$$

The α-cuts of A are

$$A_1 = \{60\}, \quad A_{0.8} = \{50, 60, 70\}$$

$$A_{0.6} = \{40, 50, 60, 70, 80\}, \quad A_{0.4} = \{30, 40, 50, 60, 70, 80, 90\} \tag{12a}$$

$$A_{0.3} = \{20, 30, 40, 50, 60, 70, 80, 90\}$$

and

$$A = 0.3A_{0.3} + 0.4A_{0.4} + 0.6A_{0.6} + 0.8A_{0.8} + A_1 \tag{12b}$$

A *fuzzy number* F in a continuous universe U is a fuzzy set F that is normal and convex:

Normal: $\quad \max_{u \in U} \mu_F(u) = 1.$ $\tag{13}$

Convex: $\quad \mu_F(\lambda u_1 + (1 - \lambda)u_2) \geq \min[\mu_F(u_1), \mu_F(u_2)]$ $\tag{14}$

$u_1, u_2 \in U, \quad \lambda \in [0, 1]$

2. Fuzzy Set Operations

Let A and B be two fuzzy sets defined in the universe of disclosure U with membership function μ_A and μ_B.

Union: The membership function of the union of fuzzy sets A and B, $A \cup B$, is defined as

$$
\begin{aligned}
\mu_{A \cup B}(u) &= \mu_A \vee \mu_B \\
&= \max(\mu_A(u), \mu_B(u)), \qquad \forall u \in U
\end{aligned} \tag{15}
$$

where "$\vee$" indicates the max operation.

Intersection: The membership function of the intersection of fuzzy sets A and B, $A \cap B$, is defined as

$$
\begin{aligned}
\mu_{A \cap B}(u) &= \mu_A(u) \wedge \mu_B(u) \\
&= \min(\mu_A(u), \mu_B(u)), \qquad \forall u \in U
\end{aligned} \tag{16}
$$

where "$\wedge$" indicates the min operation.

Maximum and minimum are not the unique choices for union and intersection operations of fuzzy sets. There are other choices based on different theories [3, 7].

Complement: The complement of A, denoted by $\overline{A}$, is defined by its membership function as

$$
\mu_{\overline{A}}(u) = 1 - \mu_A(u), \qquad \forall u \in U \tag{17}
$$

Equality: A and B are equal if and only if

$$
\mu_A(u) = \mu_B(u), \qquad \forall u \in U \tag{18}
$$

Cartesian product: Let $A_1, A_2, \ldots, A_n$ be fuzzy sets defined in universes $U_1, U_2, \ldots, U_n$, respectively. Their Cartesian product is also a fuzzy set in the space $U_1 \times U_2 \times \cdots \times U_n$, and its membership function is defined as

$$
\begin{aligned}
\mu_{A_1 \times A_2 \times \cdots \times A_n}&(u_1, u_2, \ldots, u_n) \\
&= \min[\mu_{A_1}(u_1), \mu_{A_2}(u_2), \ldots, \mu_{A_n}(u_n)]
\end{aligned} \tag{19}
$$

where $u_1 \in U_1, u_2 \in U_2, \ldots, u_n \in U_n$.

Extension Principle   The extension principle provides a way to generalize crisp mathematical concepts to operate on fuzzy sets. Any mathematical function that operates on nonfuzzy data can be extended to deal with fuzzy sets by the extension principle.

Let $A_1, A_2, \ldots, A_n, B$ be n fuzzy sets defined in universes $X_1, X_2, \ldots, X_n, Y$, respectively; and let X be the Cartesian product $X_1 \times X_2 \times \cdots \times X_n$. Let $f(\cdot)$ be a function that maps an n-tuple $(x_1, x_2, \ldots, x_n) \in X$ to a point $y \in Y$:

$$
y = f(x_1, x_2, \ldots, x_n) \tag{20}
$$

The extension principle generalizes $f(\cdot)$ to operate on n fuzzy sets A_1, $A_2, \ldots, A_n$, such that $B = f(A_1, A_2, \ldots, A_n)$; B is the fuzzy mapping image of $A_1, A_2, \ldots, A_n$ through $f(\cdot)$ and is defined by

$$B = \{(y, \mu_B(y)) \mid y = f(x_1, x_2, \ldots, x_n), (x_1, x_2, \ldots, x_n) \in X\}$$

(21)

and

$$\mu_B(y) = \max_{\substack{(x_1,x_2,\ldots,x_n) \in X \\ y = f(x_1,x_2,\ldots,x_n)}} \{\min[\mu_{A_1}(x_1), \mu_{A_2}(x_2), \ldots, \mu_{A_n}(x_n)]\}$$

(22)

Example 4: We are given a function $f(x_1, x_2) = x_1 x_2$ and two fuzzy sets A_1 and A_2, where A_1, A_2, x_1, x_2, and y are all defined on the universe of real space:

$$A_1 = \frac{0.1}{1} + \frac{0.2}{2} + \frac{0.5}{3}, \qquad A_2 = \frac{0.4}{1} + \frac{0.3}{2} + \frac{1}{4}$$

(23)

By the extension principle, f can also operate on fuzzy sets, A_1 and A_2:

$$B = f(A_1, A_2)$$

$$+ \frac{(0.1 \wedge 0.4)}{1} + \frac{(0.1 \wedge 0.3) \vee (0.2 \wedge 0.4)}{2} + \frac{(0.5 \wedge 0.4)}{3}$$

$$= \frac{(0.1 \wedge 0.4) \vee (0.2 \wedge 0.3)}{4} + \frac{(0.5 \wedge 0.3)}{6} + \frac{(0.5 \wedge 0.1)}{12}$$

(24)

$$= \frac{0.1}{1} + \frac{0.2}{2} + \frac{0.4}{3} + \frac{0.2}{4} + \frac{0.3}{6} + \frac{0.5}{12}$$

B. Membership Functions

Defining membership functions for fuzzy sets is subjective and context dependent. Experimental construction of the membership functions usually involves two steps. The first step is the experimental acquisition of the degrees of the membership functions. The second is to construct the membership function by curve fitting the collected data to a certain membership function. Several methods have appeared in the literature and are summarized as the following [8].

1. Methods for Acquisition of the Degrees of the Membership Functions

Direct Rating Method A randomly selected element x from the universal set U is presented to the subject and the question such as "How

A is x?" or "To what degree is A x?" is asked. A is the concerned fuzzy set. The answer is a value y in the range of $[y_{max}, y_{min}]$, which can be normalized to [0, 1]. The same question is asked n times for the same value of x, but to avoid memorization, questions are randomly presented with others concerning different x values. Denote the value of the answer by y/x, where x is the value presented in the question. The result of all the observed y/x is a conditional distribution that can be assumed to be Gaussian distribution $N(m, \sigma^2)$. Its mean value $m_{y/x}$ and variance σ of the estimated degree of the membership function $\mu_A(x)$ are

$$\mu_A(x) = m_{y/x} = \frac{\sum_{i=1}^{n} (y_i/x)}{n}, \qquad \sigma^2_{y/x} = \frac{\sum_{i=1}^{n} (y_i/x - m_{y/x})^2}{n-1} \qquad (25)$$

Polling Method   The values of membership functions are found by randomly and repeatedly presenting the questions such as "Do you agree that x is A?" answered either "yes" or "no." A is the concerned fuzzy set, and x is an element defined in the universal set U. The degree of membership function μ_A at x is derived by the proportion of respondents answering "yes":

$$\mu_A(x) = \frac{\text{Total number of "yes" responses for } x}{\text{Total number of "yes" and "no" responses for } x} \qquad (26)$$

Set-Valued Statistics Method A_i is the set-valued observation, m_i is the frequency of observation of A_i, and A_{α_i} is the α-cut of the fuzzy set A. The fuzzy set A can be approximated by

$$\mu_{A_i}(x) = \sum_i m_i, \qquad x \in A_{\alpha_i} \qquad (27)$$

Other types of approximations are also available [8].

Reverse Rating Method   The question asked in this method is "What value of x possesses degree y of membership in the fuzzy set A?" Questions with different values of y are randomly presented n times, as in direct rating. Let the value of x given in the answer be denoted x/y. The conditional distribution of x/y is assumed to be a Gaussian distribution $N(m, \sigma^2)$. The mean value $m_{x/y}$ and variance σ of x are

$$m_{x/y} = \frac{\sum_{i=1}^{n} (x_i/y)}{n}, \qquad \sigma^2_{x/y} = \frac{\sum_{i=1}^{n} (x_i/y - m_{x/y})^2}{n-1} \qquad (28)$$

2.   Methods for Construction of the Membership Function

Once we acquire the values of the membership function, we can construct the membership function by curve-fitting to a predefined function with undetermined parameters. Two of such functions are shown in the following:

S-Shaped Function

$$\mu(x) = (1 + e^{a-bx})^{-1} \tag{29}$$

Linear Filter Function

$$m(x) = \begin{cases} 0, & \text{if } x \le p - w \\ x - p + w, & \text{if } p - w < x < p + w \\ 1, & \text{if } x \ge p + w \end{cases} \tag{30}$$

where a, b, p, and w are parameters of the functions.

C. Fuzzy Relations

Fuzzy relations are extensions of ordinary relations. The concept is important in many application areas such as fuzzy modeling, fuzzy control, and fuzzy expert systems. It provides a way to express the ambiguous relations such as "x and y are *almost the same*" and "x is *much smaller than y*" from natural language.

The *fuzzy relation* among X_1, X_2, . . . , X_n is a fuzzy set defined on the Cartesian product space $X_1 \times X_2 \times \cdots \times X_n$; the strength of this relation is represented by its membership function μ_R as

$$R = \{(x_1, x_2, \ldots, x_n), \mu_R(x_1, x_2, \ldots, x_n) \,|\, (x_1, x_2, \ldots, x_n) \tag{31}$$

$$\in X_1 \times X_2 \times \cdots \times X_n\}$$

$$\mu_R: X_1 \times X_2 \times \cdots \times X_n \to [0, 1] \tag{32}$$

Example 5: x_1, x_2 are defined on universes of real space, X_1, X_2, respectively. The membership function of the fuzzy relation R representing "x_1 is very close to x_2" could be defined, for example, as

$$\mu_R(x_1, x_2) = \begin{cases} \sqrt{x_2/x_1}, & x_1 > x_2 \\ \sqrt{x_1/x_2}, & x_1 < x_2 \\ 1, & x_1 = x_2 \end{cases} \tag{33}$$

A *fuzzy matrix* is an alternative representation of the fuzzy relation. If the support of two universal sets X and Y are discrete and countable,

$$X = \{x_1, x_2, \ldots, x_n\}, \qquad Y = \{y_1, y_2, \ldots, y_m\} \tag{34}$$

then a fuzzy relation R of two fuzzy sets A, B defined in X, Y, respectively, can be expressed by a matrix M of size $n \times m$, called a fuzzy matrix. Its element at the ith row and the jth column is

$$M(i, j) = \mu_A(x_i, y_j)$$
$$= \min[\mu_A(x_i), \mu_B(y_j)], \quad i \in \{1, 2, \ldots, n\}, j \in \{1, 2, \ldots, m\}$$

$$(35)$$

1. Composition of Two Fuzzy Relations

Fuzzy relations of different spaces can be combined into a new fuzzy relation by *fuzzy composition*. Different versions of composition rules have been suggested [3, 7]. The most often used one is max-* composition, defined as follows. Let $R_1(x, y)$, $R_2(y, z)$ be two fuzzy relations, $(x, y) \in X \times Y$, $(y, z) \in Y \times Z$; the composition of fuzzy relations R_1 and R_2 is a new fuzzy relation, $R_1 \circ R_2$, and its membership function is defined as

$$\mu_{R_1 \circ R_2}(x, z) = \max_{(x,z)} \{\mu_{R_1}(x, y)^* \mu_{R_2}(y, z)\} \quad (36)$$

In this type of composition rule, "*" can stand for different fuzzy operations, yielding different composition rules. Two special cases are

Max–min composition: $\mu_{R_1 \circ R_2}(x, z) = \max_{(x,z)} \{\min_y [\mu_{R_1}(x, y), \mu_{R_2}(y, z)]\}$

$$(37)$$

Max–product composition: $\mu_{R_1 \circ R_2}(x, z) = \max_{(x,z)} \{\mu_{R_1}(x, y) \mu_{R_2}(y, z)\}$

$$(38)$$

Example 6: Consider two fuzzy relations, R_1: "x is very close to y" and R_2: "y is very close to z." Let X, Y, Z be finite discrete sets, and $x \in X$, $y \in Y$, $z \in Z$, where $X = [x_1, x_2, x_3]$, $Y = [y_1, y_2, y_3, y_4]$, $Z = [z_1, z_2, z_3]$. R_1 and R_2 are represented by fuzzy matrices M_{R_1} and M_{R_2}:

$$M_{R_1} = \begin{bmatrix} 1 & 0.58 & 0.45 & 0.32 \\ 0.71 & 0.82 & 0.63 & 0.45 \\ 0.58 & 1 & 0.77 & 0.55 \end{bmatrix},$$

$$M_{R_2} = \begin{bmatrix} 0.45 & 0.71 & 0.71 \\ 0.77 & 0.82 & 0.41 \\ 1 & 0.63 & 0.32 \\ 0.71 & 0.45 & 0.22 \end{bmatrix} \quad (39)$$

The max-min composition of R_1 and R_2, $R_1 \circ R_2$, can be calculated by Eq. (36); its fuzzy matrix is

$$M_{R_1 \circ R_2} = \begin{bmatrix} 0.58 & 0.71 & 0.71 \\ 0.77 & 0.82 & 0.71 \\ 0.77 & 0.82 & 0.58 \end{bmatrix} \qquad (40)$$

D. Fuzzy Logic and Approximate Reasoning

Fuzzy logic can represent the imprecise meaning of natural language, and it is able to perform the imprecise reasoning that plays an important role in human decision making under uncertainty and imprecise information.

1. Fuzzy Logic

Fuzzy logic is the basis of the approximate reasoning. A proposition P is a sentence that can be expressed in the canonical form

$$P: \text{``}x \text{ is } A\text{''} \qquad (41)$$

where x is the subject and A is a predicate that characterizes the subject x. For example, in the proposition "Vitamins in food are heat sensitive," "Vitamins in food" is the subject, which is characterized by the predicate "heat sensitive."

In classical logic, the predicates are nonfuzzy and the truth values of the propositions are either true or false. The most frequently used logic operations between two sets, A and B, are conjunction ($\wedge$), disjunction ($\vee$), implication ($\Rightarrow$), and equivalence ($\Leftrightarrow$). They are interpreted as "A and B," "A or B," "if A then B," and "A if and only if B," respectively. The truth value of the these logic operations on propositions A and B can be defined by a truth table. In fuzzy logic, however, we could have many different ways to express the vagueness. For example, the truth value of proposition could be a fuzzy set, the predicate in the proposition might contain a fuzzy predicate, some fuzzy quantifier other than existence and universe might be used to modify the proposition, or fuzzy qualifiers can be used to modify the proposition [9]. These possibilities are explained in more detail in the following.

Fuzzy Truth Value In fuzzy logic the truth values of propositions are not limited to the choices "true" and "false." Instead, the truth value of a fuzzy proposition P can be expressed by a linguistic truth value denoted by $v(P)$ and defined as an fuzzy set in the interval of [0, 1] such as "very true," "almost true," "not true," and so on.

Let P_1, P_2 be two propositions and $v(P_1), v(P_2)$ be their truth values, respectively:

$$v(P_1) = \sum_i \frac{\alpha_i}{x_i}, \qquad v(P_2) = \sum_j \frac{\beta_j}{y_j} \qquad (42)$$

Their logical operation can be derived by applying the extension principle given in Eqs. (21) and (22):

$$v(P_1 \text{ and } P_2) = \sum_{i,j} \frac{\min(\alpha_i, \beta_j)}{\min(x_i, y_j)}, \qquad v(P_1 \text{ or } P_2) = \sum_{i,j} \frac{\max(\alpha_i, \beta_j)}{\max(x_i, y_j)}$$

$$v(\text{not } P_1) = \sum_i \frac{\alpha_i}{1 - x_i}, \tag{43}$$

$$v(P_1 \Rightarrow P_2) = \sum_{i,j} \frac{\max(\alpha_i, \beta_j)}{\max[1 - x_i, \min(x_i, y_j)]}$$

Example 7: Consider two truth values of two propositions P_1, P_2. Their fuzzy truth values are $v(P_1) =$ "true" and $v(P_2) =$ "very true," respectively:

$$v_1(P) = \frac{0.4}{0.6} + \frac{0.5}{0.7} + \frac{0.6}{0.8} + \frac{0.7}{0.9} + \frac{1}{1} \qquad \text{and}$$

$$v_2(P) = \frac{0.7}{0.6} + \frac{0.8}{0.7} + \frac{0.9}{0.8} + \frac{1}{0.9} + \frac{1}{1}$$

The truth value of $(P_1 \text{ or } P_2)$ could be derived using Eq. (43) as

$$v(P_1 \vee P_2) = v(P_1) \vee v(P_1)$$

$$= \frac{0.4 \vee 0.7}{0.6} + \frac{0.5 \vee 0.8}{0.7} + \frac{0.6 \vee 0.9}{0.8} + \frac{0.7 \vee 1}{0.9} + \frac{1 \vee 1}{1}$$

$$= \frac{0.7}{0.6} + \frac{0.8}{0.7} + \frac{0.9}{0.8} + \frac{1}{0.9} + \frac{1}{1}$$

$$\tag{44}$$

Fuzzy Propositions   A proposition is called fuzzy proposition if it contains a fuzzy predicate or is modified by fuzzy quantifiers or fuzzy qualifiers.

   i.   Fuzzy predicates.   The predicates of the propositions are fuzzy sets. For example, in the proposition "The concentration is more or less high," "more or less high" is a fuzzy predicate.

   ii. Fuzzy quantifiers.   The quantifiers of fuzzy propositions in fuzzy logic are not only existential and universal quantifiers but also include other terms. Terms such as "most," "several," "few," "frequently," "about hundreds," and so on can be used to modify propositions. For example, "Most sterilized products are overprocessed." "Most" is a fuzzy quantifier.

   iii. Fuzzy qualifiers.   A proposition can also be qualified by associating the given proposition with fuzzy truth value qualifiers, possibility qualifiers, necessity qualifiers, probability qualifiers, or usuality qualifiers. Some examples of these are "very true" as a fuzzy truth qualifier, "very likely" as a fuzzy probability qualifier, "almost impossible" as a possibility qualifier, and "usually" as a usuality qualifier.

Fuzzy Linguistic Variables   A linguistic variable is a variable whose values are words or sentences in natural language. Those values can be created from some basic terms and augmented by some grammars.

A more formal definition of fuzzy linguistic variable: A linguistic variable is characterized by $(u, T(u), U, G, M)$, where u is the name of variable, $T(u)$ is the term set that contains linguistic values (names) of variables, U is the universe of disclosure of variable u, G is the syntactic rule for generating the values in term set $T(u)$, and M is the semantic rule for associating each linguistic value of u with a proper meaning.

Example 8: Consider a linguistic variable "Moisture content of corn" in the percentage range of 0–100%.

The linguistic variable u = "Moisture content of corn."
Universe of disclosure U = [0, 100]%.
The term set $T(u)$ is defined as {very low, medium low, low, normal, high, medium high, very high}.
The syntactic rule G for generating the term set $T(u)$ is based on three basic terms: "low," "normal," "high," and two modifiers: "very," "medium."
The semantic rules M are

M(very high) = the fuzzy set for "moisture above 90%,"
M(medium high) = the fuzzy set for "moisture ranged from 80% to 90%," and so on.

Linguistic Hedge (Modifier)   A linguistic hedge (modifier) is an operator used to modify a term of a fuzzy set to generate a new term. In Example 8, "very" and "medium" are the linguistic hedges that modify the terms representing "high" and "low" to create terms representing "very high," "medium low," etc. Three often used linguistic hedges are the following:

1. Concentration: con(A),

$$\mu_{con(A)} = [\mu_A(x)]^2 \tag{45}$$

2. Dilation: dil(A),

$$\mu_{dil(A)} = [\mu_A(x)]^{1/2} \tag{46}$$

3. Intensification: intens(A),

$$\mu_{\text{intens}(A)} = \begin{cases} 2[\mu_A(x)]^2, & \mu_A(x) \in [0, 0.5] \\ 1 - 2[1 - \mu_A(x)]^2, & \text{otherwise} \end{cases} \qquad (47)$$

2. Meaning Representation of Fuzzy Propositions

Natural language can be viewed as a collection of propositions that give constraints on the values of a collection of variables. Those propositions are called the elastic constraints. Most constraints are usually expressed implicitly rather than explicitly in our natural language. The process of making constraints and variables in fuzzy propositions explicit is called meaning representation [9–11]. It is important in constructing a knowledge base that consists of a collection of fuzzy propositions.

   a. Fuzzy Predicate Proposition   The fuzzy logic proposition P can be represented in a canonical form: "x is A." It implies that the possibility distribution of x is equal to the membership function of the fuzzy set A in the universe set U,

$$\text{poss}(x = u) = \pi_x(u) = \mu_A(u), \qquad u \in U \qquad (48)$$

where $\text{poss}(x = u)$ denotes the possibility that x's value is u, and $\pi_x(u)$ is the possibility distribution function of x in U. The possible value of x is constrained by the fuzzy proposition P by Eq. (48).

   If a fuzzy proposition's predicate is modified by a linguistic hedge h, "x is hA," then the meaning representation can be obtained from $\pi_x(u) = \mu_{A'}(u)$, where A' is a modification of fuzzy set A, $A' = hA$. $\pi_x(u) = \mu_A(u)$ for the proposition "x is A."

   b. Fuzzy Conditional Propositions   The canonical form is expressed as, if "x is A" then "y is B." It implies that proposition P is a conditional possibility distribution of y given x:

$$\mu_P(u, v) = \pi_{y|x}(u, v). \qquad (49)$$

Given two propositions P_1 and P_2:

$$P_1: \text{``}x \text{ is } A,\text{''} \qquad \pi_x(u) = \mu_A(u) \qquad (50)$$

$$P_2: \text{``}y \text{ is } B,\text{''} \qquad \pi_y(v) = \mu_B(v) \qquad (51)$$

Fuzzy conditional propositions imply the fuzzy relation between two fuzzy propositions; therefore, they are also called fuzzy implications. About 40 different methods for fuzzy implications have been described in the literature [7, 12]. Three often seen fuzzy implications are defined as follows:

1. Mamdani's minimum-operation, R_M

$$\mu_{A \rightarrow B}(u, v) = \min[\mu_A(u), \mu_B(v)] \tag{52}$$

2. Larsen's product-operation, R_L

$$\mu_{A \rightarrow B}(u, v) = \mu_A(u)\mu_B(v) \tag{53}$$

3. Lakasiewicz's implication, R_A

$$\mu_{A \rightarrow B}(u, v) = \max_{u,v} [1, 1 - \mu_A(u) + \mu_B(v)] \tag{54}$$

The knowledge base in fuzzy control systems or fuzzy models is usually composed of linguistic statements representing the decision rules in the form of fuzzy conditional propositions. For example, "If the temperature of the product is much lower than the default temperature, then open the steam valve wider," "If the color of the banana is yellow, then it is ripe," etc.

Fuzzy Quantification Propositions Proportions like "Q A's are B's" are called quantification rules. Here Q is a fuzzy quantifier, A, B are fuzzy predicates, and the universe set $U = \{u_1, u_2, \ldots, u_m\}$. The rule could be interpreted as the proportion of B in A or the relative carnality of B in A; it is defined as

$$\text{prop}(B/A) = \sum \text{count}(B/A)$$

$$= \frac{\sum \text{count}(A \cap B)}{\sum \text{count}(A)} = \frac{\sum_i [\mu_A(u_i) \wedge \mu_B(u_i)]}{\sum_i \mu_A(u_i)}, \tag{55}$$

$$j = 1, 2, \ldots, m$$

where $\sum \text{count}(B/A)$ is the relative carnality of a B in A. The fuzzy quantification proposition could then be interpreted by a fuzzy predicate proposition:

$$\text{"prop}(B/A) \text{ is } Q\text{"} \tag{56}$$

and from Eq. (48), it can be represented by

$$\pi_{\text{prop}(B/A)}(u) = \mu_Q(u) \tag{57}$$

For example, the quantification proposition "Most ripe tomatoes are soft" could be interpreted as "prop(soft tomato/ripe tomato) is MOST."

Fuzzy Qualification Propositions When propositions are modified by qualifiers such as fuzzy truth qualifiers, fuzzy probability qualifiers, fuzzy possibility qualifiers, or fuzzy usuality qualifiers, they are called fuzzy qualification propositions. Two of their meaning representations are shown here:

The proposition "x is A" is known, and $\pi_x(u) = \mu_A(u)$ as in Eq. (48).

1. *Fuzzy truth qualification propositions*

   P: "(x is A) is true" $\hspace{4cm}$ (58)

   The proposition P could be represented by $\mu_P(v) = \mu_T[\mu_A(u)]$, where $\mu_T(v)$ is the fuzzy set representing the linguistic truth value T.

2. *Fuzzy usuality qualification propositions*

   P: "Usually (x is A)" $\hspace{4cm}$ (59)

   The proposition is interpreted as a quantification proposition: "Most u are A." Here x is taken as a sequence of values u_1, $u_2, \ldots, u_n$ in the universe U. Therefore proposition P defines a possibility distribution of the constrained variable ($\sum$ count(A)/n) by

   $$\pi\left[\frac{\sum \text{count}(A)}{n}\right] = \mu_{\text{MOST}}\left[\frac{\sum \text{count}(A)}{n}\right] \hspace{2cm} (60)$$

   where MOST is the fuzzy set representing the fuzzy quantifier MOST, and $\sum$ count(A) is the carnality of A.

Example 9: Consider the meaning representation of the proposition, "Over the last few months the new product A has sold more than most of the similar products in the company." The following information is obtained from the database of the company:

1. The quantity of product j that has been sold in the last ith month, Q_{ij}.
2. The membership function $\mu_{\text{last few months}}(j)$ represents the degree that the jth month preceding belonged to the description of "last few months."
3. The membership function $\mu_{\text{sold much more}}(i, j)$ represents the degree that the i quantity has sold much more than the j quantity.
4. The membership function $\mu_{\text{similar}}(i, j)$ represents the degree that product i is similar to product j.
5. The membership function μ_{MOST} represents the fuzzy quantifier, MOST.

From this data, the quantity of product A sold out in the last few months is

$$T_A = \sum_i \mu_{\text{last few months}}(j)Q_{Ai} \tag{61}$$

and quality of product j sold in last few months is

$$T_j = \sum_i \mu_{\text{last few months}}(j)Q_{ji} \tag{62}$$

The membership function of fuzzy set, X, representing product A selling much more than the other product j is

$$\mu_X(u_j) = \mu_{\text{sold much more}}(T_A, T_j) \tag{63}$$

The membership function of the fuzzy set, Y, representing that product is similar to product A is

$$\mu_Y(u_j) = \mu_{\text{similar}}(j, A) \tag{64}$$

The original proposition could be represented by "Most X's are Y's," and this quantification proposition could be represented as in Eqs. (55)–(57):

$$\sum \text{count}(Y/X) = \frac{\sum \text{count}(X \cap Y)}{\sum \text{count}(Y)} = \frac{\sum_j \min[\mu_X(u_j), \mu_Y(u_j)]}{\sum_j \mu_X(u_j)} \tag{65}$$

$$\pi\left[\sum \text{count}(Y/X)\right] = \mu_{\text{MOST}}\left[\sum \text{count}(Y/X)\right] \tag{66}$$

3. Approximate Reasoning

In classic logic some important inference rules used for inference are

Modus ponens	$(A \wedge (A \Rightarrow B)) \Rightarrow B$	(67)
Modus tollens	$(\overline{B} \wedge (A \Rightarrow B)) \Rightarrow \overline{A}$	(68)
Hypothetical syllogism	$((A \Rightarrow B) \wedge (B \Rightarrow C)) \Rightarrow (A \Rightarrow C)$	(69)

These inference rules, called tautologies, are always true regardless of the truth value of proposition A, B, and C. The modus ponens is related to the forward data-driven inference, often used in control systems. The modus tollens is related to backward goal-driven inference, mainly used in expert systems.

There are many kinds of different reasoning in fuzzy logic. According to the existence of fuzzy modifiers, these can be categorized into three types of reasoning: categorical, syllogistic, and dispositional reasoning. The premises and conclusions can be expressed in their canonical forms

in the reasoning. Therefore, each premise is a constraint on some variables, and the conclusion is the induced constraint for those variables
[9–11].

Categorical Reasoning   In categorical reasoning the premises contain
no fuzzy quantifiers nor fuzzy qualifiers. The premises are in the canonical
form "x is A" or in the conditional canonical form "if x is A then Y is
B." Let A, B, C be fuzzy predicates, and let x, y, z be variables in universes
U, V, W.

1. Conjunction rule of inference

    x is A

    $$\frac{y \text{ is } B}{x \text{ is } A \cap B} \tag{70}$$

2. Cartesian product rule of inference

    x is A

    $$\frac{y \text{ is } B}{(x, y) \text{ is } A \times B} \tag{71}$$

3. Entailment rule of inference

    x is A

    $$\frac{A \subset B}{x \text{ is } B} \tag{72}$$

4. Composition rule of inference

    x is A

    $$\frac{(x, y) \text{ is } R}{y \text{ is } A \circ R} \tag{73}$$

5. Generalized modus ponens (GMP) of inference

    x is A

    $$\frac{\text{if } x \text{ is } B \text{ then } y \text{ is } C}{y \text{ is } A \circ R_{A \to B}} \tag{74}$$

For example,

The color of the banana is very yellow

$$\frac{\text{If the color of the banana is very yellow, then banana is very ripe}}{\text{The banana is very ripe}}$$

The GMP is a special case of the compositional rule of inference. It is related to the partial matching of premises of rules between "x is A" and "x is B."

If the Mamdani's min-operation of fuzzy implication in Eq. (52) and max-min fuzzy composition in Eq. (37) are used for fuzzy inference, then in the previous example, the membership function of fuzzy inference would be obtained as

$$\mu_{C'}(y) = \max_{y} \min[\mu_A(x), \min\{\mu_B(x)\mu_C(y)\}] \tag{75}$$

If the Larsen's product-operation in Eq. (53) and max-product fuzzy composition in Eq. (38) are used, then the result of inference would be

$$\mu_{C'}(y) = \max_{y} \min\{\mu_A(x), \mu_B(x)\mu_C(y)\} \tag{76}$$

where $C' = A \circ (B \to C)$

Fuzzy Syllogistic Reasoning In fuzzy syllogistic reasoning, the propositions in premises contain fuzzy quantifiers. They could be expressed as

Q_1 A's are B's

$$\frac{Q_2 \ C\text{'s are } D\text{'s}}{Q_3 \ E\text{'s are } F\text{'s}} \tag{77}$$

where A, B, C, D, E, F are fuzzy predicates. Q_1, Q_2 are given fuzzy quantifiers, and Q_3 a quantifier to be decided. Two of the basic fuzzy syllogisms are described:

1. Intersection syllogism

   Q_1 A's are B's

   $$\frac{Q_2 \ (A \text{ and } B)\text{'s are } D\text{'s}}{(Q_1 \ (\cdot) \ Q_2) \ A\text{'s are } (B \text{ and } C)\text{'s}}$$

   where $(Q_1 \ (\cdot) \ Q_2)$ denotes the product of fuzzy numbers Q_1 and Q_2.

2. Consequent conjunction syllogism

   Q_1 A's are B's

   $$\frac{Q_2 \ B\text{'s are } C\text{'s}}{Q_3 \ A\text{'s are } (B \text{ and } C)\text{'s}} \tag{79}$$

   where Q_3 is defined by $0 \ (\vee) \ (Q_1 \ (+) \ Q_2 \ (-) \ 1) < Q_3 < Q_1 \ (\wedge) \ Q_2$, and $(+)$, $(-)$, $(\vee)$, $(\wedge)$ are addition, subtraction, maximum, and minimum operators on fuzzy numbers.

Dispositional Reasoning In dispositional reasoning, the premises contain explicitly or implicitly the fuzzy quantifier "usually." Therefore, the propositions are preponderantly but not necessarily always true. For example, "Tomato is a red color," "The shape of an orange is spherical." The importance of dispositional logic is that most commonsense knowledge can be viewed as a collection of dispositional propositions, and we can infer from this common sense. Three forms of dispositional reasoning are introduced in the following:

1. Dispositional entailment rule of inference

   (usually) x is A

   $$\frac{\text{(usually) } A \subset B}{\text{(usually) } x \text{ is } B} \tag{80}$$

   or (usually)^2 x is B

2. Dispositional modus ponens

   (usually) x is A

   $$\frac{\text{(usually) if } x \text{ is } A \text{ then } y \text{ is } B}{\text{(usually)}^2 \ y \text{ is } B} \tag{81}$$

3. Dispositional consequent conjunction syllogism

   (usually) A's are B's

   $$\frac{\text{(usually) } B\text{'s are } C\text{'s}}{(2 \text{ usually } - \ 1)(A\text{'s are } (B \text{ and } C)\text{'s})} \tag{82}$$

E. Fuzzification and Defuzzification

Fuzzification is the encoding of observed nonfuzzy data into fuzzy sets defined in the universes of input variables. Most data measured by sensors in processes are crisp data such as temperature, flow rate, and pressure. A simple, intuitive method of fuzzification is to convert a crisp datum into a fuzzy singleton. This method also greatly simplifies the operation in fuzzy reasoning; thus it is often used in fuzzy control applications.

Defuzzification is a mapping from the space of a fuzzy set to a space of crisp values. For most applications in process control, a final nonfuzzy output is required to actuate the process, but the result of fuzzy inference is a fuzzy set. Therefore, defuzzification is required to decode the fuzzy control output into crisp control output. Two methods often used in fuzzy control are center of area (COA) method, mean of maximum (MOM) method [13].

Assume q fuzzy sets $C_1, C_2, \ldots, C_q$ of an output variable y are obtained after inference from q fuzzy rules. (Fig. 2):

1. Center of area (COA) method: The defuzzification of the output variable is the center of area of the possible distribution of the composite fuzzy set C':

$$C' = C_1 \vee C_2 \vee \cdots \vee C_q \tag{83}$$

$$y_{COA} = \frac{\sum_{j=1}^{n} y_j \mu_{C'}(y_j)}{\sum_{j=1}^{n} \mu_{C'}(y_j)} \tag{84}$$

where the universe set U of the output variable is quantified as $\{y_1, y_2, \ldots, y_n\}$; n is the number of quantization levels; y_j is the amount of control output at the jth quantization level; $\mu_{C'}(y_j)$ is the value of the membership function of fuzzy set C' at y_j.

2. Mean of maximum (MOM) method: The output w_j is the mean value of outputs that have the maximum degree of the membership function for the jth fuzzy set C_j, and the final output y is the average of the w_j:

$$z_{MOM} = \sum_{j=1}^{q} \frac{w_j}{q} \tag{85}$$

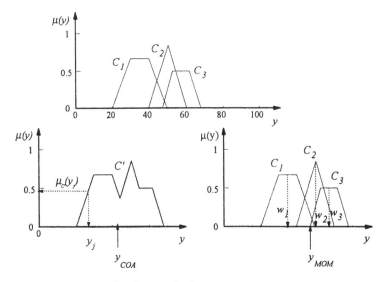

FIGURE 2 Defuzzification methods.

III. NEURAL NETWORK PRINCIPLES

A. Introduction to Artificial Neural Networks

An artificial neural network (ANN) is a computational structure inspired by biological neural systems. The biological system can handle complicated tasks such as image and speech recognition, classification, generalization, adaptive learning, and so on. Those tasks are difficult or impossible to accomplish by conventional computing methods. Therefore, the motivation for ANNs is to achieve many of those desirable abilities of the biological neural systems.

All ANN models attempt to achieve good performance through dense interconnection of simple computation elements. They are specified by their net topology, node characteristics, and learning rules. The basic processing unit is called a neuron (node), which performs the simple functions of summing inputs and nonlinear mapping. Each connection comes with a numerical value, called a weight, expressing the strength of connection. The network responds to the input data collectively and simultaneously.

In contrast to conventional methods, the designed models are usually analyzed and built for specific systems. The ANN is developed in a different procedure: choosing the proper network architecture, assigning initial weights of connections, selecting a proper training algorithm, and feeding the proper data set for training. The ANN has potential advantages that we attempt to exploit [14, 15]:

1. *Adaptation and learning ability*: A priori knowledge of the system is not needed for constructing the ANN, because the ANN will learn its internal representation from the input/output data of its environment and response. It can continue adapting itself after training.
2. *Robust, or fault tolerance*: The ANN is more tolerant of noisy and incomplete data, because the information is distributed in the massive processing nodes and connections. Minor damage to parameters in the network will not degrade overall performance significantly.
3. *High computational speed*: The ANN is an inherently parallel architecture. The result comes from the collective behavior of a large number of simple parallel processing units.

B. Biological Neural Networks

Although ANN is inspired by the biological neural system, no ANN has successfully duplicated the performance of the human brain. A simple understanding of biological neurons, however, will give insight into the ANN [16].

The neuron is the fundamental unit in the neural system. A schematic diagram of a neuron is shown in Fig. 3. There are many different types of neurons, but a typical neuron contains three major parts: soma (cell body), axon, and dendrites. Dendrites are fine brushes of fibers attached on the cell body. Dendrites receive signals from other neurons through their axons, which carry impulses from neurons. A long filament extending out from the other end of the cell body is the axon, which has many branches at its end. At the end of the axon, it contacts dendrites of neighboring neurons at a special contact organ, called the synapse, where the signals are passed between neurons. The signals are transmitted electrically and affected by chemical transmitters released at the synapse. The chemical transmitters also affect the response of the neuron that receives the signals.

The incoming signals from neighboring neurons are in an excitatory state if they cause firing, or an inhibitory state if they hinder firing of response. The condition for firing or not firing is decided by the state of aggregation of impulses. If it is in excitatory mode and exceeds a certain level, called the threshold value, then the neuron will generate a pulse response and send it out through its axon. Thus the activation depends on the number of signals received, the strength of the incoming signals, and the synaptic strength of the connections. The magnitude of signals is not significantly different among biological neurons; therefore, we can treat the neuron as passing information by means of binary signals. The foregoing explanation is a greatly simplified consideration of a biological neuron model.

Therefore a neuron can be considered as a simple signal processing unit with multiple inputs from other neurons and only a single output that is distributed to other neurons. The neuron aggregates the incoming

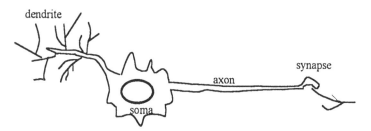

FIGURE 3 The biological neuron model.

signals, and when the signal exceeds a certain threshold level, it will produce an output signal and pass it to other neurons. Each neuron performs only a simple function, but when neurons are massively connected together, they can perform complicated tasks.

C. Basic Concepts of Artificial Neural Networks

1. Artificial Neurons

The artificial neurons are simple processing units similar to the biological neurons: they receive multiple inputs from other neurons but generate only a single output. The generated output may be propagated to several other neurons.

The first artificial neuron model proposed by McCulloch and Pitts (Fig. 4a) is based on the simplified consideration of the biological model [2]. The elementary computing neuron functions as an arithmetic logic computing element. The binary inputs of the neurons are $x_1, x_2, \ldots, x_n$. Zero represents absence, and one represents existence. The weights of connections between the ith input x_i and the neuron are represented by w_i. When $w_i > 0$, the input is excitatory. When $w_i < 0$, it is inhibitory.

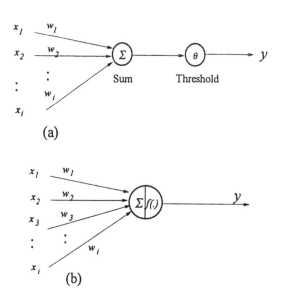

FIGURE 4 Artificial neuron models: (a) McCulloch and Pitts model [2]; (b) general neuron model.

The net summation of inputs weighted by the synaptic strength w_i at connection i is

$$net = \sum_{i=1}^{n} w_i x_i \tag{86}$$

The *net* value is then mapped through an activation function $f(\cdot)$ of the neuron output. The activation function used in the model is a threshold function:

$$y = f(net) \tag{87}$$

$$f(x) = \begin{cases} 1, & x > \theta \\ 0, & \text{otherwise} \end{cases} \tag{88}$$

where θ is the threshold value.

The neuron models used in current neural networks are constructed in a more general way. The input and output signals are not limited to binary data, and the activation function can be any continuous function other than the threshold function used in the earlier model (Fig. 4b). The activation function is typically a monotonic nondecreasing nonlinear function. Some of the often used activation functions are (where α denotes the parameter, and θ denotes the threshold value)

1. Threshold function:

$$f(x) = \begin{cases} 1, & x > \theta \\ 0, & x < \theta \end{cases} \tag{89}$$

2. Sigmoid function:

$$f(x) = \frac{1}{1 + e^{-\alpha x}} \tag{90}$$

3. Hyperbolic function:

$$f(x) = \tanh(\alpha x) = \frac{e^{\alpha x} - e^{-\alpha x}}{e^{\alpha x} + e^{-\alpha x}} \tag{91}$$

4. Linear threshold:

$$f(x) = \begin{cases} 1, & x \geq \theta \\ x/\theta, & 0 < x < \theta \\ 0, & x \leq 0 \end{cases} \tag{92}$$

5. Gaussian function:

$$f(x) = e^{-\alpha x^2} \tag{93}$$

2. Network Structures

The usual structure of an ANN is to form layers of neurons and interconnect them together as in Fig. 5a. According to the location of the layers of the neuron, there are an input layer, hidden layers, and an output layer. The input layer receives input signals directly. The output signal is sent out through the output layer in the end. Those layers not directly connected to nodes at the input and output layers are called hidden layers.

There are different ways of interconnections between neurons, shown in Figs. 5a–e:

a. *Feed-forward connections*: The connections are linked in the direction from input layer toward the output layer.
b. *Feedback connections*: The connections are linked in the reverse direction to feed-forward connections.
c. *Lateral connections*: The connections are between neurons in the same layer.
d. *Time-delayed connections*: The connections have time-delayed effects.
e. *Recurrent network*: The nodes in the output layers are connected to the nodes in the input layers.

3. Training and Testing

There are two phases in building an ANN: training and testing. In training, a set of training data is fed into the network to determine the parameters of networks, synaptic weights, and thresholds. It often involves optimizing some "energy function." In testing, another set of data is applied to the well-trained ANN network to test its generalization.

4. Supervised and Unsupervised Learning

In supervised learning, the training data set consists of pairs of input and desired output data. Each time an input is given, the desired output (response) is also given by the training data. The error signal is generated as the difference between the actual output and the desired output. Error signals are then used to update weights and thresholds of networks. Examples of this situation are perception learning and back propagation learning. In unsupervised learning, only input data are fed into network, because the desired output is not known, and thus no explicit error information is given. The learning is based on the observation of a response to inputs to with minimal a priori information available. Examples of this type of learning are winner-take-all learning and ART1 learning. In the following sections, some of the important learning rules and the corresponding networks are introduced.

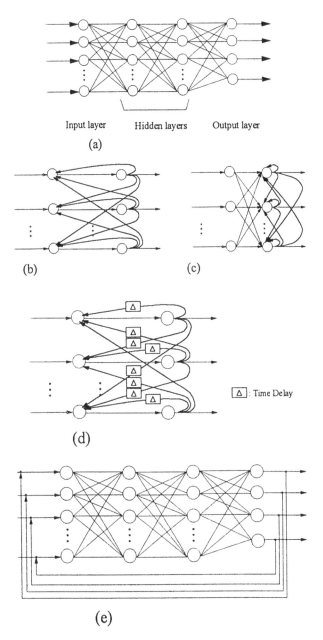

Figure 5 Structures of artificial neural networks: (a) feed-forward connections; (b) feedback connections; (c) lateral connections; (d) time-delayed connections; (e) recurrent network.

D. Important Learning Rules

1. The Perceptron Learning Rule

A perceptron was developed by Rosenblatt [17] as a simple two-class classifier. A single perceptron can learn to recognize the output target pattern when the input vector is presented (Fig. 6a). Given are n-tuple input vectors $X = [x_1, x_2, \ldots, x_n]$. Those variables belong to either class A or B. The single node computes the weighted sum of input elements as

$$net = \sum_{i=1}^{n} w_i x_i \tag{94}$$

and then passes through a threshold function:

$$f(net) = \begin{cases} 1, & net > \theta \\ -1, & \text{otherwise} \end{cases} \tag{95}$$

$$y = f(net) = f\left(\sum_{i=1}^{n} w_i x_i\right) \tag{96}$$

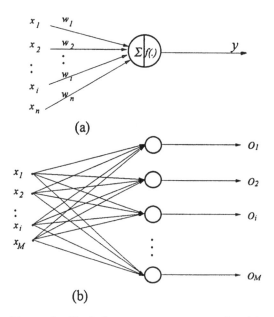

(a)

(b)

FIGURE 6   Single-layer perceptron networks: (a) a single perceptron; (b) M-class classifier.

where y is the output of perceptron and θ is the threshold value. If the perceptron output is 1, the input vector is class A. If output is -1, the input vector is class B. Therefore, a single perceptron forms a hyperplane that separates the input space into two regions representing classes A and B.

The connecting weight is updated by the perceptron learning rule. The error signal r between the actual output y and desired output d is defined as

$$r = d - y \tag{97}$$

The weights are updated by

$$w_i(t + 1) = w_i(t) + \Delta w_i(t) \tag{98}$$

$$\Delta w_i(t) = \alpha r x_i = \alpha(d - y)x_i \tag{99}$$

where α is a constant learning rate.

The learning procedure is summarized in the following:

1. Initialize the weight w_i and threshold θ with small random numbers.
2. Present a new input vector $[x_1, x_2, \ldots, x_n]$ and the desired output d to the network.
3. Calculate actual output of the perceptron by Eqs. (94)–(96).
4. Update weights by Eqs. (97)–(99).
5. If not all input vectors are presented, go to step 2.
6. Test convergence. If it is not satisfied, go back to step 2 and assume that none of the input vectors has been presented.

The M-classes of the classifier could be constructed by M perceptrons as shown in Fig. 6b. Each output node represents one of the classes. Only one output node has value 1, and the other nodes are -1 for the corresponding class. The weights of connections linked to the nodes that have the incorrectly classified class are updated in the same way as in the single perceptron (Eqs. 97–99). Another modification of perceptron learning rule is the least mean square (LMS) or Widrow–Hoff [18] learning rule that replaces the threshold activation function in Eq. (99) by a linear activation function:

$$\Delta w_i(t) = \alpha(d - net)x_i \tag{100}$$

2. Back Propagation Learning Rule

The back propagation learning rule is the generalization of the perceptron learning rule as shown in Fig. 5a for the multilayers of a neural network by Rumelhart et. al. [19]. The back propagation learning rule overcomes the past difficulty of training multilayered networks and gives new computing potential to layered networks. The pairs of input and desired output data are given to the network for training. The goal is for the neural network be able to have actual output that matches the desired output when the corresponding input is given.

The notations of the network are explained first before introducing the learning rule: The layer number is denoted by m. At input layer $m = 0$, and at output layer $m = n_f$:

$net_m(k)$ represents the net input to the kth neuron at the mth layer.
$o_m(k)$ represents the output of the kth neuron at the mth layer.
N_m is the number of neurons at the mth layer.
W_m is the weight matrix of the mth stage, between the mth and $(m - 1)$th layers.
$W_m(i, j)$ represents the element of W_m at (i, j) connecting the jth node of the $(m - 1)$th layer with the ith node of the mth layer.

Each node performs two functions in the feed-forward calculations (Fig. 7a): aggregating the weighted sum of outputs of the previous layer,

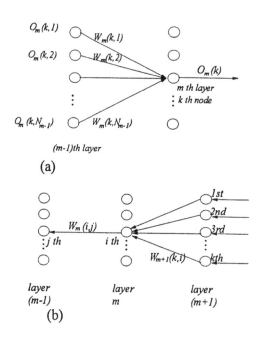

FIGURE 7   Back propagation learning: (a) feed-forward calculation; (b) error back propagation.

and mapping the sum through an activation function to get the output of this node:

$$net_m(k) = \sum_{j=0}^{N_{m-1}-1} W_n(k, j)o_{m-1}(j) \qquad (101)$$

$$o_m(k) = f[net_m(k)] \qquad (102)$$

where $f(\cdot)$ is an invertable differentiable nonlinear function.

The gradient descent method is applied to adjust the connection weights of the network toward the negative direction of an error function. At the beginning of training, the weights are usually randomly chosen values. The updating of weights is performed by moving in the direction of negative gradient along the multidimensional surface of the error function:

$$\Delta W_m(i, j) = -\alpha \frac{\partial E}{\partial W_m(i, j)} \qquad (103)$$

The energy function E_T is defined as

$$E_T = \frac{1}{2} \sum_{l=0}^{L-1} \sum_{k=0}^{N_{n_f}-1} [d_l(k) - o_{n_f}(k)]^2 \qquad (104)$$

where α is the learning rate, $d_l(k)$ is the desired output at the kth node of the output layer for the lth training data, and L is the total number of training data. For convenience we commonly update weights by considering only the error function of one training datum each time instead of all the data. Therefore, the energy function is defined as

$$E = E_l = \frac{1}{2} \sum_{k=0}^{N_{n_f}-1} [d_l(k) - o_{n_f}(k)]^2 \qquad (105)$$

Applying the chain rule to Eqs. (101) and (103),

$$\frac{\partial E}{\partial W_m(i, j)} = \frac{\partial E}{\partial\ net_m(i)} \frac{\partial\ net_m(i)}{\partial W_m(i, j)} \qquad (106)$$

The $\partial\ net_m(i)/\partial W_m(i, j)$ term here could be derived by substituting Eqs. (101) and (102) into the feed-forward calculation. Therefore,

$$\frac{\partial\ net_m(i)}{\partial W_m(i, j)} = o_{m-1}(j) \qquad (107)$$

The $\partial E/\partial\ net_m(i)$ terms are derived differently for the output layer and the hidden layers.

1. *At the output layer:* $m = n_f$

$$\delta_m(i) = \delta_{n_f}(i) \equiv -\frac{\partial E}{\partial\ net_{n_f}(i)}$$

$$= -\frac{\partial E}{\partial o_{n_f}(i)} \frac{\partial o_{n_f}(i)}{\partial\ net_{n_f}(i)} \tag{108}$$

$$= [d(i) - o_{n_f}(i)]f'[net_{n_f}(i)]$$

where $\delta_m(i)$ is called delta error signal produced at the ith node of output layer. Substitute Eqs. (107) and (108) into (106). The updating equations could be obtained as

$$\Delta W_{n_f}(i, j) = \alpha\ \delta_{n_f}(i)o_{n_f-1}(j)$$

$$= \alpha[d(i) - o_{n_f}(i)]f'[net_{n_f}(i)]o_{n_f}(j) \tag{109}$$

2. *At the hidden layer* m: The delta error signal at the ith neuron of the mth layer is defined as

$$\delta_m(i) \equiv -\frac{\partial E}{\partial\ net_m(i)}$$

$$= -\frac{\partial E}{\partial o_m(i)} \frac{\partial o_m(i)}{\partial\ net_m(i)} \tag{110}$$

$$= -\frac{\partial E}{\partial o_m(i)}\ f'[net_m(i)]$$

The error signal at hidden layers could not be obtained directly in the hidden layer as in the output layer, but it can be derived from the error signal of the next layer:

$$-\frac{\partial E}{\partial o_m(i)} = \sum_{k=0}^{N_{m+1}-1} \left[-\frac{\partial E}{\partial\ net_{m+1}(k)} \frac{\partial\ net_{m+1}(k)}{\partial o_m(i)} \right] \tag{111}$$

$$= \sum_{k=0}^{N_{m+1}-1} [\delta_{m+1}(k)\ W_{m+1}(k, i)]$$

where $\delta_{m+1}(k)$ is the error signal at the $(m + 1)$th layer and $\partial net_{m+1}(k)/\partial o_m(i) = W_{m+1}(k, i)$.

Therefore, the weight updating equations at the mth layer are

$$\Delta W_m(i, j) = \alpha\ \delta_m(i)\ o_{m-1}(j) \tag{112}$$

$$\delta_m(i) = f_i[net_m(i)] \sum_{k=0}^{N_{m+1}-1} [\delta_{m+1}(k)\ W_{m+1}(k, i)] \tag{113}$$

Because the error signal is propagated from the output layer back to the input layer as shown in Fig. 7b, it is called error back propagation learning.

The procedure for back propagation learning is summarized in the following:

1. Initialize the weights w_i with small random numbers.
2. Present a new input vector $[x_1, x_2, \ldots, x_{N_0}]$ and the desired output $[y_1, y_2, \ldots, y_{N_{n_f}}]$

$$[x_1, x_2, \ldots, x_{N_0}] = [o_0(1), o_0(2), \ldots, o_0(N_0)] \qquad (114)$$

$$[y_1, y_2, \ldots, y_{N_{n_f}}] = [o_{n_f}(1), o_{n_f}(2), \ldots, o_{n_f}(N_{n_f})] \qquad (115)$$

3. Calculate outputs of all nodes from the first layer to the output layer by forward calculation as in Eqs. (101) and (102).
4. Calculate the delta error signals of all nodes from the output layer backward to the input layer by Eq. (108) for the output layer, Eq. (110) for the input layer.
5. Adapt weights by Eq. (109) for the output layer, Eq. (112) for the hidden layers.
6. If not all input vectors are presented, go to step 2.
7. Test convergence. If it is not satisfied, go to step 2 and assume none of inputs has been presented.

Back propagation learning always leads to the nearest local minimum. Thus whether a good minimum is obtained depends on the initial values of the weights and on the selection of the error measurement. It is possible to obtain input–output relationships of a nonlinear system using a three-layered back propagation neural network [19].

3. Hebbian Learning Rule

The Hebbian learning rule is for the neural networks that function as an associative system [20]. An associative system can give the associated output when a given input is presented. The training phase is to record the associative input–output exemplars in the network. An associative system functions as a content addressable memory.

To examine the Hebbian learning rule, we consider a single-layer neuron. The binary input vector is denoted by $[x_1, x_2, \ldots, x_n]$, and binary output is denoted by $[o_1, o_2, \ldots, o_m]$; w_{ij} is the weight of connection between the ith output node and the jth input node. The Hebbian learning rule assumes that if two neurons were active at the same time, the strength of connection between them should be increased. If correlation term $o_i x_j$ between input and output nodes is positive, then increase the weight w_{ij}; otherwise, decrease the weight:

$$o_i = f\left(\sum_{j=1}^{n} w_{ij}x_j\right), \qquad f(net) = \begin{cases} 1, & net > 0 \\ 0, & \text{otherwise} \end{cases} \tag{116}$$

$$\Delta w_{ij} = \alpha o_i x_j \tag{117}$$

The procedure of Hebbian learning rule is as follows:

1. Initialize the weights w_i with small random numbers around zero.
2. Present a new input vector $[x_1, x_2, \ldots, x_n]$ and the desired output d.
3. Calculate the actual output of the perceptron by Eq. (116).
4. Adapt weights by Eq. (117).
5. If not all input vectors are presented, go to step 2.
6. Test convergence. If it is not satisfied, go to step 2 and assume none of inputs has been presented.

4. Hamming Net and MAXNET Learning Rule

The procedure of Hamming net and MAXNET is an unsupervised learning classifier for binary input vectors. It consists of two layers as shown in Fig. 8. The first layer is a Hamming network, which is a feed-forward type network and selects the stored class that is closest to the input vector. The Hamming distance (HD) is defined as the number of different bits between two vectors. The minimum HD between the input and the exemplar is indicated by the output response of the neuron at the first layer. The second layer is a MAXNET, which is a recurrent network. Its function is to suppress the values of MAXNET outputs that are not maximal at the beginning [14].

Assume the input vector is an n-tuple $[x_1, x_2, \ldots, x_n]$ and there exist m exemplars representing p classes, $C^{(m)}$, $m = 1, 2, \ldots, p$:

$$C^{(m)} = [c_1^{(m)}, c_2^{(m)}, \ldots, c_p^{(m)}] \tag{118}$$

For a Hamming network, the weights w_{ij} of connections from input i to node j and thresholds θ_j of node j are assigned values:

$$w_{ij} = \frac{c_i^{\,j}}{2}, \qquad \theta_{ij} = \frac{n}{2} \tag{119}$$

The net sum of input at the mth node, net_i, is

$$net_i = \sum_{j=1}^{n} w_{ij}x_j + \frac{n}{2} \tag{120}$$

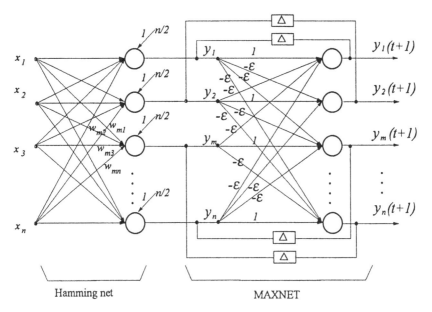

FIGURE 8   Hamming net and MAXNET.

The activation function of the Hamming net is a linear function:

$$f_H(\text{net}) = \frac{\text{net}}{n} \tag{121}$$

The output of the closest class to the input vectors will have the largest output value (or the smallest Hamming distance).

For a MAXNET network, a recurrent network is used to enhance the output value of the dominate neuron at the output of the Hamming net and suppress the values of the other nodes. The weight is assigned by

$$b_{ij} = \begin{cases} 1, & \text{if } i = j \\ -\epsilon, & \text{if } i \neq j, \, 0 < \epsilon < 1/p \end{cases} \tag{122}$$

The activation function of MAXNET is

$$f_M(\text{net}) = \begin{cases} 0, & \text{net} < 0 \\ \text{net}, & \text{net} \geq 0 \end{cases} \tag{123}$$

The output of MAXNET is

$$
\begin{aligned}
y_j(t + 1) &= f_M \left[\sum_{i=1}^{n} b_{ij} y_i(t) \right] \\
&= f_M \left[y_j(t) - \epsilon \sum_{k \neq j} y_k(t) \right]
\end{aligned}
\tag{124}
$$

Equation (124) is iterated until the convergence criterion is satisfied. Ultimately there will be only one positive output node of MAXNET.

The procedure of Hamming net and MAXNET is as follows:

1. Initialize the weights w_i and threshold θ_j by Eq. (119).
2. Present a new input vector $[x_1, x_2, \ldots, x_n]$.
3. Calculate output of the Hamming net by Eqs. (120) and (121).
4. Calculate output of MAXNET by Eq. (124).
5. If more than one output of MAXNET is positive, go to step 4.

5. Winner-Take-All Learning Rule

Winter-take-all is an unsupervised learning rule [21], and the network is called a Kohonen network [22]. The network can classify input vectors into one of the P classes. Assume the input vector is an n-tuple vector X and there exist P classes. The number of output nodes is the same as the number of classes in the network, as shown in Fig. 9.

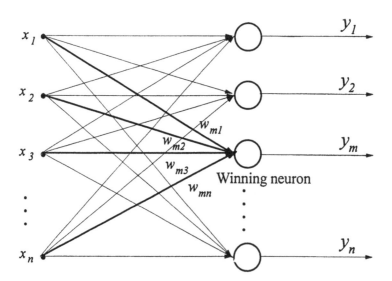

FIGURE 9 The winner-take-all learning rule.

The initial weights w_{ij} for connection from input node j to output node i are initialized at normalized random values. The winning neuron is selected as that having the maximum response due to the input X. If we assume the mth neuron is the winning neuron, i.e.,

$$\sum_{j=1}^{n} w_{mj}x_j = \max_{i=1,2,\ldots,P} \left(\sum_{j=1}^{n} w_{ij}x_j\right) \tag{125}$$

then only the weights of connections that link to the wining neuron (fan-in weights) are updated by

$$\Delta w_{mj} = \alpha(x_j - w_{mj}), \quad \text{for } j = 1, 2, \ldots, n \tag{126}$$

where α is the constant learning rate.

6. ART1 Learning Rule

The Adaptive Resonance Theory 1 (ART1) network [14, 23] is shown in Fig. 10. It can recognize vectors from previously learned categories while creating a new class in response to the presentation of the new vectors. It can form new clusters and can be trained without supervision. It can also incorporate the new clusters without affecting the storage clusters that already have been learned. The network generates clusters by itself, if the presented input is identified as not belonging to some existing cluster. The stored cluster information will be updated.

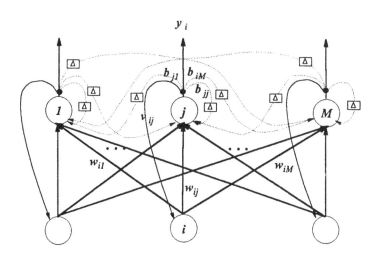

FIGURE 10 The ART1 network.

The first cluster is created with the first input pattern vector received, and then a second cluster is created if the distance between the presented data later and the first cluster exceeds a threshold value. Therefore, the presented input either follows and accepts the old cluster or creates a new cluster.

The weights of bottom-up connection w_{ij} and top-down connection $m_{ij}(t)$ connect the input node i and output node j. A binary unipolar input vector $[x_1, x_2, \ldots, x_n]$ is presented to the input of network. The output of the jth output neuron in the beginning is y_j^0:

$$y_j^0 = \sum_{i=1}^{n} (w_{ij}x_i), \quad \text{for } m = 1, 2, \ldots, M \tag{127}$$

M is the number of output nodes, which is the maximum possible number of cluster values. The value of y_m^0 represents an initial matching score that measures the similarity between the input vector and the existing cluster. The best matching existing cluster is m, and

$$y_m^0 = \max_{m=1, 2, \ldots, M} y_m^0 \tag{128}$$

is selected by the MAXNET learning method as in Eqs. (122)–(124).

The top-down part of the network is to check the similarity of the input vector to the selected stored cluster m. The vigilance test is then performed as

$$y_m^0 = \frac{\sum_{i=1}^{n} (b_{ij}x_i)}{\sum_{i=1}^{n} |x_i|} > \rho \tag{129}$$

where ρ is the vigilance threshold. If the test is good, the new input belongs to one of the existing clusters j. The weight connections related to output node j are updated by

$$w_{ij}(t + 1) = \frac{b_{ij}(t)x_i}{0.5 + \sum_{i=1}^{n} b_{ij}(t)x_i} \tag{130}$$

$$b_{ij}(t + 1) = x_i b_{ij}(t) \tag{131}$$

If the test is failed, the node j is disabled by setting y_j to zero temporarily, and then we go back to find the best matching scores. The algorithm will attempt to establish a new cluster different from the pattern j under test.

The procedure for ART1 learning:

1. Initialize weights and set vigilance threshold value:

$$w_{ij} = 1, \quad b_{ij} = \frac{1}{1 + n}, \quad 0 \le \rho \le 1 \tag{132}$$

2. Present a binary unipolar input to network.
3. Compute the matching scores by Eq. (127) and select the best one by Eq. (128).
4. Test similarity by the vigilance test of Eq. (129).
5. If the test is good, update the weights by Eqs. (126) and (127). If the test is not good, disable the selected best matching y_j temporarily and go back to step 3. It will create a new cluster.
6. If input data are not all presented, go back to step 2.

IV. IMPLEMENTATION OF NEURO-FUZZY TECHNOLOGY FOR PROCESS AUTOMATION

In this section some applications of fuzzy logic and neural networks for food process automation, such as process modeling, controls, sensors, and classification are briefly reviewed. The basic concepts and theories for implementing those techniques are explained. Furthermore, some new applications of neural fuzzy control systems are also introduced.

A. Process Modeling

For process automation, we must be able to predict process behavior. Therefore process modeling is essential for food process automation. Conventional methods for process modeling are either deterministic or stochastic [24]. Deterministic modeling describes and analyzes the process by governing equations. However, those equations are often not available for complicated processes. Stochastic modeling uses the past data to infer the probability distributions of future events. Fuzzy modeling and neural network modeling are two different modeling methods from these conventional ones.

1. Fuzzy Modeling

The fuzzy modeling method applies fuzzy logic and approximate reasoning to describe the relations among the process variables and deduce the response of output variables from the given values of input variables. Fuzzy modeling is suitable for food processes because many food processes have a complex nature, insufficient and inaccurate information, and much uncertainty [25]. Fuzzy logic can deal with uncertainty due to vagueness or fuzziness rather than randomness. Therefore, we can use data at certain levels of inconsistency and minimize the loss of valuable information during modeling [26]. Furthermore, food qualities are language oriented and difficult to describe and measure. Fuzzy logic can better handle this situation than deterministic or stochastic modeling.

A fuzzy model is constructed by a set of fuzzy conditional propositions describing the relations R_i among process variables [25, 26]. The process model, which is also the knowledge base of the process R_{KB}, is described by

$$R_{KB}: \text{``}R_1 \text{ also } R_2 \text{ also } R_3 \ldots \text{ also } R_n\text{''} \tag{133}$$

$$R_{KB} = R_1 + R_2 + \cdots + R_n \tag{134}$$

$$R_i: \text{``if } (x_1 \text{ is } A_{i,1}) \text{ and } (x_2 \text{ is } A_{i,2}) \text{ and } \ldots \text{ and } (x_n \text{ is } A_{i,n}), \text{ then } (y \text{ is } B_i)\text{''} \tag{135}$$

where x_i is the ith state variable, y is the output variable, A and B are the linguistic values of the corresponding process variables, "also" is represented by the maximum operator " $+$ ", and R_i is the ith fuzzy proposition for describing the relations among process variables.

Consider an n-dimensional query for the process output y in the form

$$Q: \text{``}(x_1 \text{ is } Q_1) \text{ and } (x_2 \text{ is } Q_2) \text{ and } \ldots \text{ and } (x_n \text{ is } Q_n)\text{''} \tag{136}$$

$$Q = Q_1 \wedge Q_2 \wedge \cdots \wedge Q_n \tag{137}$$

The answer to the query can be obtained using two methods:

1. Apply the composition rule of inference by Eq. (74) on the query Q and the fuzzy model of the process R_{KB}:

    $$\begin{aligned} y &= Q \circ R_{KB} \\ &= Q \circ (R_1 \wedge R_2 \wedge \cdots \wedge R_n) \\ &= (Q \circ R_1) \wedge (Q \circ R_2) \wedge \cdots \wedge (Q \circ R_n) \end{aligned} \tag{138}$$

2. Consider the degree of consistency (or similarity) between the query Q and the antecedent of fuzzy proposition "$(x_1 \text{ is } A_{i,1})$ and $(x_2 \text{ is } A_{i,2})$ and $\ldots$ and $(x_n \text{ is } A_{i,n})$":

    $$A_i = A_{i1} \wedge A_{i2} \wedge \cdots \wedge A_{in} \tag{139}$$

    Usually there is no exact match between Q and A_i. The degree of consistency between Q_i and A_{ij} is defined as

    $$\gamma_{ij} = \vee(Q_j \wedge A_{ij}) = \bigvee_{v_j} [\mu_{Q_j}(u_j) \wedge \mu_{A_{ij}}(u_j)] \tag{140}$$

    The overall degree of consistency for the antecedent of the ith rule γ_i is

    $$\gamma_i = \gamma_{i1} \wedge \gamma_{i2} \wedge \cdots \wedge \gamma_{in} \tag{141}$$

    If all $\gamma_i = 0$, the knowledge base has no information for the query at all. The values of γ_i also represent the weighted coefficient for the rule consequent B_i. Therefore, the desired output y is

$$y = \gamma_1 \wedge B_1 + \gamma_2 \wedge B_2 + \cdots + \gamma_{im} \wedge B_m \tag{142}$$

If a crisp output is required, then the COA and MOM defuzzification methods in Eqs. (84) and (85) can be applied.

Here is an example of food-related fuzzy modeling. A fuzzy prediction model is developed to predict the corn breakage level during the drying process [27–30]. The prediction model maps the weighted factors of process state variables into linguistic values of an output variable. Let $x_1, x_2, \ldots, x_m$ denote m process state variables of a drying process, such as drying temperature, initial moisture content of product, equilibrium moisture content of product, and so on. The values of y denote the process output variable, i.e., the corn breakage level.

The linguistic variables can be described by n linguistic values l_1, $l_2, \ldots, l_n$, such as [very high, high, medium, low, very low, low]:

$$A(x_i) = [l_1, l_2, \ldots, l_n], \qquad T(y) = [l_1, l_2, \ldots, l_n] \tag{143}$$

Let g_i represent the degree that the breakage level y is l_i, and let v_i represent the weighting factor of the measured input variable x_i in one of $A(x_i)$.

The fuzzy matrix R describes the relation between the weighting factor g_i of the measured input variable and the breakage level. R is a $m \times n$ matrix:

$$R = \begin{bmatrix} r_{11} & r_{12} & \cdots & r_{1n} \\ r_{21} & r_{22} & \cdots & r_{2n} \\ \vdots & \vdots & \ddots & \vdots \\ r_{m1} & r_{m2} & \cdots & r_{mn} \end{bmatrix} \tag{144}$$

Element r_{ij} represents the degree of membership that state variable x_i will lead to the jth linguistic value l_j of the process output variable y.

Let $G = [g_1, g_2, \ldots, g_n]$ and $V = [v_1, v_2, \ldots, v_n]$; then

$$G = V \circ R \tag{145}$$

$$g_j = \bigvee_{i=1}^{n} (v_i \wedge r_{ij}) \tag{146}$$

The values of R and V can be obtained experimentally by the polling methods described in Eq. (26):

$$r_{ij} = \frac{O[A(x_i, l_j)]}{O[A(x_i)]}, \qquad v_i = \frac{O[A(x_i, \sigma)]}{O[A(x_i)]} \tag{147}$$

where $O[A(x_i)]$ is the number of samples belonging to $A(x_i)$ of state variable x_i, $O[A(x_i, l_j)]$ is the number of samples belonging to $A(x_i)$ of variable x_i and jth linguistic value l_j of output variable y, and $O[A(x_i, \sigma)]$ is the number of samples within the standard deviation of the average of x_i. Once we get the values of R and the weighting vector V, we can predict the degree of membership of each output variable in its linguistic value (the corn breakage level) by Eq. (145).

Fuzzy modeling and fractal analysis were applied to revitalize food process information in [26]. The meat chilling and malt modification processes were studied to demonstrate that fuzzy modeling can handle data with uncertainty and increase reasoning ability because of less information loss.

2. Neural Network Modeling

A neural network can model a process without much a priori knowledge of the process because the network is taught from exemplar training data sets. It is especially good for modeling some food processes that are ill-defined, not well-known, nonlinear, multivariate, or involve handling massive data. The modeling method is to identify the input and output variables of the process, select the proper neural network structure and learning rules, and train the network by a set of training data in supervised learning or its own output response in unsupervised learning.

A simple, and the most often used, neural network for process modeling is the multilayer network with back propagation learning rule. The process input and output variables are monitored and recorded as the input and desired output of the neural network. The network can be trained off line or on line.

A three-layer feed-forward neural network training with back propagation learning has been used to model extrusion cooking processes. Moisture, feed rate, and screw speed were nodes at the input layer, and expansion, motor current, die pressure, and density were nodes at the output layer. The network was trained by experimental data sets. The reverse calculation of the network output was suggested to be a suitable value for extrusion cooker control [31].

Hoop stress of tomato fruit, which is the key factor for crack occurrence, was predicted by neural network models. The inputs of model were ambient temperature changes, relative humidity changes, and physical properties of tomatoes. The hoop stress was obtained by finite element methods because it was difficult to measure directly. The trained neural network can be considered as the mapping of finite element models. The simulation result was consistent with field experiment results for predicting the critical ambient water potential change that causes severe tomato cracking [32].

The cucumber fermentation process was modeled by a back propagation neural network and compared with the model of response surface methodology (RSM). The results showed that the back propagation networks were as good as or better than the RSM in modeling food processes [33].

A time-lagged recurrent neural network was used to model a food dehydration process. The recurrent network could be viewed as a number of feed-forward neural networks cascaded together for each time step, and the training is similar to the back propagation through time [24]. A dynamic neural network modeling for a snack food frying process was proposed in [34]. The networks were started with a sufficiently high order of structure, and the unimportant nodes were deleted to achieve an optimal process model with minimum acceptable size of the network model.

B. Process Control

1. Fuzzy Logic Control

The earliest application of fuzzy logic control was by Mamdani and Assilian [35]. Fuzzy logic control is one of the most successful applications of fuzzy sets, fuzzy logic, and fuzzy reasoning [36]. Fuzzy logic control can incorporate the rules of thumb of operators and the process knowledge of experts. It is beneficial for control of those food processes that are difficult to analyze by mathematical models but can be accomplished by experienced operators. It offers a better man–machine interface for process control by expressing the control strategies and food properties in the humanlike language. It is also robust with respect to undesirable perturbations and system aging because of its generalization and redundancy induced by its parallel architecture of rules. Even when some rules are corrupt the system still functions well.

Properly chosen state variables and control variables are essential to the performance of a fuzzy control system. They are usually chosen based on the expert's experience or engineer's knowledge. The input and output variables of a fuzzy logic controller are linguistic variables, as defined in Section II.D. Consider a multi-input, single-output process control. Let x_i represent an input state linguistic variable, y the output control linguistic variable:

$$\{x_i, \{T^1_{x_i}, T^2_{x_i}, \ldots, T^{g_i}_{x_i}\}, U_i, \{M^1_{x_i}, M^2_{x_i}, \ldots, M^{g_i}_{x_i}\}\}$$
$$\{y, \{T_y^{\,1}, T_y^{\,2}, \ldots, T_y^{\,h_i}\}, V_i, \{M_y^{\,1}, M_y^{\,2}, \ldots, M_y^{\,g_i}\}\}$$

$$(148)$$

$T^{g_i}_{x_i}$, $T_y^{\,h_i}$ are the associated term sets of the input variable x_i and output variable y, respectively. For example, these could be Positive Big (PB),

Approximate Zero (AZ), Negative Medium (NM), etc. $M^j_{x_i}$, M^j_y are the membership functions of variables x_i, y, respectively. The triangular-shaped, trapezoidal-shaped, and bell-shaped functions are often chosen for fuzzy control applications.

The typical architecture of a fuzzy logic control system consists of fuzzification interface, knowledge base, fuzzy inference unit, and defuzzification interface, as shown in Fig. 11. The fuzzification interface transforms the crisp measured data into a fuzzy set representing the suitable linguistic value. The knowledge base stores empirical knowledge of process operation, such as fuzzy rules in a rule base, and membership functions in a database. The fuzzy inference unit, similarly to human decision making, performs approximate reasoning. The defuzzification interface gets a nonfuzzy decision or control action from an inferred fuzzy control action [4, 7, 13].

There are different types of fuzzy rules. The state evaluation type of fuzzy control rule is most often used and has the form

R_i: "if $(x_1$ is $A_{i,1})$ and $(x_2$ is $A_{i,2})$ and . . . and $(x_n$ is $A_{i,n})$, then $(y$ is $C_i)$"

(149)

where $A_{i,j}$ and C_i are the fuzzy term sets of the corresponding linguistic variables for the ith rule. The state variables are evaluated to determine the degree of contribution of each rule. An example involving oven temperature control is that "if (oven temperature is about the set temperature and the gas oven temperature rises a little) then (turn down the gas a

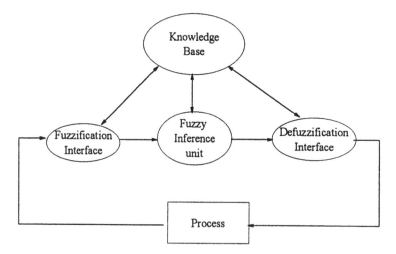

FIGURE 11 The structure of a fuzzy logic control system.

little).'' The rule base could be derived from expert experience, control engineering knowledge, an operator's control actions, a fuzzy model of a process, or an adaptive control rule [37].

The rule base contains a collection of linguistic statements describing the strategies for process control or decision making:

$$R_{KB}: \text{``} R_1 \text{ also } R_2 \text{ also } R_3 \ldots \text{ also } R_n \text{''} \tag{150}$$

$$R_{KB} = R_1 + R_2 + \cdots + R_n \tag{151}$$

where "also" is the sentence connective term and is represented by the maximum operator " + ". The state variables are observed as

$$R: \text{``} (x_1 \text{ is } A'_{i,1}) \text{ and } (x_2 \text{ is } A'_{i,2}) \text{ and } \ldots \text{ and } (x_n \text{ is } A'_{i,n}) \tag{152}$$

The most important part of fuzzy logic control is the fuzzy inference unit, which is based on fuzzy logic and approximate reasoning. This procedure is similar to a human decision-making procedure based on the observed process data and vague control strategies to make a decision for the next control action. The generalized modus ponens inference is applied to derive the output:

$$y = R \circ R_{KB} \tag{153}$$
$$= R \circ (R_1 \vee R_2 \vee \cdots \vee R_n)$$

When the input variables are crisp data and fuzzified as fuzzy singletons, the inference result simplifies to a very simple result.

1. If we choose R_M as a fuzzy implication, max-min composition as a compositional operation "$\circ$", fuzzy intersection as a sentence connective "and," and fuzzy union as a sentence connective "also," then the membership function of the fuzzy inference can be shown to be

$$\mu_M(z) = \bigcup_{i=1}^{n} \{\min[\min(\mu_{A_i}(x_0), \mu_{B_i}(y_0)), \mu_{C_i}(z)]\} \tag{154}$$
$$= \max\{\min[\alpha_i, \mu_{C_i}(z)]\}$$

2. If R_L is used instead of R_M in (1), then the membership function of the fuzzy inference is

$$\mu_L(z) = \bigcup_{i=1}^{n} \{\min[\mu_{A_i}(x_0), \mu_{B_i}(y_0)] \mu_{C_i}(z)\} \tag{155}$$
$$= \max\{\alpha_i \mu_{C_i}(z)\}$$

Here $\alpha_i = \min\{\mu_{A_i}(x_0), \mu_{B_i}(y_0)\}$ is called the firing strength, representing the degree of contribution for each control rule.

There are many other fuzzy inference methods that can be used [5, 7].

Some applications of fuzzy logic control in food processing are:

A fuzzy controller is designed and simulated for an extrusion process [25]. The water feed rate is inferred based on the feed moisture and mass feed rate. The fuzzy rules are derived by changing the water feed rate proportional to the mass feed rate and the difference between the optimal and the current moisture levels.

The temperature of a High-Temperature-Short-Time (HTST) heat exchanger has been controlled by a fuzzy logic control system [38]. Its rule base consisted of only five fuzzy control rules that were used for inference of the steam valve openness based on the observation of product temperature error and hot water error. The final control action was then defuzzified by the center of area method. The fuzzy controller was built in a discrete framework; therefore, the fuzzy rule could be put in the form of a rule table to shorten the time for inference. The result was not very successful in some cases, because of difficulty in deriving better fuzzy rules and selecting proper membership functions.

A sake-brewing fuzzy control system derived from sake-brewing experts was used to regulate the fermentation temperature for sake brewing, which needs skilled operators [39]. Sake mash's temperature is controlled according to the specific gravity and alcohol concentration. The membership functions of the control system were changeable for different brewing periods. The simulation result showed that the estimated fermentation temperature by the fuzzy control system was closer to the expert's suggestion than the estimated temperature obtained from another adaptive controller.

A coke oven gas cooling plant has been controlled by a fuzzy logic control system to regulate the cooling water flow rate and maintenance schedule [40]. The coal is burned to produce the raw gas, and gas is sent to the plant to be cooled down to the desired temperature and also for removal of some impurities. The gas cooling process is difficult to control because of the constantly changing system parameters and uneven gas flow rate. The impurities accumulate very easily on the pipe, interfering with the cooling. Therefore, maintenance is scheduled based on the "goodness" of the cooling tower by fuzzy logic inference. The system has been operating for over a year, and the result has demonstrated the robustness of the fuzzy controller to varying atmospheric temperature changes, changing configuration, and uneven flow rate.

Fuzzy controllers were used to deal with the control of cultivation of microorganisms in a fermentation process [41]. It is difficult to control the process using a classical controller, because the biological mechanism is not completely understood and on-line measurements of some variables are not reliable or accurate.

A fuzzy expert system, which is similar to a fuzzy control system, has been applied to a decision system for transfer of cows from high to low feeding groups to obtain the optimal individual or cow herd performance. The fuzzy expert system is based on information of body weight, milk production rate, etc. The membership functions are constructed by analyzing empirical data, and the rule base is derived from field experts. The data from 40 cows was analyzed by both the fuzzy expert system and the field experts. In 70% of the cases, the fuzzy expert's decisions were within one week of the experts'. Ten percent of the decisions from the fuzzy expert system were considered superior by the experts [42].

2. Neural Network Control

Neural networks can also be applied in process control. They can used in process identification as in process modeling and then combined with other controllers or used as a controller directly. Several types of neural control systems have been introduced [16, 43].

Direct Inverse Control The neural network is connected with the plant as in Fig. 12a, for process identification first. The plant output y is used as network input, and the error signal for training the network is computed as the difference between plant input and network output, $x - o$. The neural network performs the mapping of the control actions taken at the plant in inverse order. After training, it is connected with the plant as in Fig. 12a, for feed-forward control. The problem of the inverse control is when the plant inverse is not uniquely defined or the mapping is many-to-one between the input and output of plant. A simulation system for a flat bread extrusion process was developed in [44]. The control actions were taken as the inverse of the calculations of the neural networks.

Indirect Inverse Control Two neural networks A and B are connected as in Fig. 12b for process control. Network A is connected as the plant inverse identification and trained during the process. Network B is the exact copy of A. The error signal for training is the difference between the outputs of both neural networks, $x - o$. Because A and B are the same neural network, desired plant output d and actual plant output y are the same (zero error signal) when their inputs are the same.

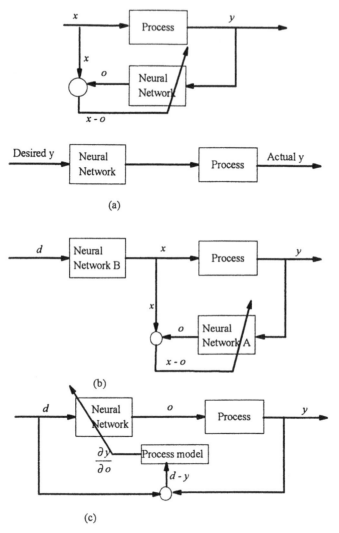

FIGURE 12 Neural network control: (a) direct inverse control; (b) indirect inverse control; (c) specialized on-line learning control.

Specialized On-Line Learning Control The neural network functions as a controller in the control system, as in Fig. 12c. The error at the plant output $(d - y)$ is used to train the network by back propagation. The error $(d - y)$ must be propagated to the plant input first before training the neural network. Let the error function of plant output be defined as

$$E = \frac{1}{2} \sum_{i=1}^{m} (d_i - y_i)^2 \tag{156}$$

The output error E is propagated back to the plant inputs by

$$-\frac{\partial E}{\partial o_i} = \left(d_1 - y_1\right)\frac{\partial y_1}{\partial o_i} + \left(d_2 - y_2\right)\frac{\partial y_2}{\partial o_i} + \cdots + \left(d_m - y_m\right)\frac{\partial y_m}{\partial o_i} \tag{157}$$

If the plant's analytical form at the operational point is known, the $\partial y_i / \partial o_i$ term can be easily obtained. If it is not, the perturbation method can be used to measure approximate values of $\partial y_j / \partial o_i$ by deviating plant's input i by Δo_i from the operational point and then measuring the resulting change at the output j, Δy_j.

C. Intelligent Sensors

One of the major problems for food process automation is a lack of on-line sensors, because product quality can be determined only in a limited way on-line [45–47]. The food quality usually depends nonlinearly on many factors and is more language oriented. Therefore, more than direct measurement at the transducer level is required to derive a higher level of information by integration of data from multiple sensors, called sensor fusion [48]. Intelligent sensors also can provide more functions than simple measurement, such as diagnosis, scaling, and calibration. The potential benefits of intelligent sensors include obtaining a higher level information, which is more than the direct gathering of the information from individual sensors; providing redundancy and reducing the overall uncertainty; robustness for the failure of some sensors; reducing the load of the controller; ability to adapt itself to the environment; and providing humanlike sensors.

Intelligent sensors can integrate information at three different levels [47]: integration at the sensor level, the measured signals of which are processed first individually by the transducers and then sent for integration; integration at the central level, which feeds the measured signals directly into a fusion processing unit; or hybrid integration, the combination the two preceding types of integration. Multiple sensors used could be similar sensors, dissimilar sensors, spatial sensors, or temporal sensors. The strategies for combining information from different sensors could involve competitive, complementary, or independent strategy.

Fuzzy logic can also analyze multiple sources of data with the rule base's inference. The input signals are inferred to reach a conclusion. This is suitable for measuring food properties that can be represented by abstract knowledge symbolically with a limited number of rules. It is also good for describing food qualities that often involve uncertain and vague information interpretation. Its rule base nature makes it easy to apply with rule-based expert systems for diagnosis and defect estimation. For example, problems could be to classify the vegetative cell or sporiferous cell, or to estimate the number of defective parts based on the measured data [45].

Neural networks can model complex and nonlinear empirical transfer functions for sensor fusion. They are appropriate for developing ill-defined relations between food qualities and measured signals. Neural networks are good for classification of data patterns and tolerate noise well. For example, they are useful for color classification based on spectral patterns [45], and the applications described in the following section on pattern classification.

D. Pattern Classification

The sorting of agricultural products still relies largely on manual operation, because most qualities in the process have no direct measurement. To automate quality sorting, a decision-making unit is required to transfer multiple quantitative data from multiple sensors into a decision on the overall product quality for classification. Pattern classification is a process for assigning each element of a data set to one of the finite set of classes. It compresses the massive quantity of data while retaining the inherent information in the data. The system usually consists of three components: transducer, feature extractor, and classifier [48, 49].

1. Fuzzy Classification

Fuzzy classification groups individual samples into classes that do not have sharply defined boundaries, as compared with conventional classification techniques. Fuzzy C-mean classifications are introduced in the following to show how fuzzy logic concepts are used in hard-limit C-mean classification [50, 51].

C-means classification is an algorithm looking for clusters in feature space by successively adding vectors into the closest class that has already been established and then updating the mean of the cluster center of all feature vectors assigned to this class. Consider n data vectors X_1,

$X_2, \ldots, X_n$ classified into m classes. Each data vector has f components representing f data features: $X_i = [x_{i1}, x_{i2}, \ldots, x_{if}]$

An $n \times f$ membership matrix M is used to describe the degree of membership that a data vector has in certain classes: $M = [m_{ij}]_{n \times f}$. m_{ij} satisfies the following conditions:

1. The sum of memberships for a individual datum across all classes is unity:

$$\sum_{j=1}^{f} m_{ij} = 1, \qquad 1 \leq i \leq n \tag{158}$$

2. The memberships of classes are positive:

$$\sum_{j=1}^{f} m_{ij} > 0, \qquad 1 \leq j \leq f \tag{159}$$

3. For the crisp theory class, a data vector either belongs absolutely or not at all to each class:

$$m_{ij} \in \{0, 1\}, \qquad 1 \leq i \leq n; \;\; 1 \leq j \leq f$$

$$m_{ij} = 1, \qquad \text{if the } i\text{th data belongs to class } j \tag{160}$$

$$m_{ij} = 0, \qquad \text{otherwise}$$

For fuzzy classification theory, the third condition is relaxed to having class membership values in $[0, 1]$,

$$m_{ij} \in [0, 1], \qquad 1 \leq i \leq n; \;\; 1 \leq j \leq c \tag{161}$$

The fuzzy C-mean classification is defined by minimizing the objective function [49, 50]:

$$J = \sum_{i=1}^{n} \sum_{j=1}^{f} m_{ij}^{\Phi} d^2(x_{iv}, \mu_{jv}), \qquad \Phi \in (1, \infty) \tag{162}$$

where

x_{iv} represents the value of the vth class in the ith data vector
μ_{jv} represents the centroid of class j for variable v
$d^2(x_{iv}, \mu_{jv})$ is the square distance between x_{iv} and c_{jv}
Φ is the fuzziness exponent

Equation (162) can be solved by Lagrangian differentiation using Picard iteration to update the class centroids and membership matrix:

$$m_{ij} = \frac{d^{-2/(\Phi-1)}(x_{iv}, \mu_{jv})}{\sum_{k=1}^{f} d_{ik}^{-2/(\Phi-1)}(x_{iv}, \mu_{jv})}, \qquad 1 \leq i \leq n; \;\; 1 \leq j \leq f \tag{163}$$

$$\mu_j = \frac{\sum_{i=1}^{n} m_{ij}^{\Phi} x_i}{\sum_{i=1}^{n} m_{ij}^{\Phi}}, \qquad 1 \leq j \leq f \tag{164}$$

For solving data points that are outliers, i.e., outside the classes (not in between classes) and inadequately represented, modified theories have been studied [49, 51, 52].

A fuzzy classification was applied to recognize the doneness of beef steak, which was determined by its internal temperature and classified as K subsets. The red, green, and blue data of the image of steak were acquired and converted to hue, intensity, and saturation as the features of steaks. The fuzzy C-means and fuzzy K-nearest classification were applied to classify the steaks based on their features [48]. The fuzzy classification methods were equal or better than the crisp classification methods. The characteristic benefit of fuzzy classification is that the degree of membership functions can provide more information about the confidence of the class assignment.

2. Neural Network Classification

For product classification, we need a relation function to transform physical properties to quality factors. Most neural networks, supervised or nonsupervised, can associate the input patterns and generate the new patterns for the corresponding output class. And the classification does not need detailed knowledge underlying the relationships between those features and the classes for building classifiers.

Some applications of neural networks in classifying food products are summarized:

A multilayer neural network was developed to recognize three barley seed varieties. The geometry features of seeds were extracted by the ZR Fourier descriptor method. Those features were fed as the input for the neural network. A three-layer classifier could reach a total recognition accuracy of 80.4% using 20 harmonics of the ZR Fourier descriptor spectra as inputs [53]. Grading color and firmness of tomatoes based on the mechanical properties such as drop impact, cyclic deformation, and resonance frequency of tomatoes using a neural network as classifier has been performed [54].

A back propagation neural network and a Kohonen self-organizing neural network were developed for a robust method to measure the amount of intramuscular fat in cuts of beef [55]. The fat percentage of meat was related to its palatability. Ultrasound was used to noninvasively characterize the amount of fat in the lean tissue. The radiofrequency signals received were transformed by digital Fourier transform. Those features were extracted from the frequency domain for the basis of classification and used as neural network input.

The maturity levels of green tomatoes at harvest were determined by a back propagation network and a Carpenter and Grossberg network. X-ray computation tomography was converted into 64-gray-level histograms. Fourteen gray levels were chosen to be the features to be correlated with maturity by neural networks [56].

E. Neuro-Fuzzy Control Systems

Recently, attention has been given to combining fuzzy logic and neural networks to gain the benefits from both techniques. One type of combination is to use neural networks for process modeling and fuzzy logic for the control system. For example, the quality factors of a bread-baking process, such as volume and browning, were modeled by multilayer perceptron neural networks. Those quality factors were also the state variables in fuzzy logic control of the baking process. These neural networks actually functioned as sensors relating the measurement of quality factors to the control system [57].

Another combination is to provide the fuzzy system with the learning or adaptive ability of a neural network to overcome the difficulty and lack of reliability in verbal elicitation of a knowledge base from humans. Many different types of neuro-fuzzy control systems have been proposed. One straightforward system is to construct the fuzzy logic control system in the architecture, isomorphic to multilayer neural network [58–60]. After the network structure is built, it can be applied with the existing neural network learning rules. One simple neural fuzzy control that applies Takagi and Sugeno's type of [37] rules is introduced in the following:

The rule is in the form

$$R^i: \text{``if } (x_1 \text{ is } A_1{}^j \text{ and } x_2 \text{ is } A_2{}^j \text{ and } \ldots) \text{ then } (y \text{ is } f(x_1, x_2, \ldots))\text{''}.$$
(165)

The membership function in the rule consequent is a nonfuzzy function of input variables. $A_1{}^j, A_2{}^j, \ldots$ are the membership functions of the input variables $x_1, x_2, \ldots$. Let's consider $f(\cdot)$ only as a constant with parameter c_i in this example, i.e., $f(x_1, x_2, \ldots) = w_j$.

The structure of the neural fuzzy control system (NFCS) is shown in Fig. 13. It is a three-layer feed-forward network containing defuzzification layer, inference layer, and defuzzification layer.

1. Fuzzification Layer

The input data is first fuzzified in this layer to perform the fuzzy inference in the next layer, and so each node corresponds to one of the membership

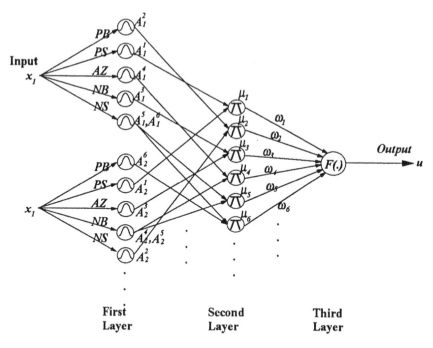

First   Second   Third
Layer   Layer   Layer

FIGURE 13 Neural fuzzy control system.

functions of the input variables. The node number of the first layer is equal to the number of fuzzy term sets of all linguistic input variables. Consider choosing the bell-shaped function for the input variables:

$$A_i^j(x_i) = \exp\left[-\left(\frac{x_i - a_{ij}}{b_{ij}}\right)^2\right] \tag{166}$$

where a_{ij} and b_{ij} are parameters that determine the locations and widths of the bell-shaped functions.

2. Inference Layer

The output of each node in this layer represents the firing strength of a rule. The firing strength (or the degree of matching) μ_j of each rule is calculated by

$$m_j = A_1^j(x_1)A_2^j(x_2) \cdots A_i^j(x_i) \cdots \qquad \text{for rule } j \tag{167}$$

The node in the second layer is connected to the nodes of the first layer according to the rule antecedent. The total number of nodes is equal to the number of fuzzy rules.

3. Defuzzification Layer

The defuzzification of Takagi and Sugenos's method has a simple form, as in Eq. (165). The control output is the weighted combination of the firing strength:

$$u = F(w_1, w_2, \ldots, m_1, m_2, \ldots) = \frac{\sum_k w_k m_k}{\sum_k m_k} \tag{168}$$

The rules and membership functions of the neural fuzzy control system can be tuned by changing the parameters of the membership functions. The error back propagation (or gradient descent) method could be used as a learning rule. This training data set could be obtained by recording the states of input and output variables of the process while it is operated by experienced operators or experts.

Update the parameters of network toward the direction of steepest gradient descent of error functions as applied in ANN learning. The error function E is defined as

$$E = \frac{1}{2}(u_d - u)^2 \tag{169}$$

where u is the actual output and u_d is the desired output from training data. The delta error Δu of output u is defined as

$$\Delta u \equiv u_d - u \tag{170}$$

1. Tuning w_i in the rule consequent is by

$$w_i(t + 1) = w_i(t) - \beta_w \frac{\partial E}{\partial w_i}$$
$$= w_i(t) + \beta_w \Delta u \frac{\mu_i}{\sum_i \mu_i} \tag{171}$$

where β_w is the learning rate for w_i.

2. Tuning parameters a_{ij} and b_{ij} of the bell-shaped fuzzy membership functions in rule antecedent is by

$$a_{ij}(t + 1) = a_{ij}(t) - \beta_a \frac{\partial E}{\partial a_{ij}} \tag{172}$$
$$= a_{ij}(t) - \beta_a \Delta u \frac{w_i - u}{\sum_i \mu_i} \mu_i \frac{2 \ln(A_i^j)}{x_i - a_{ij}}$$

$$b_{ij}(t + 1) = b_{ij}(t) - \beta_b \frac{\partial E}{\partial b_{ij}} \tag{173}$$

$$= b_{ij}(t) - \beta_b \, \Delta u \, \frac{w_i - u}{\sum_i \mu_i} \, \mu_i \, \frac{2 \ln(A_i^j)}{b_{ij}}$$

where β_a, β_b are learning rates for a_{ij}, b_{ij}, respectively.

There are variations of the foregoing neuro-fuzzy control system. For example, some systems use minimum-product fuzzy inference, or noncrisp values of fuzzy membership functions in the rule consequent. Because a continuous function is required for back propagation learning, some approximation in calculation is required such as using soft-computing in the action selection network [61], considering the existing derivative only [62], or defuzzification by approximate center of area using sum operation instead of maximum operation, etc. [62, 63].

After the neuro-fuzzy system is constructed, we can substitute the neural fuzzy system for the neural network in most applications of neural network control or modeling [64]. The advantages of the foregoing neuro-fuzzy system lies in its computational efficiency and the small number of adjusting parameters, due to the use of expert knowledge before learning. A recent application of a neural fuzzy controller for multivariable process control was to a snack food frying process [65].

There are also some neuro-fuzzy systems using learning rules other than back propagation. Kohonen's feature-maps algorithms are used to determine the input–output space division of the membership function from the clustering of the training data [63, 64]. A self-learning fuzzy controller using a credit assignment method to build an intelligent rule-based system has been proposed [65]. The system comes with two special neurons: associative critic neuron (ACN) and associative learning neuron (ALN). The ACN is to criticize from the system state and environment, and ALN is a content addressable memory system continuously updating the position of membership functions in the rule premise from the internal reinforcement signal generated by ACN and the states before a failure signal occurs. Generalized-approximate-reasoning-based-intelligent-control-architecture (GARIC) [66] consists of three components: the action selection network (ASN), the action evaluation network (AEN), and the stochastic action modifier (SAM). The ASN is similar to the general neural structured fuzzy inference system and is used to decide the control action. The AEN maps the state vector and failure signal to a score, which indicates the state goodness and product internal reinforcement. SAM modifies the control action from ASN with the internal reinforcement signal from AEN stochastically to acquire the final control action.

V. SUMMARY

In this chapter, some basic principles of fuzzy logic and artificial neural networks have been described. Their applications in the automation of food processes, such as process modeling, process control, intelligent sensors, and pattern recognition, were reviewed.

Fuzzy logic and artificial neural networks are very promising for use in certain food processes that are difficult or impossible to control adequately by conventional methods. These methods complement, not replace, conventional methods. Research in both fuzzy logic and neural networks still under development. The new trend is to apply a combination of both fuzzy logic and neural networks in process automation.

REFERENCES

1. L. A. Zadeh, Fuzzy sets, *Information and Control 8*:338 (1965).
2. W. S. McCulloch and W. Pitts, A logical calculus of the ideas imminent in nervous activity, *Bulletin of Mathematical Biophysics 5*:115 (1943).
3. G. J. Klir and T. A. Folger, *Fuzzy Sets, Uncertainty, and Information*, Prentice Hall, Englewood Cliffs, New Jersey, 1988.
4. W. Predrycz, *Fuzzy Control and Fuzzy System*, Wiley, New York, 1993.
5. T. Terano, K. Asai, and M. Sugeno, *Fuzzy Systems Theory and Its Applications*, Academic Press, San Diego, 1991.
6. H. J. Zimmermann, *Fuzzy Set Theory and Its Applications*, Kluwer Academic Publishers, Boston, 1985.
7. C. C. Lee, Fuzzy logic in control systems: Fuzzy logic controller—Parts I, II, *IEEE Trans. on System Man and Cybernetics 21*(2):404 (1990).
8. J. B. Turksen, Measurement of membership functions and their acquisition, *Fuzzy Sets and Systems 40*:5 (1991).
9. L. A. Zadeh, Fuzzy logic, *Computer 21*(4):83 (1988).
10. L. A. Zadeh, A computational approach to fuzzy quantifiers in nature languages, *Computers and Mathematics 9*:149 (1983).
11. L. A. Zadeh, Test-score semantics as a basis for a computational approach to the representation of meaning, *Literary and Linguistic Computing 1*(1): 24 (1986).
12. M. Mizumoto, Fuzzy controls under various fuzzy reasoning methods, *Information Science 45*:129 (1988).
13. R. K. Singh and F. Ou-Yang, Knowledge-based fuzzy control of aseptic processing, *Food Technology 48*(6):155 (1994).
14. R. P. Lippmann, An introduction to computing with neural nets, *IEEE Magazine on Acoustics, Signal and Speech Processing 4*(2):4 (1987).
15. G. Wells, An introduction to neural networks, *Application of Artificial Intelligence in Process control* (L. Boullart, A. Krijgsman, and R. A. Vingerhoeds, eds), Pergamon Press, New York, 1992, p. 144.

16. J. M. Zurada, *Introduction to Artificial Neural Systems*, West Publishing, St. Paul, Minnesota, 1992.
17. R. Rosenblatt, *Principles of Neurodynamics*, Spartan Books, New York, 1959.
18. B. Widrow and M. E. Hoff, Adaptive switching circuits, 1960 IRE WESCON Conv. Record, Los Angeles, California, Part 4, p. 96.
19. D. E. Rumelhart, G. E. Hinton, and R. J. Williams, Learning internal representations by error propagation, *Parallel Distributed Processing: Explorations in the Microstructure of Cognition, Vol. 1: Foundations* (D. E. Rumelart and J. L. McClelland, eds.), MIT Press, Cambridge, Massachusetts, 1986, p. 319.
20. D. O. Hebb, *The Organization of Behavior: A Neuropsychological Theory*, John Wiley, New York, 1949.
21. R. Hecht-Nielsen, Counterpropagation networks, *Appl. Opt. 26*(23):4979 (1987).
22. T. Kohonen, The self-organizing map, *ITTT Computer 27*(3):11 (1990).
23. G. A. Carpenter and S. Grossberg, Neural dynamics of category learning and recognition: Attention, memory consolidation and amnesia, *Brain Structure, Learning and Memory* (J. Davis, I. Newburgh, and I. Wegman, eds.) AAAS Symp. Series, Westview Press, Boulder, Colorado, 1988.
24. D. C. Bullock and D. Whittaker, Continuous prediction of a food process with a stationary neural model, ASAE Paper No. 933508, ASAE International Winter Meeting, Chicago, December, 1993.
25. T. Eerikäinen, S. Linko, and P. Linko, The potential of fuzzy logic in optimization and control: Fuzzy reasoning in extrusion cooker control, *Automatic Control and Optimization of Food Processes* (M. Renard and J. J. Bimbenet, eds.), Elsevier Applied Science, London, 1988, p. 183.
26. M. Dohnal, J. Vystreil, J. Dohnalova, K. Marecek, M. Krapilik, and P. Bures, Fuzzy food engineering, *J. of Food Eng. 19*:171 (1993).
27. Q. Zhang and J. B. Litchfield, Fuzzy expert system: A prototype for control of corn breakage during drying, *J. of Food Process Eng. 12*:259 (1990).
28. Q. Zhang, J. B. Litchfield, and J. Bentsman, Fuzzy predictive control system for corn quality control during drying, Proceedings of the Food Processing Automation II Conference, Lexington, Kentucky, 1992, p. 313.
29. Q. Zhang, J. B. Litchfield, and J. Bentsman, Fuzzy prediction of maize breakage, *J. Agric. Engng. Res. 52*:77 (1992).
30. Q. Zhang and J. B. Litchfield, Fuzzy logic control for a continuous crossflow grain dryer, *J. of Food Process Eng. 16*:59 (1993).
31. P. Linko, K. Uemura, and T. Eerikäinen, Neural networks in fuzzy extrusion control, *Food Engineering in a Computer Climate*, Inst. of Chemical Engineering, Symposium Series 12 Rugby, U.K. 1992, p. 401.
32. H. Murase, Y. Nishiura, N. Honami, and N. Kondo, Neural network model for tomato fruit cracking, ASAE Paper No. 923593, ASAE International Winter Meeting, Nashville, Tennessee, December 1992.
33. V. Gnanasekharan and J. D. Floros, Comparison of back propagation net-

work (BPN) performance and response surface methodology (RSM) for modeling food processing, Proceedings of the Rutgers' Conference on Computer Integrated Manufacturing in the Process Industries, New Brunswick, New Jersey, 1994, p. 763.

34. Y. Huang and A. D. Whittaker, Input–output modeling of biological product processing through neural networks, ASAE Paper No. 933507, ASAE International Winter Meeting, Chicago, December 1993.

35. E. H. Mamdani and S. Assilian. An experiment in linguistic synthesis with a fuzzy logic controller, *International J. of Man–Machine Studies 7*:1 (1975).

36. M. Sugeno, *Industrial Application of Fuzzy Control*, Elsevier Science Publishers Amsterdam, 1985.

37. T. Takagi and M. Sugeno, Derivation of fuzzy control rules from human operator's control actions, Proc. of the IFAC Symp. on Fuzzy Information, Knowledge Representation and Decision Analysis, Marseilles, France, July 1983, p. 55.

38. J. S. Shieh, M. C. Chen, and L. H. Ferng, Fuzzy logic control of HTST heat exchanger, *Food Control 3*:91 (1992).

39. K. Oishi, M. Tominaga, A. Kawato, Y. Abe, S. Imayasu, and A. Nanba, Application of fuzzy control theory to the sake brewing process, *J. of Ferment. and Bioeng. 72*(2):115 (1991).

40. T. Tobi, T. Hanafusa, S. Ito, and N. Kashiwagi, The application of fuzzy control to a coke oven gas cooling plant, *Fuzzy Sets and Systems 46*:373 (1992).

41. E. Czogala and T. Rawlik, Modeling of a fuzzy controller with application to the control of biological processes, *Fuzzy Sets and Systems 31*:13 (1989).

42. Y. Edan, P. Grinspan, E. Maltz, and H. Kahn, Fuzzy logic applications in the dairy industry, ASAE Paper No. 923600, ASAE International Winter Meeting, Nashville, Tennessee, December 1992.

43. G. Lightbody, Q. H. Wu, and G. W. Irwin, Control applications for feed forward networks, *Neural Network for Control Systems* (K. Warwick, G. W. Irwin, and K. J. Hunt, eds.), Peter Peregrinus Ltd., Herts, U.K., 1988, p. 51.

44. R. Zbikowski and P. J. Gawthrop, A survey for neural networks for control, *Neural Network for Control Systems* (K. Warwick, G. W. Irwin, and K. J. Hunt, eds.), Peter Peregrinus Ltd., Herts, U.K., 1988, p. 31.

45. Q. Zhang, J. B. Litchfield, and J. F. Ried, Smart sensor technologies for food and bioprocess automation, Proceedings of the Food Processing Automation III Conference, Orlando, Florida, 1993, p. 289.

46. J. Giese, On-line sensors for food processing, *Food Technology 47*:88 (1993).

47. J. M. Fides and A. Cindar, Sensor fusion and intelligent control for food processing, Proceedings of the Food Processing Automation II Conference, Lexington, Kentucky, 1992, p. 72.

48. K. Unklesbay, J. Keller, N. Unklesbay, and D. Subhangkasen, Determination of doneness of beef steaks using fuzzy pattern recognition, *J. of Food Eng. 8*:79 (1988).

49. I. O. A. Odeh, A. B. McBratney, and D. J. Chittleborough, Soil pattern recognition with fuzzy-c-means: Application to classification and soil-land-form interrelationships, *Soil Sci. Soc. Am. J. 56*:505 (1992).

50. J. C. Bezdek, *Pattern Recognition with Fuzzy Objective Function Algorithms*, Plenum Press, New York, 1981.

51. J. Hartigan, *Clustering Algorithms*, John Wiley and Sons, New York, 1975.

52. Y. Ohashi, Fuzzy clustering and robust estimation, Int. Meet. SAS Users Group 9th, Hollywood Beach, Florida, March 1984, p. 18.

53. M. Romaniuk, S. Sokhansanj, and H. C. Wood, Barley seed recognition using a multi-layer neural network, ASAE Paper No. 93-6569, ASAE International Winter Meeting, Chicago, December 1993.

54. Y. Edan, I. Shmulevich, D. Rachmani, E. Fallik, and S. Grinberg, Neural networks for quality grading of tomatoes based on mechanical properties, Proceedings of the Food Processing Automation III Conference, Orlando, Florida, 1994, p. 347.

55. A. D. Whittaker, B. S. Park, and J. M. McCauley, Ultrasonic signal classification for beef quality grading through neural networks, Automated Agriculture for the 21st Century, Proceedings of the 1991 Symposium, Chicago, p. 116.

56. C. N. Thai, J. N. Pease, and E. W. Tollner, Determination of green tomato maturity from x-ray computed tomography images, Automated Agriculture for the 21st Century, Proceedings of the 1991 Symposium, Chicago, p. 134.

57. S. I. Cho and S. C. Kim, Neural network modeling and fuzzy control of baking process, ASAE Paper No. 946502, ASAE International Winter Meeting, Chicago, December 1993.

58. J. R. Jang, Self-learning fuzzy controllers based on temporal back propagation, *IEEE Trans. on Neural Networks 3*(5):714 (1992).

59. S. Horikawa, T. Furuhashi, and Y. Uchikawa, On fuzzy modeling using fuzzy neural networks with the back-propagation algorithm, *IEEE Trans. on Neural Networks 3*(5):801 (1992).

60. S. Horikawa, T. Furuhashi, and Y. Uchikawa, On identification of structures in premises of a fuzzy model using a fuzzy neural network, 2nd IEEE Int. Conf. on Fuzzy Systems, San Francisco, 1993, p. 661.

61. H. R. Berenji, Learning and tuning fuzzy logic controllers through reinforcements, *IEEE Trans. on Neural Networks 3*(5):724 (1992).

62. C. T. Lin and C. S. G. Lee, Neural-network-based fuzzy logic control and decision system, *IEEE Trans. on Computers 40*(12):1323 (1991).

63. J. Bruske, *Neural Fuzzy Decision Systems*, Diploma thesis, University of Kaiserslautern, Germany, 1993.

64. P. J. Werbos, Neurocontrol and fuzzy logic: Connections and designs, *Int. J. of Approximate Reasoning 6*:185 (1992).

65. Y. S. Choi, A. D. Whittaker, and D. C. Bullock, Predictive neuro-fuzzy controller for multivariable process control, ASAE Paper No. 933509, ASAE International Winter Meeting, Chicago, December 1993.

66. R. B. Hamid and P. Khedkar, Learning and tuning fuzzy logic controllers through reinforcements, *IEEE Trans. on Neutral Networks 3*(5):724 (1992).

6

Fuzzy Control for Food Processes

V. J. Davidson
University of Guelph, Guelph, Ontario, Canada

I. INTRODUCTION

A. Context for Fuzzy Control in Food Processing

Fuzzy logic is one of the cornerstones of soft computing, a term which Zadeh [1] uses to describe methodologies that "aim to exploit the tolerance for imprecision and uncertainty to achieve tractability, robustness and low-solution cost." Traditional "hard" computing tools demand a degree of precision that is difficult to achieve when dealing with complex systems. Precision also incurs substantial cost in control systems in hardware requirements (e.g., microprocessor and instrumentation), computing time, and maintenance resources. Soft computing techniques are successful when reducing the level of precision does not compromise satisfactory control.

The tolerance for imprecision can be illustrated by a simple control problem in the context of everyday food processing. A chef can consistently produce an outstanding sauce or dish by tasting during the preparation and adjusting for different ingredients or quality variations in ingredients. Adjustments are based on the chef's palate, which has been well trained by experience but which does not discern the precise level of flavor components, simply approximate values of key flavor, aroma, and textural characteristics. There are analytical techniques to determine physical and chemical properties that contribute to sensory perceptions. However, if chefs relied on precise analysis in cooking, preparation time would be considerably longer and the end result would undoubtedly be less satisfactory.

Many food-processing operations continue to be controlled by experienced operators. Operators frequently make observations and judgments, particularly about sensory attributes, that are difficult (or impossible) to reproduce using on-line sensors. As a result, it has been difficult to implement computer-based control strategies in many food-manufacturing operations. Fuzzy techniques make it possible to use operators' observations and process knowledge in computer-based control systems. As a result, fuzzy control systems offer considerable potential for increasing automatic control in the food industry.

Fuzzy control systems can be based entirely on fuzzy techniques or combined with conventional control techniques in hybrid systems. Fuzzy tools offer ways to span the spectrum that now exists from conventional, analytic methods to artificial intelligence and expert-system approaches for computer-based control.

B. Fuzzy Control Literature

The control literature, both academic research and popular press, contains many articles, conference proceedings, and books dealing with fuzzy control for a wide range of applications. For example, a recent search on the INSPEC database (4000 physics, electronics, and computing journals) based on title words "fuzzy control" retrieved 1410 citations for the period 1990 to 1995. This points to the maturing of fuzzy control technology as well as the daunting task of attempting to summarize concisely the current state of the art.

Within the constraint of a single chapter, it is not possible to discuss all of the aspects related to the design of fuzzy control systems. In addition to the information presented in this chapter and in Chapter 5, references [2, 3] are strongly recommended for readers with interests in theory as well as practical design information for fuzzy control systems. The textbook on fuzzy control [2] is comprehensive and can be understood by readers from a wide range of backgrounds (engineers and computer scientists; students and industrial practitioners; novices and expert control system designers). The handbook [3] emphasizes practical applications of fuzzy systems for a wide range of problems (financial decision making as well as control).

II. FUZZY CONTROL SYSTEMS

Zadeh [1, 4] has coined the expression *fuzzy dependency and command language (FDCL)* to include the components of fuzzy set theory that are used to represent and manipulate fuzzy dependencies. FDCL is the basis for fuzzy control systems. Some of the basic concepts of fuzzy sets and set operations have been introduced in Chapter 5 (Section II). In this chapter, the emphasis is on specific components of FDCL that are important for understanding the design and performance of fuzzy control systems for food processes.

FDCL includes representations for linguistic variables and fuzzy rules as well a mathematics for working with fuzzy rules and graphs. It is a calculus rather than a logic in the sense that inferences are not based on rigid, absolute rules but on set manipulations chosen by the system designer.

In their control text, Driankov et al. [2] use the term "approximate reasoning" in a similar sense. Approximate reasoning describes computations with fuzzy sets that represent the meaning of fuzzy propositions or word statements of relations. Conclusions depend on the meanings (semantics) that are attached to fuzzy values and propositions and on the manipulations that are applied to the fuzzy sets to make inferences. This flexibility has been the source of considerable academic debate, but practical applications in different control systems have demonstrated that more than one inference method gives satisfactory performance.

A. Data Representation

1. Fuzzy Sets and Linguistic Variables

Zadeh [5] introduced the theory of classes with unsharp or gradual boundaries and the concept of a *fuzzy set*. Fuzzy sets are extremely useful for delimiting values that have overlap. In the real world, it is often difficult and erroneous to define absolute demarcations in properties. But it is important to recognize a distinction between "fuzzy" in the sense of a gradual but defined boundary and "vague" in the sense of lacking sufficient information to describe a property.* The former connotation is one reason that fuzzy representations are so useful in control systems. The latter meaning is one that detractors, in their misunderstanding of the term "fuzzy," use to argue against fuzzy control but which would also make it impossible to implement a fuzzy control system.

---

* L. A. Zadeh, Lectures CS294-1, "Fuzzy logic, neural networks and soft computing," January 1995.

In Section II.A.1 of Chapter 5, the concept of membership functions to represent partial belonging of set elements is defined. In most control applications, membership functions provide mappings between the real number domain and a fuzzy value or set of real numbers. For convenience, the fuzzy set is often given a linguistic label or term from natural language that describes the characteristic that is common to the set elements. If a process variable such as temperature is defined in terms of fuzzy sets with linguistic values like cold, warm, and hot, it is referred to as a *linguistic variable*, to indicate that it takes linguistic values. Membership functions must be defined to provide a mapping from a numerical domain of temperature values to a linguistic domain. These mappings give linguistic labels substantial expressive power in a form that can be used in computer algorithms for mathematical computations and provide one means of "computing with words."*

The use of fuzzy sets and linguistic values for data representations offers several important advantages in control systems, particularly for food systems:

Fuzzy Quantization of Variables   Zadeh [1, 6] points out that linguistic values are a form of data compression. It is simple and parsimonious to represent a variable such as temperature by a few linguistic values (e.g., cold, warm, and hot). All values that can be resolved by a measurement system (such as a thermocouple) may not be significantly different in terms of the need to actuate control. In other words, there is a tolerance for imprecision and data compression. Fuzzy quantization is appropriate in defining many aspects of food systems (physical, chemical, sensory properties) because it recognizes the inherent variability in food materials. Fairly precise data may be available for individual components, but most food processes involve composite materials of variable composition and their characteristics cannot be defined by precise values.

Smooth Transitions   Membership functions for inputs or outputs in control systems are defined with some degree of overlap. Thus input and output states can have partial membership in more than one fuzzy value or level. With proper design, this results in smooth transitions in control actions. It also contributes to robustness in the control system because the ultimate action is reached by combining a range of possible actions.

Operator Interaction   The use of linguistic variables with natural language labels results in easy interaction between operators and the control system. Information about the process state is simplified, and operators

---

* L. A. Zadeh, Lectures CS294-1, "Fuzzy logic, neural networks and soft computing," January 1995.

can make observations that are used as "natural language" inputs to control algorithms.

In food processes, this is a substantial advantage because operators can make observations about sensory attributes that may be difficult or impossible to measure on-line or in a quality control laboratory. Furthermore, human assessments can integrate several different components to make an inference about a more complex attribute. For example, experts can make rapid and consistent assessments about the texture, aroma, and flavor of the product that cannot be duplicated by instrumental analysis. Methodologies for sensory evaluation are well developed in the food industry, and there are excellent training procedures to ensure that terminology is used in a consistent way. FDCL also provides a means of understanding subtle modifications of meaning through manipulation of membership functions. Hedges or modifiers are discussed in Section II.D.1.d of Chapter 5 as well as by Cox [3, Chapter 5].

2. Membership Functions for Control System Inputs/Outputs

The mapping into the linguistic domain is commonly called "fuzzification." The most common examples of this data transformation in control systems are

1. From a single numerical value (singleton) to a linguistic value (or values, because partial association with overlapping membership functions is possible) as shown in Fig. 1a
2. From a set of real numbers into a fuzzy set (Fig. 1b)

The first mapping is usually necessary when a sensor input is a single numerical value. A singleton value may represent the current reading from the sensor or may be an average over multiple values. The second mapping may be used when a measured value is noisy and it is advantageous to capture characteristics of the noise in the fuzzy value(s). The inverse transform, defuzzification, is used to calculate the appropriate level (i.e., a single value) for an actuator. Fuzzification and defuzzification are advantageous data transformations for control analysis, just as Laplace and Fourier transforms are effective in conventional control analysis. Both operations are closely linked to the membership functions that are defined for inputs and outputs.

There are several important characteristics for input and output membership functions in control systems.

Compact Representation of Membership Functions For simplicity and ease of performing computations with membership functions, simple

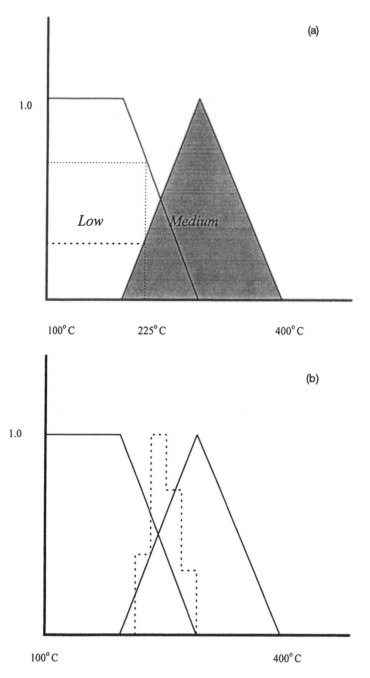

FIGURE 1 (a) Fuzzy mapping of temperature value of 225°C. μ_{low} = 0.7; μ_{medium} = 0.3. (b) Fuzzy mapping of noisy temperature value (dashed line).

shapes that can be described by a few parameters (e.g., triangles, trapezoids, Gaussian distributions) are sufficiently flexible and the most practical.

Overlap of Membership Functions for Neighboring Values The membership functions for neighboring fuzzy inputs should cross to ensure that all real values of measurements can be mapped into the fuzzy domain. Crosspoints for output values are required for smooth transitions in output levels.

Sum of Membership Grades for a Singleton Input (or "Condition Width" [2, p. 122]) A singleton input may belong to more than one fuzzy value. In control systems, it is desirable that membership grades for all possible fuzzy values of a singleton input sum to unity. This is achieved by overlap of the membership functions for neighboring fuzzy values. The effect of different overlaps on control output is illustrated in Fig. 2. In this example, the outputs associated with each of the fuzzy input levels are three singleton values (0.2, 0.5, and 0.8). For singleton inputs less than 0, condition width is not satisfied. As a result, the control output is not a smooth function and has abrupt changes. The output is constant (0.2) for input values between -1.0 and -0.5 (i.e., in this range, degree of membership in the lowest fuzzy value has no effect on the defuzzified output). The control output is also constant for input values between -0.3 and 0.0 (output $= 0.5$) because of the symmetry of the membership function defined for the middle input value. For inputs between 0 and 1, condition width is satisfied and the output is a smooth, linear function.

Bounds for Fuzzy Ranges Inputs and outputs are defined on domains of real values that represent normal process ranges for each variable. At the limits of each domain, membership functions should be unity to satisfy point (c), because no neighboring fuzzy values are defined. It should also be recognized that a control output only reaches saturation (or values of -1.0, 0.0, or 1.0 in a normalized domains) if singleton values for the maximum or minimum output are defined.

Normalized Input Domains Some control inputs (e.g., error, change in error) are suited to scaling between $[-1, +1]$ or $[0, 1]$ as appropriate for the variable. This is accomplished by defining scaling factors based on a priori knowledge of the range of values for each input. Scaling of inputs makes the control representation more general and unencumbered by dimensions. In fuzzy controllers that resemble PID (proportional–integral–derivative) algorithms, the scaling factors are similar to a gain coefficient. However, scaling is not always necessary or sound judgment.

Number of levels for fuzzy input/output variables: Membership functions are defined for each level of a fuzzy variable. The appropriate num-

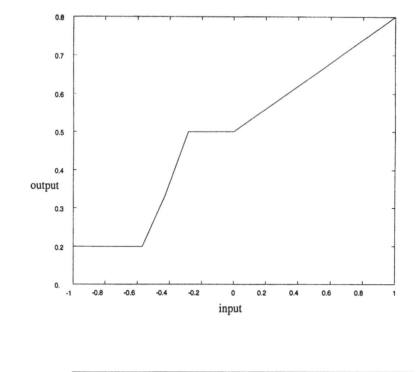

FIGURE 2 Defuzzified output for single-input system. Membership functions for three input values are shown below input axis.

ber of levels (or coarseness of the fuzzy quantization) is defined by the control system designer. The decision requires some understanding of the system's behavior and judgment to weigh the advantages of soft computing (low cost and simplicity) against the need for precision. In some ways, designing the coarseness of the granulation is analogous to deciding the appropriate number of significant figures in numerical computations. However, most scientists and engineers are less comfortable with semantic truncation and, initially, are tempted to define many levels for fuzzy variables.

Continuous versus discretized membership functions: Either continuous (e.g., membership functions in Fig. 1a) or discretized (e.g., membership function for the noisy signal Fig. 1b) representations of membership functions can be used. In discretized representations, the real number range is divided into a convenient number of intervals (e.g., 21 intervals in fuzzy floating point data representation*; 256 intervals [3]). Some designers consider a discrete representation too coarse, but again the question is, "what is the tolerance for imprecision in the control system?" If fuzzy control is being considered, then there is some tolerance for coarse quantization in the membership function.

A discretized representation has advantages in ease of computations with membership functions. Inference procedures result in aggregated fuzzy values with complex geometries. Aggregation and defuzzification (components of interpolation, discussed in Section II.C) are straightforward computations with discretized membership functions.

B. Representation of Control Knowledge

1. Fuzzy Rules and Graphs

Most fuzzy control systems are based on a set of rules or conditional statements. A fuzzy conditional statement is a convenient way to express a relationship between a set of process observations and a control action that is required to maintain the process at its desired operating state or to maintain the desired process trajectory in a batch process. Fuzzy control rules span a spectrum from rules that share common characteristics with conventional PID control algorithms to rules that form an expert-system type of controller. The entire spectrum has relevance to food process control systems, but it is important to understand the range of the spectrum and some of the implications of moving from one end to the other.

Fuzzy Rules that Resemble PID Algorithms   The desired operating state is expressed as a setpoint or target value for a process output. Error (the difference between setpoint and actual values), change of error, and sum of past errors are used, either as individual or combined conditions, to judge the need for control action. In a fuzzy control system these conditions are expressed as fuzzy values, but the structure and behavior of the controller has much in common with conventional PID control. In [2, pp. 107–114], there is thorough coverage of fuzzy PID-like control, including the effects of membership function characteristics on controller perfor-

---

* G. L. Hayward, Lectures 050612, "Fuzzy logic control" January 1992.

mance. An advantage of fuzzy PID-like control for food processes is the ability to develop simple, nonlinear controllers.

Fuzzy Rules that Resemble Expert Systems   A single measurement or set of measurements is used to assess the process state. Although not stated explicitly in the fuzzy rule, the observed set of conditions is not the desired operating state, and a control action is recommended. The relationship between the observations (conditions) and the control action can be derived from an experienced operator's knowledge or from a model of the process dynamics. Fuzzy values are often consistent with the expert's basis for judgment, and rules are often more flexible and simpler than mathematical equations. However, expert-system rules are specific to the application, and development and design time can be considerable.

Some examples of fuzzy rules:

Simple rule (single-input, single-output, SISO), e.g.,
"If error is large, then valve position is wide open."

Compound rule (multiple-input, single-output, MISO), e.g.,
"If air temperature is high and mass loading on belt is low, then dryer belt speed is fast."

Interpretation of a Fuzzy Rule   Fuzzy control rules express a causal relationship between antecedent(s) and consequence(s) (i.e., an observed set of process conditions results in a specific control action). However, one must remember that this type of fuzzy rule does not express any knowledge of the underlying cause(s) of the observed process state or a diagnosis to remedy a particular problem. It simply recognizes that a particular variable can be manipulated to counteract disturbance(s) creating the undesirable state. Thus it should be recognized as a special type of conditional statement and not considered as a general implication. The success or failure of the conditional statement (in terms of invoking the appropriate control action) depends on the skill of the designer in choosing the manipulated variable and in assigning meaning to the fuzzy conditions (inputs and control outputs) through membership functions.

There have been different interpretations of the conditional statement as it is used in fuzzy control. One of the original researchers in the area of fuzzy control, Mamdani, interpreted a fuzzy rule as an implication, but this requires a special implication operation to be defined (i.e., a Mamdani implication or minimum operator). Zadeh [1, 4] points out that a fuzzy control rule can be considered as the application of fuzzy constraints (which are elastic) on the process conditions (X) and the manipulated variable(s) (Y). In this interpretation, the simple conditional statement

"If X is A then Y is B"   (i.e., SISO control rule)

is translated as the Cartesian product of fuzzy sets A and B defined on X and Y, respectively. Applying the general definition of Cartesian product (Chapter 5, II.A.2), the resulting membership function for the SISO fuzzy rule is a two-dimensional surface defined by

$$\mu_{A \times B}(u, v) = \mu_A(u) \wedge \mu_B(v), \qquad u \in U, v \in V$$

where μ_A and μ_B are the membership functions of A and B, respectively; U and V are the universes of discourse of X and Y and $\wedge$ is the conjunction operator, which is usually defined as a minimum operation. In fact the Mamdani implication operator is also defined as minimum, and so the operations are identical with slightly different interpretations (elastic constraints versus implication).

The input/output region of an SISO fuzzy rule can be represented on a two-dimensional graph (Fig. 3a). Zadeh refers to this representation of the extreme limits of the rule as a granule, and the rule is calibrated by defining the membership functions μ_A and μ_B. The Cartesian product (minimum operation) creates a surface in a third dimension, orthogonal to the plane of the granule, which defines the output over all possible values of the fuzzy antecedent (Fig. 3b). In the case of a more complex rule, such as an MISO case, the operations that connect the antecedents (minimum for "AND," maximum for "OR") produce a membership function surface for the compound antecedent. The Cartesian product representing the rule cannot be visualized in three-dimensional geometry.

Interpretation of a Set of Rules   Rules represent the knowledge base for controlling the process, and a set of rules is usually required to cover the entire operating range of the process. The set of fuzzy rules can be viewed as the superposition of the granules representing individual rules to produce a fuzzy graph [1, 4]. This is shown in Fig. 4 for an SISO control system. Superpositioning of granules is equivalent to a union operation on all rules defined in the relevant operating space. In the limiting case, the granules can be reduced to points, without overlap, and the graph appears as a conventional mathematical function. The heavy lines outline the effective operating region when Cartesian product or Mamdani-type implication is used [7].

A fuzzy rule set is a useful approximation method in control when mathematical function(s) are too precise. It is a parsimonious representation. With fuzzy data compression, many control systems can be defined with a small number of rules. The output characteristics of the fuzzy control system can be modified (if adaptation is required) by adjusting the membership functions (the equivalent of tuning in conventional control systems).

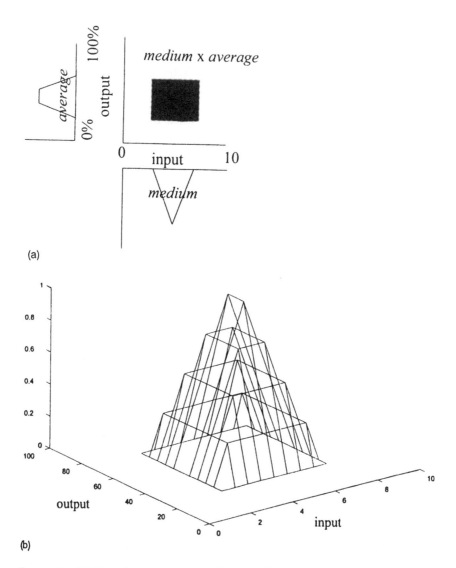

(a)

(b)

FIGURE 3   (a) Two-dimensional and (b) three-dimensional representations of a fuzzy granule.

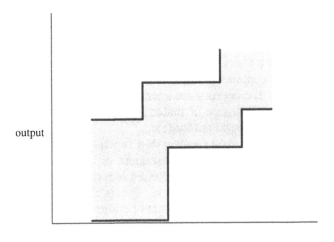

input

FIGURE 4 Fuzzy graph showing superposition of three granules.

In control applications one must define a complete rule set (i.e., a rule set that defines relationships between inputs and control outputs for all possible operating states). Initially this may be difficult, but as operating experience is gained, membership function domains may be expanded and rules may be modified, added, or deleted.

C. Interpolation or Inference

A fuzzy rule set is a set of conditional statements relating fuzzy inputs/ observations to control outputs. Actual process observations are not necessarily exact matches for rule antecedents (i.e., there can be partial belonging to more than one fuzzy input value). The inference procedure to determine the appropriate control output is a process of interpolation within the granules of the fuzzy graph.

Again there are several approaches to the interpolation procedure. One of the most common is to consider every rule at each inference. Driankov et al. [2, pp. 129–131] refer to this as individual-rule based inference and discuss it in comparison with composition-based inference. One advantage of individual-rule based inferencing is that the procedure is simple in terms of computational steps. The basic steps for the inference are as follows:

1. *Fuzzification and connective operations*: Actual input(s) are mapped into the fuzzy domain to determine the degree of match with rule antecedents. Some inputs (e.g., operator observations) may be intrinsically fuzzy values and no transformation is required. If the rule contains multiple antecedents, connected by "AND" or "OR," the appropriate operations are performed to determine the overall degree of match (i.e., new membership function) for the compound antecedent.

2. *Clipping or firing*: The new membership function for the rule antecedent is applied to the rule consequent or control output because the degree of match for the consequent must match the membership function of the antecedent. This is also referred to as "firing" the rule and is the minimum operation of the Cartesian product or Mamdani implication.

3. *Aggregation and transformation*: All the rule outputs are combined by union (or maximum operations) to define an aggregate fuzzy output value. This fuzzy output must be transformed into the appropriate form for the control action. Most commonly, this is complete defuzzification, since actuators require crisp

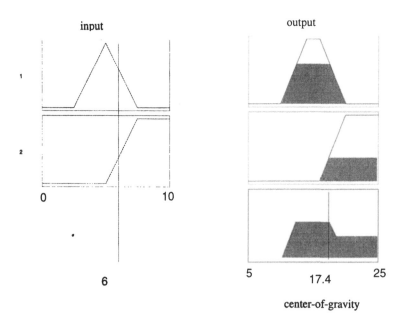

Figure 5 Fuzzy inference with two rules and center-of-gravity defuzzification.

output values. Several defuzzification methods are reasonable for control outputs, and these are defined in Chapter 5 (Section II.E) as center-of-area (Eq. 84) and mean-of-maximum (Eq. 84). However, for operator assistance (diagnosis and troubleshooting) it may be appropriate to present the aggregated or partially defuzzified output. The center-of-gravity inference procedure is shown step by step in Fig. 5. The mean-of-maximum output for the same rule set is 15.

III. SPECIFIC APPLICATIONS OF FUZZY CONTROL TO FOOD PROCESSES

In the previous section, basic elements of fuzzy control systems were discussed with emphasis on several key advantages of FDCL for food processes: simple representations for values and relations, tolerance for imprecision, and linguistic variables. The academic literature includes research papers on applications of fuzzy control to food operations and a reference list is provided for readers with specific interests: extrusion [8], fermentation processes [9–13], drying [14, 15], aseptic processing [16], snack food manufacture [17, 18], and a general review [19].

The discussion that follows in this section focuses on the use of fuzzy techniques in two specific areas of application: drying and fermentation. Typical industrial control problems within these applications are outlined, and some of the implementation issues associated with fuzzy control strategies are discussed.

A. Drying

Dryers are used across the entire spectrum of food processing, from post-harvest operations to the final stages of manufacturing. The primary purpose is to stabilize water activity food products; however, the thermal processing may be intentionally pushed to the point of toasting or roasting to impart specific flavor and textural characteristics.

There are many dryer configurations and different modes of operation (batch versus continuous, manual versus automatic), but for all modes of operation there is a common motivation for good process control. The objective is to achieve a final moisture specification but to minimize damage to thermally sensitive components (e.g., flavor, enzymes) and, in some cases, to produce desired functional properties such as rehydration characteristics and texture in the dried product. Overdrying is a cost to a food manufacturer in lost production or shrinkage losses as well as the potential for quality deterioration due to oxidative changes, unnecessary flavor loss, or stress cracking. Underdrying is also a problem. Rejected material may be reworked, but this incurs costs in warehousing, reduction in productivity, and the potential to lower the quality of new production. The need to balance quality factors (some subjective) with optimal conditions for moisture removal adds complexity to the dehydration control strategy.

Many dryers in the food industry are controlled manually by experienced operators. Automatic control has been introduced for some types of food dryers; examples are discussed in Chapter 10. However, computer-based control systems for industrial food dryers must satisfy some generic requirements. These requirements are outlined briefly with comments on the potential of fuzzy techniques to satisfy these needs:

1. *A relatively simple control model that is tractable in data and computational requirements.* Even for simple dryer configurations, models of the fundamental heat- and mass-transfer processes are complex mathematically and not suitable for real-time control purposes. There are many approaches to simplifying the control model, including fuzzy techniques. Fuzzy rules that are based on models of the dryer dynamics can be used to determine control actions based on inputs of current operating conditions. Fuzzy relational models that are developed from historic operating data or identification tests are also used for dryer control [20].

2. *On-line inputs.* Time delays resulting from off-line measurements cause problems in any control system. There are a number of reliable on-line sensors that estimate moisture content through indirect methods (e.g., capacitance, infrared reflectance, or transmittance). However, there are additional quality parameters that are difficult or impossible to measure on-line (e.g., stress cracking, checking, flavor components) but which an experienced operator can judge quickly and consistently. Fuzzy linguistic values are natural and appropriate for these types of control inputs.

3. *Adaptive capability.* The drying characteristics of agricultural materials can vary from season to season or as ingredients in prepared foods change with formulations or suppliers. This creates a need for adaptive control strategies. In fuzzy strategies, this can be achieved by adjusting membership functions or by updating relational models using on-line operating data.

Fuzzy control systems for grain and food dryers have been developed to the pilot plant stage [14, 15], and work on industrial-scale systems is in progress [20]. Two examples are used to demonstrate the application of different fuzzy methods and to provide a basis for comparison with other control methods.

Example 1. Fuzzy rules to simplify drying model for a deep-bed process.

Background: Many drying operations for food materials are carried out in deep beds. Hot air is passed through a bed of wet material to provide thermal energy and mass-transfer conditions for moisture removal. In our research, we have modeled the heat- and mass-transfer processes in deep beds for roasting [21] and drying [15] operations.

Because of computational requirements, it was not practical to use the numerical simulations of the heat- and mass-transfer processes directly in the real-time control systems that we have developed. However, the fundamental process models had several benefits in off-line analysis. Simulations were used for sensitivity analysis to identify the process variables that have the greatest effects on heating time (peanut roasting) and drying time (soybean drying). Furthermore, it was relatively easy to modify the model for different materials or dryer configurations (e.g., changes in direction of air flow) and to use this as a basis for modifying the control system.

Figure 6 shows a graphical output from the soybean drying simulation for a range of air temperatures and feed moisture contents [15]. Solid contours connect equal times for drying soybeans to an outlet moisture content of 12.3% (wb). Martineau [15] used this graph and others (at different relative humidity conditions) to develop a fuzzy rule base for estimat-

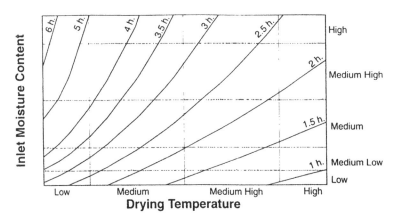

FIGURE 6 Contours of equal drying times for soybeans. Relative humidity of drying air is low. (From Ref. 15.)

ing drying time in a feed-forward controller. A similar approach has been used in control systems for extrusion [8] and for peanut roasting [17].

Fuzzy solution: The inputs, air temperature and feed moisture content, were partitioned into fuzzy intervals based on the simulation output and common sense judgment. The grids show the fuzzy intervals (without indicating the overlap between neighboring values) that were chosen for air temperature and feed moisture content. In defining membership functions, the recommendations outlined in Section II.B were followed (e.g., triangular membership functions that satisfied condition width for inputs).

Rules were created to define relationships between process inputs and drying time. After defuzzification, the rule outputs were essentially the same as the mathematical simulation for soybean drying. When the rule base was tested on a pilot-scale dryer, however, tuning was required to reduce the drying time estimates. The initial membership functions for drying time were modified based on the experimental results. It must be admitted that this approach was time consuming. It took several months to develop the simulation, including experimental work to determine the drying characteristics of the soybeans. However, the experimental work was also used to estimate the effect of drying conditions on soybean viability (the ability of the soybeans to germinate). In the control system, this quality constraint was used in addition to the final moisture constraint to determine the appropriate operating conditions for drying [15].

Comparison with other approaches: There are other approaches to developing simple control models for deep-bed drying. Empirical models with parameters that are determined by statistical techniques are currently used in control systems for industrial food dryers (e.g., [22]). It is often difficult, however, to interpret the significance of empirical parameters in terms of process variables. For example, a single parameter may combine effects due to product characteristics with effects of process variables such as air temperature and relative humidity. If the drying characteristics change (e.g., seasonal variation), the entire model must be recalibrated (i.e., new estimates of parameter values). In the fuzzy rule approach outlined in Example 1, there are two possible approaches to this problem. New drying characteristics can be determined (by experiment) and incorporated in the numerical simulation. However, these results must be translated into the fuzzy rule base, either as new rules or modified membership functions. Another approach is to use adaptive fuzzy techniques such as gradient descent [2, p. 207]. This works well with simple rule bases.

Example 2. Fuzzy relational model for drying control.

Background: Postlethwaite [23] has worked out an alternative approach for developing a fuzzy control system and is currently applying it to a dryer for spent grain and yeast in a distillery [20]. Material is conveyed through the dryer by a screw operating at constant revolutions per minute (RPM), and indirect steam heating supplies the energy required for moisture removal. In the industrial operation, dryer loading is variable and load disturbances (based on measurement of motor amperage) are compensated for by adjusting steam pressure. The new control system includes an on-line moisture analyzer at the dryer discharge. Previously, moisture was determined off-line, but the two-hour delay was about the same magnitude as the residence time and was unacceptable.

Fuzzy solution: A fuzzy relational model defines relationships between possible combinations of process variables and output values (i.e., for the distillery dryer, relationships between fuzzy values for steam pressure, motor amperage, and outlet moisture). This is different from fuzzy relations or rules that relate inputs to a control action. The relational model can be represented as a fuzzy graph or fuzzy rules, but an array format is also convenient. The example below shows an array format for a simple SISO model. The elements of the array (membership grades) represent the strength of the relationship between a fuzzy value of an input and the output. If this array was converted to rule format, consequents would take multiple fuzzy values.

u = [very low, low, medium, high, very high] (fuzzy values of process input)

y = [low, medium, high] (fuzzy values of process output)

$$R = u \times y \text{ (relational array)} = \begin{bmatrix} 0.0 & 0.8 & 1.0 \\ 0.2 & 0.5 & 0.7 \\ 0.5 & 0.4 & 0.2 \\ 0.8 & 0.5 & 0.0 \\ 1.0 & 0.0 & 0.0 \end{bmatrix}$$

Past observations of process inputs and outputs are used to determine the elements of the array (i.e., identify the relational model). The simplest method is to use one input/output observation at a time to update the array, and this method is outlined elsewhere [2, p.235]. The technique can be extended to update the process model during on-line operation so that the fuzzy model can be adaptive.

The current value of the process output is determined by a composition operation

$$y'(k) = R \circ y(k - 1) \circ u_1(k - \tau_1) \circ \cdots \circ u_n(k - \tau_n)$$

where

y' = array of memberships of the output reference sets (e.g., levels of outlet moisture)

R = relational array, which defines fuzzy relationships between inputs (e.g., steam pressure and motor amperage values) and outputs

u_i = array of memberships of the ith input set (e.g., steam pressure may be partitioned into fuzzy values such as low, medium, high, and the current process value represented by an array [0.8, 0.2, 0.0])

τ_i = a time delay between a change in an input variable and the observed effect on the output variable

○ = fuzzy compositional operator (defined in Chapter 5)

This representation is compact, but the composition operation is more complex computationally than making inferences by individual-rule firing.

In conventional model-based control, the control action can be determined by inverting the process model to determine the value of the manipulated variable that satisfies the control objective(s). The problem with a fuzzy relational model is that it is more difficult to invert to obtain the controller equation (i.e., given a target moisture range and value(s) of motor amperage and past value(s) of outlet moisture, what is the appropriate value for steam pressure?). This problem was solved [23] by developing a hybrid controller with a fuzzy relational model of the process and an optimizer that determines the control output based on defuzzified values from the fuzzy model. Postlethwaite [23] has demonstrated reasonable results with this approach for simulations and lab-scale experiments with nonlinear control problems (e.g., heat-exchanger problem).

Implementation issues: The work required to develop and implement this type of controller includes

1. Identification of the process model. In the distillery this involved several months of on-line data logging to capture process information, followed by computations to determine the relational array R.
2. Verification of the process model. This can be done during on-line operation or by using a set of historical process data that was not used in model identification.
3. Controller formulation and off-line verification.
4. On-line verification.

It was estimated [20] that this work could be completed in a time frame of six to seven months including one month for stage 4.

B. Fermentation Processes

Fermentation processes are also ubiquitous in the food industry, ranging

From large-scale operations for products such as beer to small-scale production of specialty ingredients

From traditional processes such as brewing and winemaking to recent developments based on advances in biotechnology

From production of high-quality foods for consumers to waste-treat-
ment processes

All of these processes rely on growth and maintenance of microorganisms
in complex reaction environments. However, the control requirements
can be quite different. In some cases, process economics are closely linked
to precise regulatory control. In others, the objectives are to produce
consistent products within some elastic constraints of production time and
product specifications. In many processes, one must identify abnormal
process trajectories quickly and, if possible, make corrections. With such
a wide range of control objectives, it is not surprising that many control
strategies are required. Some of these are discussed in Chapter 8. Here,
the focus is on applications of fuzzy techniques in fermentation control.
The following list identifies some process environments where fuzzy techniques are useful and examples of successful applications in each area are
discussed:

1. Fermentation is traditionally controlled or managed by an expert
   with many years of experience.
2. Control constraints are elastic, but operator assistance for troubleshooting and identification of process trajectory is important.
3. Nonlinear control is required.

1. Fermentation Is Traditionally Managed by an Expert with Many Years of Experience

Winemasters and brewmasters are usually trained by years of apprenticeship. Some of this expertise can be captured in expert-system type controllers. Fuzzy techniques are an advantage in these expert-system controllers
because many of the critical observations must be made by humans and
are based on subjective judgments. Some type of reasoning is usually
needed in control decisions.

Japanese sake brewing is based on more than 1000 years of tradition,
but there is now some concern that fewer young people are training to be
master sake brewers [12]. A group of researchers (University of Tokyo,
Hiroshima University, and Hiroshima Prefectural Food Industry Technology Center) collaborating with Chugoku Jozo K. K. have developed a
fuzzy expert-system for controlling the temperature profile during sake
fermentation. The master brewer regulates the maximum temperature and
rate of temperature change to control the essential enzymatic reactions
during the fermentation. This expert system is based on 13 rules (i.e.,
parsimonious) that were developed based on 70 process histories as well
as discussions with master brewers. Testing was conducted in two stages:

a comparison of two manually controlled fermentations to six fermentations under fuzzy control in pilot-scale fermentors followed by commercial-scale testing (28 batches). In the commercial tests, different yeast and types of rice were used in different batches. In spite of this variability in raw materials, the test batches were completed in the normal fermentation time (20 ± 1 days), within the normal process specifications for specific gravity (1.3 to 2.0 Baumé) and with sensory quality that was similar to conventional production.

2. Control Constraints are Somewhat Elastic but Operator Assistance Is Useful

It is difficult and costly to try to apply precise regulatory control in most industrial-scale fermentation processes. Process variables such as temperature, substrate concentration, and pH must be controlled within specific optimum ranges. However, precise control of macroparameters is not justified because there are many microfactors that vary and affect the fermentation dynamics. For example, in the sake brewery example, the normal variability in fermentation time (with master brewer control) is 1 day, not 24.0 hours. Product characteristics are also variable, within certain limits. The manufacturer wants to maintain product characteristics within certain quality control tolerances, but more precise control does not necessarily result in a higher selling price or consumer market share. There are economic advantages in computer-based control that offers consistency (i.e., more likely to satisfy elastic constraints) and operator assistance in identifying the current process state or in predicting its approximate trajectory. The economic advantages include less rejected production, reduced need to blend batches, and less reliance on expensive manpower (experts) for round-the-clock consultations. Furthermore, improved predictive capability makes it easier to schedule production operations upstream and downstream of the fermentation step.

The sake brewing example illustrates the type of elastic constraints that are common in food processing. A range of specific gravity values (1.3 to 2.0 Baumé) and alcohol contents (not specified) was with acceptable limits set by the commercial brewery, and sensory evaluation of the test batches was judged to be not greatly different from normal production. In other words, some attributes can be measured by physical and chemical analysis and must meet specifications but the ultimate criterion is sensory analysis, a subjective evaluation. The integration of numerical information from on-line measurements, quality control laboratory analysis, and sensory evaluation is easily handled by fuzzy techniques.

Many industrial fermentations are batch processes. Normal process trajectories, (i.e., changes in composition and physical parameters, such as specific gravity and temperature, with time) follow approximately the same pattern but rarely coincide. It is important in terms of process control to recognize when a fermentation is following its normal trajectory (within typical limits) or when it is starting to move in an unacceptable direction (e.g., growth rate is slowing down). In some fermentations, one must identify the correct time to add critical nutrients to achieve the desired product composition.

A fuzzy expert system was developed for on-line diagnosis and control [24]. LAexpert (Smalltalk V—Macintosh) was developed for lactic acid fermentation. A knowledge network was developed to represent facts (observations, measurements) and relations between facts. As process observations are made, an inference procedure is invoked to determine if the fermentation is in a normal or abnormal state. Diagnosing the causes of abnormal behavior requires backward chaining, which is a different inference procedure than used in most of the examples that have been cited. In most control applications, the inference proceeds from data or observations to conclusions (control actions) by forward chaining. In diagnosis, a conclusion about abnormal conditions initiates a backward search to identify possible root causes. Not surprisingly, backward chaining may point to several possible causes. In the fuzzy inference procedure, truth values (based on membership grade) are estimated, and the cause with the highest truth value is considered most likely.

The fuzzy expert system [24] was connected to a process control computer (two-way communication), and the authors mention that the inference procedure was sensitive to noise in process measurements. They remedied this by filtering the process data. No indication of development time for the knowledge network was given, but lactic acid fermentation was selected as a test case because of expert knowledge as well as available process data.

3. Nonlinear Control Required

In fermentation processes, pH control is a difficult nonlinear control problem. Since by definition pH is the negative logarithm of hydrogen ion concentration, the pH response is nonlinear with respect to the amount of neutralizing agent that is added. A number of conventional, nonlinear techniques have been used. However, even in simple systems, these methods are complex, and most food fermentations are mixtures of many organic acids and bases. A fuzzy relational model-based controller for pH control has been developed and tested for a strong base–weak acid neutralization [25]. There are some practical issues to consider in terms of the structure of the relational model and the problems of measurement noise in model identification and control action.

IV. IMPLEMENTATION ISSUES

A. Hardware Versus Software

Fuzzy control systems can be implemented through software that encodes the fuzzy data structure and set manipulations for inferences but which runs on standard computational platforms. The alternative approach is hardware implementation, and there are dedicated fuzzy coprocessors available as well as developments in fuzzy support for general-purpose processors (e.g., RISC architecture). There are advantages and disadvantages to both approaches, and a very clear discussion of these issues is presented elsewhere [26]. These authors consider issues of problem complexity (number of inputs, number of outputs, number of rules), flexibility (programmability, inference method), and speed (μs or ms) in exam-

TABLE 1 Commercial Fuzzy Coprocessors

Manufacturer/ product	Inputs, outputs, rules	Inference/ defuzzification	Precision	Inference time
Fujitsu F²RU-8	no limits	max–min centroid	8 b	100 μs–10 ms
Siemens 81C99	256 inputs, 64 outputs, 256 rules			78 μs (8 inputs, 1 output, 256 rules)
Omron FP1000, FP3000, FP5000, FZ001	varies with product; e.g., FP3000 provides 8 inputs, 4 outputs, 128 rules		FP3000: 12 b	FP3000: medium; FP5000: very fast
SGS-Thomson WARP	16 inputs, 16 outputs, 256 rules	max-dot	varies; e.g., 6–7 b i/o values, 4 b alpha values	μs range (with 40 MHz clock)
Togai Infra Logic FCA	flexible technology; up to 1024 inputs and outputs; up to 511 rules/output		10 b i/o values, 8 b membership values	depends on custom features; e.g., 69 μs for 8 inputs, 4 outputs, 5 antecedents per rule, 2 consequents per rule, 20 rules (20 MHz clock)
Toshiba T/FC150	8 inputs, 1 output		10 b i/o values, 8 b membership values	maximum is μs range with simple rules

Specifications are taken from Ref. 26.

ining hardware and software alternatives. They also observe that different solutions are appropriate at different design phases (prototyping versus production). Because most control applications in food processing are in early development stages, because the time scale for control decisions is relatively long, and because some components are numerical computations, most fuzzy control systems are software implementations on standard microprocessors. Table 1 provides some idea of the special-purpose coprocessors (requiring a general-purpose processor for the nonfuzzy parts) that are currently available and their capabilities.

B. Development Tools for Control Applications

Aptronix (San Jose, California)—FIDE Fuzzy Inference Development Environment (Windows) includes an environment for defining fuzzy values, a fuzzy inference language, debugging tools, and a real-time code generator for Motorola microcontrollers (6805, 68HC05, 68HC11).

Inform (Evanston, Illinois)—fuzzyTECH is available in several editions. The fuzzyTECH Precompiler edition is a design environment (similar to FIDE) that generates C source code that can be integrated with other software on different hardware platforms.

Togai Infra Logic (Irvine, California)—TIL Shell (Windows) also allows flexible design (definition of membership functions, inference, defuzzification) and can generate C code.

MathWorks (Natick, Massachusetts)—The Fuzzy Logic Toolbox (beta edition) works within the MATLAB computing environment. It allows flexible construction of fuzzy inference systems and easy visualization through graphical tools (3-D Rule Viewer, Surface Viewer) and can build stand-alone C programs that call fuzzy inference systems developed in MATLAB.

C. Coda

Fuzzy techniques are well established in control technology and will continue to be used in control systems for food processes. In fact, fuzzy control methods, alone or in combination with other techniques (e.g., conventional PID, neural networks), are particularly suited to the control problems in food processing and will enhance the implementation of computer-based control technology in the food industry.

Successful implementations of fuzzy control systems must take advantage of FDCL to design simple, parsimonious controllers. This will result in low-cost control systems and enhanced utilization of computer-based control.

ACKNOWLEDGMENTS

The author was a visiting scholar with BISC (Berkeley Initiative in Soft Computing) when this chapter was written. The insightful lectures by Professor L. A. Zadeh and discussions with other members of the BISC group were very helpful in completing this chapter.

REFERENCES

1. L. A. Zadeh, Soft computing and fuzzy logic, *IEEE Software 11*(6):48 (1994).
2. D. Driankov, H. Hellendoorn, and M. Reinfrank, *An Introduction to Fuzzy Control*, Springer-Verlag, New York, 1993.
3. E. Cox, *The Fuzzy Systems Handbook*, Academic Press Professional, Cambridge, Massachusetts, 1994.
4. L. A. Zadeh, The role of fuzzy logic in modeling, identification and control. *Modeling, Identification and Control 15*(3):191 (1994).
5. L. A. Zadeh, Fuzzy sets, *Information and Control 8*:338 (1965).
6. L. A. Zadeh, Fuzzy logic, neural networks and soft computing, *Commun. ACM 37*(3):77 (1994).
7. D. Dubois, and H. Prade, Basic issues on fuzzy rules and their application to fuzzy control, *Fuzzy Logic and Fuzzy Control*, IJCAI'91 Workshops on Fuzzy, Sydney, Australia, 1991, Springer-Verlag, Heidelberg, 1994, p. 3.
8. T. Eerikainen, S. Linko, and P. Linko, The potential of fuzzy logic in optimization and control: Fuzzy reasoning in extrusion cooker control, *Automatic Control and Optimization of Food Processes* (M. Renard and J. J. Bimbinet, eds), Elsevier Applied Science, London, 1988, p. 183.
9. G. P. Whitnell, V. J. Davidson, R. B. Brown, and G. L. Hayward, Fuzzy predictor for fermentation time in a commercial brewery, *Computers and Chemical Eng. 17*(10):1025 (1993).
10. C. Von Numers, M. Nakajima, T. Siimes, H. Asama, P. Linko, and I. Endo, A knowledge-based system using fuzzy inference for supervisory control of bioprocesses, *J. Biotechnol. 34*:109 (1994).
11. X.-C. Zhang, A. Visala, A. Halme, and P. Linko, Functional state modeling and fuzzy control of fed-batch aerobic baker's yeast process, *J. Biotechnol. 37*:1 (1994).
12. J. Koizumi, Fuzzy control for Japanese sake—fuzzy decision controller and fuzzy simulator for Japanese sake fermentation, *Industrial Applications of Fuzzy Technology*, Springer-Verlag, Tokyo, 1992, p. 193.
13. K. Oishi, M. Tominaga, A. Kawato, Y. Abe, S. Imayasu, and A. Nanba, Application of fuzzy control theory to the sake brewing process, *J. Fermentation and Bioeng. 72*(2):115 (1991).
14. Q. Zhang, and J. B. Litchfield, Knowledge representation in a grain drier fuzzy logic controller, *J. Agric. Eng. Res. 57*:269 (1994).
15. S. Martineau, *Quality-Based Control of Soybean Drying*, M.Sc. thesis, University of Guelph, Guelph, Ontario, Canada, 1995.

16. R. K. Singh and F. Ouyang, Knowledge-based fuzzy control of aseptic processing, *Food Technol. 48*(6):423 (1994).
17. J. J. Landman, *Modelling and Control of Colour Development in Virginia Peanuts During Cross-Flow, Dry Roasting*, Ph.D. thesis, University of Guelph, Guelph, Ontario, Canada 1994.
18. Y. S. Choi and A. D. Whittaker, Self-learning fuzzy controller for snack food frying, Proceedings of IFIS '93, Cat. No. 93TH0594-2, 1993.
19. M. Dohnal, J. Vystrcil, J. Dohnalova, K. Marecek, M. Kvapilik, and P. Bures, Fuzzy food engineering, *J. Food Eng. 19*:171 (1993).
20. H. Bremner and B. Postlethwaite, The industrial evolution of a fuzzy model based controller, Proceedings of the International Conference on Control '94, IEE, Coventry, United Kingdom, 1994, vol. 1, p. 243.
21. J. J. Landman, V. J. Davidson, R. B. Brown, G. L. Hayward, and L. Otten, Modelling of a continuous peanut roasting process, Proceedings of ACOFOP III, Paris, Elsevier, New York, 1994, p. 207.
22. J. F. Forbes, B. A. Jacobson, E. Rhodes, and G. R. Sullivan, Model based control strategies for commercial grain drying systems, *Can. J. Chem. Eng. 62*:773 (1984).
23. B. Postlethwaite, A model-based fuzzy controller, *Chem. Eng. Res. Des. 72*(1):38 (1994).
24. T. Simes, M. Nakajima, H. Yada, H. Asama, T. Nagamune, P. Linko, and I. Endo, Object-oriented fuzzy expert system for on-line diagnosing and control of bioprocesses, *Appl. Microbiol. Biotechnol. 37*:756 (1992).
25. B. Kelkar and B. Postlethwaite, Study of pH control process using fuzzy modelling, International Conference on Control '94, Coventry, United Kingdom, 1994, vol. 1, p. 272.
26. A. Costa, A. De Gloria, P. Faraboschi, A. Pagni, and G. Rizzotto, Hardware solutions for fuzzy control, *Proc. IEEE 83*(3):422 (1995).
27. The Mathworks, Inc., The Fuzzy Logic Toolbox, beta edition, February, Natick, Massachusetts, 1995.

7

Image Processing and Its Applications in Food Process Control

S. Majumdar, X. Luo, and D. S. Jayas
University of Manitoba, Winnipeg, Manitoba, Canada

I. INTRODUCTION

Embodying image acquisition, storage, analysis, and pattern recognition techniques, image processing techniques are capable of extracting or measuring various visual features (related to size, shape, color, and texture) of objects and performing task-relevant analysis and interpretation with precision, objectivity, and speed. Digital imaging operates on pictorial information commonly provided by various physical sensors (instead of human eyes) such as video cameras, X-ray machines, laser scanners, ultrasound sensors, and nuclear magnetic resonance (NMR) sensors, and a computer (instead of human brain) is used to analyze and understand the obtained images. It can be viewed as a simulation and extension of the ability of the human vision system. For this reason, image processing systems are also called as machine vision systems in many cases.

Although established about 30 years ago, image processing did not become a practical technique widely used in industries until the early 1980s, when substantial advances were made in the related areas, especially in computer and imaging techniques. Commercial image processing or machine vision systems are fairly common in many industries, such as automotive, electronics, and manufacturing [1–3]. Its applications in the food industry are limited; however, some applications for the purposes of produce inspection and process guidance (robotic vision) are becoming common. Scientific research is continuing to broaden its applications to many other food and allied industries.

Application of image processing techniques to food process control is challenging. Unlike other industrial objects of defined size, shape, color, and texture, the objects or workpieces involved in the food industry are usually of a natural variability. It is given that the image processing systems for food industry must be sufficiently flexible and robust to cope with the biological variability [4–6]. In addition, processing speed demands in most food processing applications are very high, usually requiring specialized image processing hardware and software.

Machine vision systems began to appear in the food processing industry in significant numbers in the early 1980s [7]. The initial use was limited to simple sorting and inspection tasks. This application expanded rapidly in the late 1980s. In 1989, 12% of the installed machine vision systems were being installed by food processors [8]. The expansion was not only in numbers of installed machine vision systems; the scope of the application was also expanded to almost every aspect in the food processing inspection, from raw material sorting and grading to final product and package inspection. The applications of machine vision systems have saved thousands of laborers from the tedious, inefficient manual inspection tasks in the food processing industry. To the food processors, the application of machine vision systems is no longer a luxury but a necessity to keep and increase their product competition abilities in the market. Currently the need for image processing systems in the food industry is great. According to an estimate by Nello Zuech (Vision Systems International, Yardley, Pennsylvania), the potential market for machine vision systems in the agri-food industry is about US $581 million, but only about 465 units valued at US $57 million are installed.

From a technical point of view, the application of image processing in the food industry is still in its infancy. Most of the installed machine vision systems are two-dimensional (2-D) monochromatic or black-and-white systems with a resolution of 128×128 or 512×512 pixels. They are based on PCs (personal computers) with a 80286 or 80386 processor. Limited by the computing speed of the PCs, most of the systems use very simple image processing techniques [6], and many systems use specialized hardware or chips to increase the inspection speed. These systems are successful only under constrained conditions for specific applications. As high-speed microcomputers (80486 or Pentium) with reasonable and continuously dropping prices have become commercially available since the early 1990s, many image processing functions have been increasingly moved from specific hardware to software implementations, and more complicated image processing techniques have been used. This change makes recently developed systems more flexible or generic in applications. Also, color image processing systems began to emerge in the food industry in the early 1990s because of advances in solid state color imaging sensors as well as the increased computing speed of microcomputers. Although research in these areas has grown rapidly and substantially in the recent years, the adoptions of both generic and color image processing systems to food processing are few. There are still many generic problems to be overcome [6]. Most of the developments are still being studied under laboratory conditions.

The basics of digital image processing and its applications in the food industry are covered in this chapter. Many recent books introduce the concept of image processing quite well [3, 9, 10]. The reader is referred to these and similar books for further details on the digital image processing.

II. DIGITAL IMAGE PROCESSING SYSTEMS

A. System Components

A digital image processing system is a collection of hardware and software components that can acquire, store, display, and process digital images, as shown in Fig. 1. Although these components may be physically separated in space and time, each is fundamentally necessary to complete the digital image processing system.

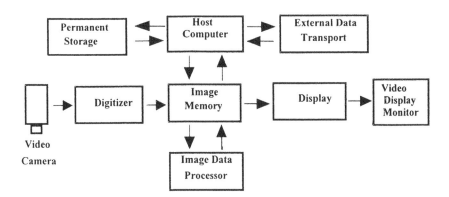

FIGURE 1 Components of a typical digital image processing system.

B. Image Acquisition

The first stage in any digital image processing system is to acquire a digital image. This is a two-step process requiring some sensor device and digitization function. The sensor device is sensitive to a band in the electromagnetic energy spectrum and produces an electrical signal output proportional to the level of energy sensed. The digitizer is a device for converting the analog electrical output of the sensor into a digital form.

Besides X-ray imaging systems, most common sensors deal with visible and infrared light. The most frequently used devices are micro densitometers, image dissectors, vidicon cameras, and solid state arrays.

The operation of vidicon cameras is based on the principle of photoconductivity. An image focused on the tube surface produces a pattern of varying conductivity that matches the distribution of brightness in the optical image. An independent, finely focused electron beam scans the rear surface of the photoconductive target, and by charge neutralization this beam creates a potential difference that produces on a collector a signal proportional to the input brightness pattern. A digital image is obtained by quantizing this signal, as well as the corresponding position of the scanning beam.

Solid state arrays are composed of discrete silicon imaging elements, called photosites, that have voltage output proportional to the intensity of the incident light. Line-scan and area sensors are two types of solid state sensors. A line-scan sensor consists of a row of photosites and produces a 2-D image by relative motion between the scene and the detector. An area sensor is composed of a matrix of photosites and therefore can capture an image in the same manner as a vidicon tube. A significant advantage of solid state array sensors is that they can be shuttered at very high speed (e.g., 1/10,000 s). This makes them ideal for applications in which freezing of motion is required.

The line-scan sensors have resolutions ranging from 256 to 4096 elements. The resolution of the area sensors ranges from 32 × 32 elements at the low end to 2048 × 2048 elements at the high end.

C. Storage

An 8-bit image of size 1024 × 1024 pixels requires about 1 Mbyte of storage. Thus providing adequate storage is usually a challenge in the design of image processing systems. One method of providing short-term storage is through computer memory. Another is by specialized boards, called frame buffers, that store one or more images and can be accessed rapidly, usually at video rates (30 images per s). On-line storage generally takes the form of magnetic disks. The magneto-optical storage has a Gbyte (one billion bytes) of storage memory on a 5.25-in. optical platter. Archival storage is characterized by massive storage requirements but infrequent need for access. Magnetic tapes and optical disks are usual media for archival applications.

D. Processing

Processing of digital images involves procedures that are usually expressed in algorithmic form. Thus, with the exception of image acquisition and display, most image processing functions can be implemented in software. The only reason for specialized image processing hardware is the need for speed in some applications or to overcome some fundamental computer limitations. Image processing is characterized by specific solutions. Hence techniques that work well in one area may be totally inadequate in another. The actual solution of a specific problem generally requires significant research and development.

E. Communication

Communication in digital image processing primarily involves local communication between components of an image processing system or between image processing systems and remote communication from one point to another, typically in connection with the transmission of image data.

F. Display

Monochrome and color TV monitors are the principal display devices used in modern image processing systems. Monitors are driven by the output(s) of a hardware image display module in the back-plane of the host computer or as part of the hardware associated with an image processor. The signals at the output of the display module can also be fed into an image recording device that produces a hard copy (slides, photographs, or transparencies) of the image being viewed on the monitor screen. Other display media include printing devices.

A host computer controls the entire system. It provides the interface to the user along with the sequencing of acquisition, storage, display, and processing. The digital image stored in memory is freely accessible for processing by the host computer. Although the host computer has the full ability to carry out any conceivable operation upon a stored image, its execution speed can be limited. To augment the host, specialized high-speed digital image processing processors are usually a part of the processing system.

This additional processing hardware can take the form of high-speed hardware circuits or secondary microprocessors, optimized to handle common digital image processing operations. For applications that must run fast enough to keep up with real-time events, such as a moving conveyer line of parts, the high-speed hardware approach is often essential.

Many off-the-shelf digital image processing software packages and function libraries are available (e.g., *Optimetric* by Optimas Corporation, Bothell, Washington; *Image-Pro Plus* by Media Cybernetics, Silver Spring, Maryland; *Global Lab Image* by Data Translation, Marlboro, Massachusetts; and *IPLab* by Signal Analytics, Vienna, Virginia). These programs contain various digital image processing operations and generally run on the host computer. Many packages also support the use of a specialized digital image processor. The processing software is often the part of the system that "glues" all other components together into a complete system.

III. DIGITAL IMAGES

A. Definition

An image is a two-dimensional (2-D) function generated by sensing a radiometric information of a scene. A scene is a collection of three-dimensional (3-D) objects with some geometrical arrangement and usually governed by the physical laws of nature. The image is represented by an image function $f(x, y)$, where the arguments of the image function—the independent variables x, y—are spatial coordinates related to physical locations in the sensor image plane, and f is the intensity or gray level at these locations. In a color image, f is a vector with three components representing hue (H), saturation (S), and intensity (I), or red (R), green (G), and blue (B).

B. Image Resolution

The quality of a digital image is directly related to the number of pixels and lines, along with the range of brightness values, in the image. These aspects are known as image resolution. The image resolution is the capability of the digital image to resolve the elements of the original scene. For digital images, the resolution characteristics can be broken down in two ways: the spatial resolution and the brightness resolution (or color resolution, if the image is in color). The number of pixels in an image is described by its spatial resolution. The more pixels in the image, the greater its spatial resolution. Every pixel in a digital image represents the intensity of the original image at the spatial location where it was sampled.

The concept of brightness resolution addresses how accurately the digital pixel's brightness can represent the intensity of the original image.

The aspect ratio is a measure of an image's rectangular form. It is calculated by dividing the image's horizontal width by its vertical height. In case of commercial broadcast television and common video equipment, images have an aspect ratio of 1.333, commonly denoted as 4:3. This means that the horizontal dimension of the image is 1.333 times as wide as the vertical dimension. An image with a 1:1 aspect ratio appears as a perfect square.

C. Color Image

If one looks very closely at a color video display screen, whether it is a cathode ray tube (CRT), liquid crystal display (LCD), or another type, one will notice individual dots of solid colors. These dots emit light in the colors of red, green, and blue. All the colors in the spectrum can be created with the primary colors of red, green, and blue (R, G, and B). This is called the additive color property, and it works for the mixing of primary colors that are emitting light. When red, green, and blue are mixed together, an entire spectrum of colors can be created, and it can be represented by a color space cube as shown in Fig. 2.

Subtractive color mixing is based on reflective colors rather than emissive colors. Instead of emitting light like a video display, subtractive colors reflect the light shined upon them. The subtractive colors, called

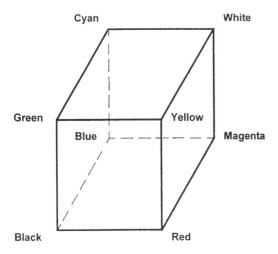

FIGURE 2 The red–green–blue (RGB) color cube.

secondary colors, are cyan, magenta, and yellow (C, M, and Y). Subtractive colors are used primarily in the printing industry.

Although RGB color space is the fundamental color space used to physically detect and generate color light, other derivative color spaces can be created to aid color image processing. The most important derivative color space is the hue, saturation, and intensity (HSI) space. This color space represents color as we perceive it. Whenever an application requires a human to interpret or control the colors of an image, HSI space is well suited. Hue indicates what color, such as green, dominates the reflected light. Saturation indicates how much of the color is there, i.e., purity of color. For example, a hue of red can have numerous saturation levels ranging from deep red (fully saturated) to pink and finally white (no saturation of red at all). Intensity indicates how bright the color is, such as light green.

IV. IMAGE ANALYSIS

Image analysis operations are used in applications that require the measurement and classification of image information. They are different from all other digital image processing operations because they almost always produce nonpictorial results. The mission of image analysis operations is to understand an image by quantifying its elements. The quantification includes such things as measures of size, indicators of shape, and descriptions of outlines. Other elements of interest can include attributes such as brightness, color, and texture.

The first step in image analysis generally is to segment the image. Segmentation subdivides an image into its constituent parts or objects. The level to which this subdivision is carried depends on the problem being solved; i.e., segmentation should stop when the objects of interest in an application have been isolated. In general, autonomous segmentation is one of the most difficult tasks in image processing. This step in the process determines the eventual success or failure of the analysis. For this reason, considerable care should be taken to improve the probability of rugged segmentation. In some situations, such as industrial inspection applications, at least some measure of control (e.g., lighting, clean environment) over the environment is possible at times.

Segmentation algorithms for monochrome images generally are based on one of two basic properties of gray-level values: discontinuity and similarity. In the first category, the approach is to partition an image based on abrupt changes in gray level. The principal areas of interest within this category are detection of isolated points and detection of lines and edges in an image. The principal approaches in the second category are based on thresholding, region growing, and region splitting and merging. The concept of segmenting an image based on discontinuity or similarity of the gray-level values of its pixels is applicable to both static and dynamic (time-varying) images. Details of different segmentation methods are described in any standard book of digital image processing [10–12].

V. FEATURE EXTRACTION

A. Use of Features

Once the image has been cleanly segmented into discrete objects of interest, the next step in the image analysis process is to measure the individual features of each object. Many features can be used to describe an object. These features are compared with the information from known objects to classify an object into one of many categories. Generally, the features that are the simplest to measure and contribute substantially toward the classification are the best to use.

B. Brightness Features

Brightness and color features of an object can be extracted by examining every pixel within the object's boundaries. The histogram of these pixels shows the brightness distribution found in the object. For color objects, the red, green, and blue pixel component values of the object can be converted to the hue, saturation, and brightness (intensity) color space. Then looking at the histogram of the hue component, image will instantly show the predominant hue of the object. For a color-sorting application, this feature alone can be all that is necessary to differentiate between two objects, such as red for Red Delicious and green for Granny Smith apples moving on a conveyer line.

Statistics of the brightness in an object can also be useful measures. The mean brightness represents the average brightness of an object. The standard deviation of brightness gives a measure of how much the object's brightness varies from the mean value. The mode brightness is the most common brightness found in the object. The sum of all pixel brightness values in an object relates to the energy, or aggregate brightness, of an object. This measure is called an object's zero-order spatial moment. The application of these statistical measures to brightness histograms or color-component histograms can help in classifying the brightness or color characteristics of an object.

C. Texture Features

The texture measure of an object can be used to discriminate between the surface finish of a smooth or a coarsely textured object. A convenient way to determine the texture of an object is to examine its spatial frequency content. A smooth object has only small variances in its brightness, which is equivalent to having primarily low spatial frequencies. A coarsely textured object has lots of minute variations, which means that it will contain high spatial frequencies. A high-pass filter will yield an image of the object's high-frequency components.

Similarly, a Fourier transform operation [10] can be used. Again, the presence of high frequencies in the frequency image represents coarse texture, and the absence of high frequencies represents a smooth texture. Some other texture features, such as gray level co-occurrence matrices [13], gray level run length matrices [14], and neighboring gray level dependence matrices [15], are also used for classification of different objects.

D. Shape Features

The most common measurements that are made on objects are those that describe shape. Shape features are physical dimensional measures that characterize the appearance of an object. Objects can have regular shapes, such as square, rectangular, circular, or elliptical, but in many cases the shape of the object is arbitrary, twisting, and turning in apparently random ways. The commonly measured shape features are briefly defined here.

Perimeter: The pixel distance around the circumference of an object. The result is a measure of the boundary length of the object.

Area: The pixel area of the interior of the object. The area is computed as the total number of pixels inside, and including, the object boundary. The result is a measure of object size.

Area to perimeter ratio: It measures the roundness of the object, given as a value between 0 and 1:

$$\text{Roundness} = \frac{4\pi \times \text{Area}}{\text{Perimeter}^2}$$

The greater the ratio, the rounder the object. If the ratio is equal to 1, the object is a perfect circle.

Major axis length: Distance between the (x, y) endpoints of the longest line that can be drawn through the object. The major axis endpoints (x_1, y_1) and (x_2, y_2) are found by computing the pixel distance between every combination of border pixels in the object boundary and finding the pair with the maximum length. The result is a measure of object length:

$$\text{Major axis length} = \sqrt{(x_2 - x_1)^2 + (y_2 - y_1)^2}$$

where (x_1, y_1) and (x_2, y_2) are the major axis endpoints.

Minor axis length: The distance between the (x, y) endpoints of the longest line that can be drawn through the object while maintaining perpendicularity with the major axis. The result is a measure of object width:

$$\text{Minor axis length} = \sqrt{(x_2 - x_1)^2 + (y_2 - y_1)^2}$$

where (x_1, y_1) and (x_2, y_2) are the minor axis endpoints.

Minor axis length to major axis length ratio: The ratio of the length of the minor axis to the length of the major axis is a measure of object elongation, given as a value between 0 and 1. If the ratio is equal to 1, the object is generally of square or circular shape.

Boundary box area: The area of the box that would entirely surround the object:

$$\text{Boundary box area} = \text{Major axis length} \times \text{Minor axis length}$$

Spatial moments: The spatial moments of an object are statistical shape measures that do not characterize the object specifically. Rather, they give statistical measures related to an object's characterizations.

The *zero-order spatial moment* is computed as the sum of the pixel brightness values in an object. For a binary image, this is simply the number of pixels in the object, because every object pixel is equal to 1 (white). Therefore, the zero-order spatial moment of a binary object is its area. For a gray level image, an object's zero-order spatial moment is the sum of the brightness values of pixels and is related to the object's energy.

The *first-order spatial moments* of an object contain two independent components, x and y. These are the x and y sums of the pixel brightness in the object, each multiplied by its respective x or y coordinate in the image. In the case of a binary image, the first-order x spatial moment is just the sum of the x coordinates of the object's pixels, because every object pixel is equal to 1. Similarly, the y spatial moment is the sum of the y coordinates of the object's pixels.

VI. OBJECT CLASSIFICATION

The next step after measuring the features of the objects is to classify the objects. The classification involves comparing the measured features of a new object with those of a known object or other known criteria and determining whether the new object belongs to a particular category of objects.

As an example, an image analysis operation is required to perform grading of salmon fish on the basis of their size. The fish is moving on a conveyer and the goal is to determine which fish belongs to which grade (say, out of three grades: grade 1, grade 2, and grade 3) so that they can be sorted into bins of corresponding grades. One approach is to measure the total image area, which generally correlates with size, and classify the fish accordingly.

VII. MEASURE INVARIANCE

In machine vision and most image interpretation applications, objects are measured against specific dimensional requirements. If an object meets the requirements, it is a member of a class (for instance, the "good" class); otherwise, it is not (it is in the "bad" class or another class). In the real imaging world, a variety of things can cause an object to appear distorted from its actual physical appearance. In particular, the object may take on geometric distortions such as rotation to a random angle, translation to a random location in the image, and scaling to appear smaller or larger. Brightness and color distortions are also common. All of these distortions can disrupt measures like object area and perimeter, and they will certainly make a shape description different. The troubling part is that because of these distortions a "good" object may be classified as a "bad" object. The worst thing that can happen in an image analysis operation is that a "bad" object is classified as a "good" object, resulting in adverse consequences to end results depending on the seriousness of the problem. A "bad" fish classified as "good" fish and processed for consumption has the potential to cause serious health concerns.

Measure invariance refers to the trait of a particular measure being insensitive to a particular variance. Let's say we have decided that an object of interest can be sufficiently discriminated as "passing" or "failing" based solely on its length and width features. We further decide that the major (length) and minor (width) axes endpoints will be used to identify these features. By examining an image of the object, we record the (x, y) endpoints of both features (Fig. 3a). We propose that every time we find an object that has the same length and width endpoints, it is a member of the "pass" class. Otherwise, it falls into the "fail" class. What if the

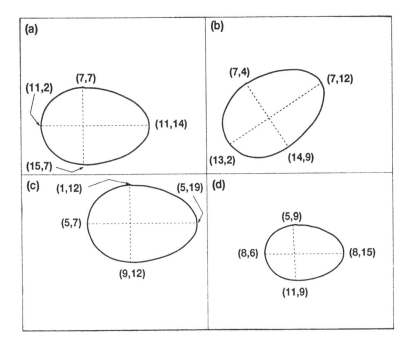

FIGURE 3 Binary (a), rotated (b), translated (c), and scaled down (d) images of an egg with major and minor axes endpoints. Note that the major and minor axes endpoints are different for each case.

identical object is rotated in the image (Fig. 3b)? What if it is translated to another part of the image (Fig. 3c)? What if it is slightly smaller or larger (Fig. 3d)? Even though the object may be a "pass" object, these variations in the way the object appears within the image will cause the length and width endpoints to vary, and the object will be misclassified as a "fail."

Instead of using the length and width endpoint measures alone, more appropriate measures can be derived that are invariant to object rotation, translation, and scaling variations. First, length and width distances can be used instead of the absolute length and width endpoints (Fig. 4a). Second, we can add scaling invariance by normalizing the length and width distance measures by a known distance measure, called a *unit distance* (Fig. 4b). This is a dynamic calibration procedure, meaning that every time the measures are made, the scaling distortion is automatically removed. Distance reference marks or an object of known dimensions in the image is used to represent a known distance.

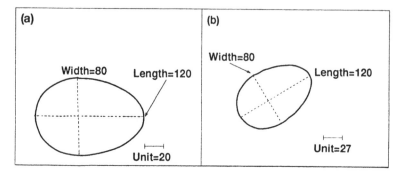

FIGURE 4 Original binary image of the egg from Fig. 3a with major and minor axes (a) and rotated and scaled down image of the same egg (b). The major and minor axes lengths are the same in both cases.

VIII. LIGHTING

Lighting and control of image brightness are two crucial aspects of a digital image processing technique for precise and accurate feature extraction, especially if gray level or color reflectance features are being used. There are mainly two types of lighting: direct and diffused. According to the type of application, different types of lighting are used. There are two basic lighting configurations for illumination of samples: front or reflected light and back-lighting with a diffusing screen (e.g., plastic) to obtain silhouette images. Back-lighting is more effective for generating high-contrast images, which subsequently helps in precise segmentation.

Regardless the type or configuration of illuminant used for lighting (e.g., incandescent, fluorescent, or halogen), there are many factors that can increase the variability in digital image brightness levels. Lamp use or burning time reduces the brightness level because of a reduction in light output. For incandescent lamps, ambient temperature variations can cause changes in light output and color values. For fluorescent lights, flickering may severely affect the light output. Halogen lamps are relatively stable, but they emit too much heat. To get a uniformity in the field of view (FOV), diffusers can be used in front of light bulbs.

Insufficient camera warm-up and other video circuitry can contribute to the error of the system. A light-compensating electronic circuit such as automatic gain control should be disabled when image brightness data are used. A video camera with gamma compensating circuits should have gamma correction factor to eliminate any error.

IX. FOOD INDUSTRY APPLICATIONS

A. Raw Material Inspection

1. Fruits and Vegetables

Produce inspection is a common need in the food industry, which involves quality verification, defect removal, process control, sorting, and grading. The manual inspection works by visual examination of produce's external features, such as morphology (size and shape), color, texture, and surface defects, which are related to the quality of the food products in one way or another. It is a tedious, inefficient operation. Combined with robotics, image processing provides a potential tool for the automation of produce inspection, which will undoubtedly bring significant economic and labor saving benefits to the food industry.

To meet the increasing demand for high-quality produce, many fruits and vegetables are sorted and graded before being sent to market or for further processing. Applying image processing to sorting and grading of fruits and vegetables has been a very active area of research in the last decade [16–18]. Most of the studies have been directed toward morphological and color sorting, and defect detection.

Morphology is an important quality indicator for many fruits and vegetables. Compared with other quality factors such as color and defects, morphological factors are easier to measure using image processing techniques. The technique of morphological sorting and grading is relatively mature, and most of the early research in monochromatic machine vision sorting and grading was in this area. An image processing technique was developed for sorting of potatoes by morphology [19, 20]. A machine vision system was used to determine orientation and shape of bell peppers [21]. Circular Hough transforms were applied to locate the stem and blossom ends, which were used to determine the orientation of the bell peppers. Circular Hough transform is generally used to identify circular or irregularly shaped objects; it is an extension to the standard Hough transform, used to identify straight lines [10]. Axial paired gradients and medial axis variance is used to describe the shape. A machine vision system was developed to determine stem–root joint for processing carrots by shape features [22]. A neural network was designed to classify carrot tips into five classes using morphological features [23]. Trained with 80 simulated carrot tips and tested with 250 fresh market carrots, the average misclassification rate was 11.5%. A machine vision system was developed to characterize and classify fresh-market carrots for forking, curvature, surface defects, and brokenness [24]. Boundary irregularity in silhouette images of asparagus was evaluated as shape features for quality assessment [25].

An image processing technique was used to analyze Florida grapefruit for shape classification [26]. Acceptable and misshaped grapefruits were analyzed for four feature vectors (e.g., area ratio, circularity, diameter range, and diameter ratio), among which the major to minor axis diameter ratio was the most significant in classification. The success rate was >96% with the Bayesian nonparametric decision model. Image processing has been used for morphological sorting of many other fruits and vegetables, including apples [27, 28], lemons [29], oranges [30], and tomatoes [31, 32].

Color is usually the most important and influential attribute of food quality [33]. Many defects of fruits and vegetables appear as discolorations on surfaces. Color sorting and grading of fruits and vegetables using image processing was previously limited by the high cost of color imaging devices and the limited computing capability of computers. Some work has been done by using monochromatic systems combined with optical color filters. As solid state color sensors, like coupled charge device (CCD) color cameras, and high-speed microcomputers, such as 80486 and Pentium, have become commercially available in the early 1990s, color sorting and grading is becoming a reality. Extensive research has been done in the last 10 years. A color image analysis technique was used to detect and classify fungal-damaged soybeans [34]. Intensity and ratios of red to blue, red to green, and green to blue were used as discriminants. Individual soybeans were correctly classified into one of five categories: healthy with 98% accuracy, and those showing symptoms of infection due to *Phomopsis sp.*, *Alternaria sp.*, *Fusarium sp.*, and *Cercospora kikuchii* with 77 to 91% accuracies. To evaluate the quality, color information from soybeans was combined with other derived morphological features [35]. Damaged soybeans were correctly classified with an accuracy of 85%. A color vision system was developed to inspect and grade fresh-market peaches [36]. Diffuse lighting and normalized luminance were used to reduce the red, green, and blue inputs to two-dimensional chromaticity coordinates. Peach color was compared with standard peach maturity colors. Machine maturity classification agreed with manual maturity classification in 54% of test samples, and was within one color standard in 88% of the tests. Bell peppers were successfully sorted using a color image analysis system [37]. Red–green–blue (RGB) pixel intensity values were mapped to one of eight possible hues. The relative hue distributions of pixels in six orthogonal views were calculated and used as color quantitative variables. Up to 96% accuracy was achieved for grading bell peppers by color. A trainable algorithm was developed on a color machine vision system for inspection of soybean seed quality [38]. The algorithm correctly classified asymptomatic soybean seeds, seeds infected by *C. kikuchii*, "seeds of other colors," and "materially damaged seeds," with 94, 97, 85, and 96% accuracy, respectively. The variables used for classification were color chromaticity coordinates and seed sphericity. A high-speed color inspection system was developed to detect foreign materials on a peanut conveyor belt in real time [39]. The system is trainable to recognize and differentiate unique color signatures or fingerprints of many types of foreign materials and food products. Other applications of color image analysis were for sorting and grading of apples, mushrooms, and potatoes [16], date fruits [40], oranges [30], kiwi fruits [41], and bananas [42].

Image processing has also been used to sort and grade fruits and vegetables for defect detection. Although defect detection is basically performed using color and morphological information about a product, it is usually more complex than sorting and grading based on color and morphological features. Defect detection requires not only the global (or average) measurements on color and morphology of a product, as color or morphological sorting and grading do, but also the local surface scrutiny. A review of the research in this area up to 1991 was reported [17]. Most recent research includes defect detections in asparagus [43], peanuts [44], prunes [45], and peaches [46].

Internal defects, such as water-core, internal breakdown, dehydration, and hollow heart, are common degrading factors in quality control of fruits and vegetables. Traditional imaging techniques cannot provide enough information for detecting internal defects [47]. Nuclear magnetic resonance (NMR) and X-ray imaging techniques have the potential for the detection of internal defects in fruits and vegetables. Since the NMR imaging technique can make almost all internal defects visible based on the changes in water content distribution caused by these defects, internal defects can be detected. The research in this area has been conducted for detecting water-core, bruises, and pits in apples; hollow heart in potatoes; and worm holes or dry regions due to frost in citrus [48, 49]. The application of NMR image processing is currently limited by the high cost and the complex operations of NMR imaging equipment. The X-ray transmission through an object is related to its density distribution; therefore, it is possible to use X-ray imaging for detecting dehydration and hollow heart in fruits and vegetables. X-ray imaging was used for detecting hollow heart in potatoes [50] and for measuring water-core in apples [51]. The application limitations are again the expensive imaging equipment and the forbidden uses in many food products.

It has been revealed by many researchers [52–56] that a single group of features (color, morphology, surface defects, and internal defects) alone is inadequate for quality inspections of many fruits and vegetables. To evaluate the overall quality of fruits and vegetables, a combination of various external and internal quality features measured by multiple sensors may be needed. Multiple sensors were investigated for tomatoes [54] and for cantaloupes [55].

2. Meat

Image processing has been used for inspection of many meat products, such as beef, pork, poultry, fish, and shrimp. An algorithm was developed to extract statistical shape features from the histograms of the three primary components of color images for quantitative measurements of the doneness in beef steaks [57]. Morphological (size and shape) features were used to segment and isolate the rib-eye and to determine marblings using images of the cut surface of beef carcasses [58]. An image analysis technique was used to locate and isolate the lion-eye area on a beef carcass from a cluttered scene [59] and to separate connected muscle tissues in beef carcass rib-eyes [60]. A machine vision system was developed to measure the areas of visible fat and lean in commercial pork cuts [61].

The potential of using color image processing to detect poultry defects such as bruises, skin tears, and systemic diseases was investigated [62]. Park and Chen [63] used six optical filters at different wavelengths to get multispectral images for poultry carcass inspection. Gray level image intensities at selected wavelengths were used to distinguish normal carcasses from "septicemic" and "cadaver" carcasses. The accuracies of classification for poultry carcasses using a neural network classifier were 92% for normal, 83.4% for "septicemic," and 96.6% for "cadaver" carcasses.

An inexpensive oyster meat grader was developed based on measuring the projected area of oyster images and relating the area to the oyster meat volume by which oysters are commercially graded [64]. Various morphological and spectral features derived from shrimp images were examined for potential in determining the optimum location for removal of shrimp heads [65, 66]. The spectral features were more effective than morphological features [65, 66]. The standard deviation of the location prediction ranged from 2.766 to 4.600 mm at a rate of three shrimps per second. Fish species were recognized by shape analysis of their images [67]. A database of fish shapes was created from the photographs of seven different types of fish.

An image processing technique was used to detect the sex of broiler chicks [68]. The sex identification of broiler chicks is currently done manually by examining the wing feathers for a sex-related feathering trait. An image analysis algorithm was developed to isolate feathers of interest from the rest of the wing.

Ultrasonic image analysis provides a tool for estimating yield and quality grade of live meat-producing animals by measuring rib-eye area, fat thickness, marblings, and other quality features. Research in this area has been reported by many researchers [69–73].

3. Eggs

The applications of image processing for egg inspection were investigated by several researchers. Cracks in stationary eggs were detected by analyzing the egg images with back-lighting [74]. An accuracy of 96% was achieved for crack detection. This research was further expanded to detect cracks in rotating eggs by using three different side images of each egg [75]. Again, back-lighting was used. A machine vision system was used for detection of blood spots and dirt stains in eggs [76]. An application of a machine vision technique was described for detecting fertility of hatching eggs [77, 78]. The images of back-lit eggs were enhanced using histogram stretch. The shape features of the histograms of egg images were used to distinguish fertile from infertile eggs. The detection accuracy was 96–100% for the four-day and 88–90% for the three-day eggs in incubation. Egg mass loss during incubation was detected by measuring the air-cell volume with a machine vision system [79].

B. Processed Food Inspection

Processed food holds the biggest portion of the food market. The quality of the processed food largely depends on the food processing technique. Inspection is needed not only for sorting and grading final food products, but also for providing quality information on products being processed for optimal food processing control. Food processing environments are usually characterized by high or low temperature, high humidity, and high speed. Manual inspection of in-process products is difficult from the viewpoint of sampling from processing lines and providing in-time quality information for process control. Machine vision is suitable for this kind of inspection task.

A typical example of inspection of in-process food is in the bakery industry. A system was designed for a multiple-lane cracker baking process [80]. Complemented with mechanical devices, the system can automatically perform selected-lane positioning and tracking, selected-lane picture-taking, and product parameter (diameter, area, and thickness) measuring. The measured data are filtered for noise, stored in the computer memory, and displayed on the computer screen for use by the operator to control the dough forming and baking process. Other applications in the bakery industry include measuring surface browning of pizza shells [81], analyzing the protein quality of simulated pizza crusts [82], and inspecting prebaked crusts used in pizza pies on a production line [4].

Image processing is widely used for final processed-food inspection. A multicamera, multiprocessor system was developed to sort dried prunes for surface defects in real time [45]. The sorting errors were 5.6% for good, 3.7% for mold, 9.1% for crack, and 16.5% for scab prunes. A machine vision system was used to grade raisins into three grades for degree of wrinkles and shape: B or better, C, and substandard [83]. The features used in the image analysis were wrinkle edge density, average gradient magnitude, and elongation. The system graded raisins with accuracy and precision comparable with the current grading methods. An X-ray inspection station was designed to automatically separate stones from almonds on a conveyor belt. Differential transmission of low-level X-rays creates an image identifying the stones, which are then removed by an air blast from a row of tubes [4]. A high-speed image processing system called ROBOSORTER was installed in 1993 in Queensland, Australia, at the world's largest ginger factory, operated by Buderim Ginger Ltd. [84]. It is being used to sort pieces of diced and sliced confectionery ginger by size and shape into up to five grades including a reject grade, at a speed of up to 45 pieces per second. Items presented for sorting are singulated onto parallel main conveyors for imaging. Images of the items are acquired, digitized, and analyzed in a pipeline fashion to extract the morphological features. These features are used in accordance with a user-defined sorting program to assign a grade to each item. Guided by the grade information, air jets then divert items into appropriate bins or onto cross-conveyors, each bearing product of a single grade. Most of the sorted items are subsequently enrobed in chocolate or sold as crystallized ginger.

Extruded food quality analysis is another example of processed food inspection using image processing. The transverse cuts of extruded biscuits were evaluated by image analysis [85]. The features of mean orientation, mean area, and standard deviation of pores were effective in discriminating two types of extruded biscuits. An image analysis technique was used to estimate internal structural parameters of extruded food [86]. The cell size distribution of puffed corn extrudates was investigated using image processing [87]. Color and texture features of an extruded food product were extracted for determining changes in the product appearance as a result of changes in the extrusion process [88]. An image processing technique was used to measure quality attributes of extruded corn products [89, 90].

C. Package Inspection

Food packaging is the last processing step in the food processing industry. Food packages (cartons, plastic, glass, and metal containers) are not only the containers but also the "beauty dresses" of the processed food. Customers get their first impression of the quality of a product from its package. Use of machine vision systems for package inspection is already a practice in many food processing plants, including label verification (that is, verifying the position, quality, and correctness of the label), seal inspection for voids, wrinkle alignment, and product contamination. A machine vision system was introduced for liquid packaging inspection (for absence of caps and labels, out-of-range locations of labels, and out-of-range levels of liquid), dry goods packaging inspection (for missing flaps, unsealed or partially sealed flaps, torn flaps, and misaligned flaps), and carton inspection (for empty-before-filling and full-before-closing checks) [91]. Seal defects were detected in food packages using image processing [92, 93]. An automated robotic vision system was introduced for pre- and post-packaging quality control [94]. A machine vision system was developed for inspection of automatic glue placement on food product box cartons [95]. Some of the fundamental issues concerning the application of machine vision in high-speed container closure manufacturing lines through application examples in the glass and metal container industries were identified [96]. An image processing system was introduced that can inspect foil-packaged food for seal integrity at a rate of up to 150 packages per minute [97].

X. FUTURE TRENDS

One of the biggest barriers in applying machine vision to the food processing industry is the system cost. Although the cost of an image processing system has been significantly reduced in the last 10 years because of the continuously dropping prices of computers, imaging devices, and other related hardware, the ratio of cost to benefit is still unacceptable in many potential applications. For example, it is difficult to justify a $250,000 fruit sorter for seasonal use. Developing generic image processing systems that can be used for multiple applications may increase acceptability [98].

Color vision is crucial in food processing inspection. Previously limited by the high prices of the color imaging devices and the extensive-computing characteristic of color image analysis, the adoption of color image processing systems in the food processing industry is limited. As high-speed microcomputers and color imaging devices are becoming commercially available at reasonable prices, the use of color image processing systems will be increasingly popular in the food processing industry.

Robustness is a basic requirement for an image processing system used in the food processing industry because of natural variations in food products. Many research results, with high accuracy levels, are reported using nicely selected samples. When naturally varying samples of biological origin are used, the accuracies drop significantly. Such types of published results make the industry skeptical of the potential use and the benefits of the image processing systems. This poses a difficult challenge to follow-up researchers. It is possible to change the industry perspective, if the research community takes a responsible role by thoroughly evaluating their algorithms. Aside from the use of specific training samples for system development, the decision-making methods used in research may be partially responsible for poor system performance. Food inspection, as formerly performed by human inspectors, is a complex decision-making process that involves many factors such as training and experience of the inspectors. This requires that an image processing system should have some humanlike abilities, such as learning and making decisions on ill-defined concepts, for the inspection task. Traditional decision-making methods, as used in many fields, are based on well-defined concepts and yes-or-no logic, and implemented in programmed procedures (computer programs), which can only handle the tasks that are predefined by the training samples. So it is inevitable that the system performance decreases dramatically when applied to samples out of the range covered by the training samples. Neural network and fuzzy logic techniques are a potential solution for this problem. Neural networks, the electronic simulations of the human brain, have self-learning and self-organizing abilities. Fuzzy logic simulates human reasoning methodology. Applying these techniques in decision making will allow an image processing system to be more robust and humanlike.

Visual information alone is usually inadequate in many inspection applications. Other than visual information, information on taste, smell, and chemical residue, which can be measured by various chemical or physical sensors, may be needed in inspections. In these cases, integrating these sensors into an image processing system to form a multisensor inspection system is necessary.

REFERENCES

1. D. H. Ballard and C. M. Brown, *Computer Vision*, Prentice-Hall, 1982, p. 523.
2. R. C. Gonzalez and R. Safabakhsh, Computer vision techniques for industrial application and robot control, *Computer 15*:17 (1982).
3. R. M. Haralick and L. Shapiro, *Computer and Robot Vision*, Addision-Wesley, Reading, Massachusetts, 1992, p. 672.
4. G. A. Kranzler, Applying digital image processing in agriculture, *Agric. Eng. 66*(3):11 (1985).

5. N. R. Sarkar, Machine vision in the food industry, *ASAE Food Engineering News*, October, St. Joseph, Michigan, 1986, p. 3.

6. R. D. Tillet, Image analysis for agricultural process: A review of potential opportunities, *J. Agric. Eng. Res. 50*:247 (1991).

7. W. E. Shaw, Machine vision for detecting defects on fruits and vegetables, Food Processing Automation I—Proceedings of the FPAC Conference, ASAE, St. Joseph, Michigan, 1990, p. 50.

8. A. R. Novini, Fundamentals of machine vision component selection, Food Processing Automation I—Proceedings of the FPAC Conference, ASAE, St. Joseph, Michigan, 1990, p. 60.

9. W. K. Pratt, *Digital Image Processing*, 2nd ed., John Wiley and Sons, New York, 1991, p. 698.

10. R. C. Gonzalez and R. E. Woods, *Digital Image Processing*, Addision-Wesley, Reading, Massachusetts, 1992, p. 716.

11. G. A. Baxes, *Digital Image Processing—Principles and Applications*, John Wiley and Sons, New York, 1994, p. 452.

12. R. J. Schalkoff, *Digital Image Processing and Computer Vision*, John Wiley and Sons, New York, 1989, p. 489.

13. R. M. Haralick, K. Shanmugam, and I. Dinstein, Textural features for image classification, *IEEE Trans. Systems, Man, and Cybernetics 3*:610 (1973).

14. M. M. Galloway, Textural analysis using gray level run lengths, *Computer Vision Graphics and Image Processing 4*:172 (1975).

15. C. Sun and W. G. Wee, Neighboring gray level dependence matrix for texture classification, *Computer Vision Graphics and Image Processing 23*:341 (1983).

16. C. T. Morrow, P. H. Heinemann, H. J. Sommer, Y. Tao, and Z. Varghese, Automated inspection of potatoes, apples, and mushrooms, Proceedings of International Advanced Robotics Programme, Avignon, France, 1990, p. 179.

17. Q. Yang, The potential for applying machine vision to defect detection in fruit and vegetable grading, *Agric. Engineer 47*(3):74 (1992).

18. V. Bellon, Tools for fruits and vegetables quality control: A review of current trends and perspectives, Food Processing Automation III—Proceedings of the FPAC III Conference, ASAE, St. Joseph, Michigan, 1994, p. 494.

19. J. E. McClure and C. T. Morrow, Computer vision sorting of potatoes, ASAE Paper No. 87-6501, ASAE, St. Joseph, Michigan, 1987, p. 26.

20. J. A. Marchant, C. M. Onyango, and M. J. Street, High speed sorting of potatoes using computer vision, ASAE Paper No. 88-3540, ASAE, St. Joseph, Michigan, 1988, p. 6.

21. R. R. Wolfe and M. Swaminathan, Determining orientation and shape of bell peppers by machine vision, *Trans. ASAE 30*(6):1853 (1987).

22. M. M. Batchelor and S. W. Searcy, Computer vision determination of stem/root joint on processing carrots, *J. Agric. Eng. Res. 43*:259 (1989).

23. J. R. Brandon, M. S. Howarth, S. W. Searcy, and N. Kehtarnavaz, A neural network for carrot tip classification, ASAE Paper No. 90-7549, ASAE, St. Joseph, Michigan, 1990, p. 13.

24. M. S. Howarth and S. W. Searcy, Inspection of fresh carrots by machine vision, Food Processing Automation II—Proceedings of the FPAC Conference, ASAE, St. Joseph, Michigan, 1992, p. 106.
25. M. P. Rigney and G. H. Brusewitz, Asparagus shape features for quality assessment, ASAE Paper No. 91-6043, ASAE, St. Joseph, Michigan, 1991, p. 13.
26. W. M. Miller, Classification analyses of Florida grapefruit based on shape parameters, Food Processing Automation II—Proceedings of the FPAC Conference, ASAE, St. Joseph, Michigan, 1992, p. 339.
27. R. W. Taylor and G. E. Rehkugler, Development of a system for automatic detection of apple bruises, Proceedings of the Agri-mation I Conference and Exposition, ASAE, St. Joseph, Michigan, February 1985, p. 53.
28. G. E. Rehkugler and J. A. Throop, Apple sorting with machine vision, *Trans. ASAE 29*(5):1388 (1986).
29. S. Chen and W. H. Chang, Machine vision guided robotic sorting of fruits, Food Processing Automation III—Proceedings of the FPAC Conference, ASAE, St. Joseph, Michigan, 1994, p. 431.
30. M. L. Harris, Machine vision for grading of oranges, *Advanced Imaging 15*: 40 (1988).
31. N. Sarkar and R. R. Wolfe, Feature extraction techniques for sorting tomatoes by computer vision, *Trans. ASAE 28*(3):970 (1985).
32. N. Sarkar and R. R. Wolfe, Computer vision based system for quality separation of fresh market tomatoes, *Trans. ASAE 28*(5):1714 (1985).
33. F. M. Clydesdale, Color perception and food quality, *J. Food Quality 14*:61 (1991).
34. W. D. Wigger, M. R. Paulsen, J. B. Litchfield, and J. B. Sinclair, Classification of fungal-damaged soybeans using color-image processing, ASAE Paper No. 88-3053, ASAE, St. Joseph, Michigan, 1988, p. 14.
35. Y. Shyy and M. K. Misra, Color image analysis for soybean quality determination, ASAE Paper No. 89-3572, ASAE, St. Joseph, Michigan, 1989, p. 12.
36. B. K. Miller and M. J. Delwiche, A color vision system for peach grading, *Trans. ASAE 32*(4):1484 (1989).
37. S. A. Shearer and F. A. Payne, Color and defect sorting of bell peppers using machine vision, *Trans. ASAE 33*(6):2045 (1990).
38. W. W. Casady, M. R. Paulsen, J. F. Reid, and J. B. Sinclair, A trainable algorithm for inspection of soybean seed quality, *Trans. ASAE 35*(6):2027 (1992).
39. K. W. White and R. J. Sellers, Foreign materials sorting by innovative real time color signatures, Food Processing Automation III—Proceedings of the FPAC Conference, ASAE, St. Joseph, Michigan, 1994, p. 29.
40. A. A. Al-Janobi and G. Kranzler, Machine vision inspection of date fruits, ASAE Paper No. 94-3575, ASAE, St. Joseph, Michigan, 1994, p. 12.
41. A. C. Roudot, Image analysis of kiwi fruit slices, *J. Food Eng. 9*(2):97 (1989).
42. K. Liao, R. P. Cavalieri, and M. J. Pitts, Hausdorff dimensional analysis and digital imaging based quality inspection, *Trans. ASAE 33*(1):298 (1990).

43. M. P. Rigney, G. H. Brusewitz, and G. A. Kranzler, Asparagus defect inspection with machine vision, *Trans. ASAE 35*(6):1873 (1992).

44. F. E. Dowell, Identifying undamaged and damaged peanut kernels using tristimulus values and spectral reflectance, *Trans. ASAE 35*(3):931 (1992).

45. M. J. Delwiche, S. Tang, and J. F. Thompson, A high-speed sorting system for dried prunes, *Trans. ASAE 36*(1):195 (1993).

46. T. G. Crowe and M. J. Delwiche, Pipeline image processing for real-time defect detection in produce, ASAE Paper No. 95-3174, ASAE, St. Joseph, Michigan, 1995, p. 28.

47. J. A. Throop, G. E. Rehlugler, and B. L. Upchurch, Application of computer vision detection watercore in apples, *Trans. ASAE 32*(6):2087 (1989).

48. S. Y. Wang, P. C. Wang, and M. Faust, Non destructive detection of watercore in apple with nuclear magnetic resonance imaging, *Scientia Horticulturae 35*:227 (1988).

49. P. Chen, M. J. McCarthy, and R. Kauten, NMR for internal evaluation of fruits and vegetables, *Trans. ASAE 32*(5):1747 (1989).

50. E. E. Finney and K. H. Norris, X-rays images of hollow heart of potatoes, *Am. Potato J. 55*:95 (1978).

51. E. W. Tollner, Y. C. Hung, B. L. Upchurch, and S. E. Prussia, Relating X-ray absorption to density and water content with apples, ASAE Paper No. 92-7056, ASAE, St. Joseph, Michigan, 1992, p. 16.

52. G. G. Dull, Nondestructive evaluation of quality of stored fruits and vegetables, *Food Technol. 40*:106 (1986).

53. K. Peleg, Multi-sensor adaptive sorting of fruits and vegetables, Proceedings of the International Workshop funded by the US–Israel Binational Agric. Research and Development Fund (BARD), ASAE, St. Joseph, Michigan, 1993, p. 111.

54. Y. Edan, H. Pasternak, D. Guedalia, N. Ozer, I. Shmulevich, D. Rachmani, E. Fallik, and S. Grinberg, Multi-sensor quality classification of tomatoes, ASAE Paper No. 94-6032, ASAE, St. Joseph, Michigan, 1994, p. 19.

55. N. Ozer, B. A. Engel, and J. E. Simon, An adaptive technique for multi-sensor fusion and classification of fruit, ASAE Paper No. 94-6025, ASAE, St. Joseph, Michigan, 1994, p. 17.

56. N. Ozer, B. A. Engel, and J. E. Simon, Neural network for quality sorting and grading of cantaloupes by multiple nondestructive sensors, ASAE Paper No. 95-3218, ASAE, St. Joseph, Michigan, 1995, p. 14.

57. J. M. Keller, N. Covavisaruch, K. Unklesbay, and N. Unklesbay, Color image analysis of food, Proceedings of IEEE Computer Society Conference on Computer Vision and Pattern Recognition, Miami Beach, Florida, June 1986, p. 619.

58. Y. R. Chen, T. P. McDonald, and J. D. Crouse, Determining percent intramuscular on ribeye surface by image processing, ASAE Paper No. 89-3009, ASAE, St. Joseph, Michigan, 1989, p. 22.

59. T. P. McDonald and Y. R. Chen, Location and isolation of beef carcass ribeye in cluttered scenes, ASAE Paper No. 89-3005, ASAE, St. Joseph, Michigan, 1989, p. 14.

60. T. P. McDonald and Y. R. Chen, Separating connected muscle tissues in images of beef carcass ribeyes, *Trans. ASAE 33*(6):2059 (1990).

61. J. Jia and A. P. Schinckel, Automatic inspection of visible pork fat and lean using machine vision, Food Processing Automation II—Proceedings of the FPAC Conference, ASAE, St. Joseph, Michigan, 1992, p. 326.

62. W. D. Daley and J. C. Thompson, Color machine vision for meat inspection, Food Processing Automation II—Proceedings of the FPAC Conference, ASAE, St. Joseph, Michigan, 1992, p. 230.

63. B. Park and Y. R. Chen, Intensified multispectral imaging system for poultry carcass inspection, Food Processing Automation III—Proceedings of the FPAC Conference, ASAE, St. Joseph, Michigan, 1994, p. 97.

64. T. W. Awa, R. K. Byler, and K. C. Diewl, Development of an inexpensive oyster meats grader, ASAE Paper No. 88-3539, ASAE, St. Joseph, Michigan, 1988, p. 17.

65. P. P. Ling, S. W. Searcy, and J. Grogan, Feature selection for a machine vision based shrimp deheading process, ASAE Paper No. 88-3542, ASAE, St. Joseph, Michigan, 1988, p. 22.

66. P. P. Ling and S. W. Searcy, Feature extraction for a vision based shrimp deheader, ASAE Paper No. 89-3571, ASAE, St. Joseph, Michigan, 1989, p. 13.

67. H. J. C. Strachan, P. Nesvadba, and A. R. Allen, Fish species recognition by shape analysis of images, *Pattern Recognition 23*(5):539 (1990).

68. M. D. Evans, Feather sexing of broiler chicks by machine vision, ASAE Paper No. 89-3008, ASAE, St. Joseph, Michigan, 1989, p. 9.

69. S. M. Berlow, D. J. Aneshansley, J. A. Throop, and J. R. Stouffer, Computer analysis of ultrasound images for grading beef, ASAE Paper No. 89-3569, ASAE, St. Joseph, Michigan, 1989, p. 20.

70. D. J. Aneshansley, J. R. Stouffer, B. Xu, Y. Liu, J. A. Throop, and S. Berlow, Ultrasonic images allow automated grading and inspection of beef and swine, Food Processing Automation I—Proceedings of the FPAC Conference, ASAE, St. Joseph, Michigan, 1990, p. 256.

71. A. D. Whittaker, B. S. Park, J. D. McCauley, and Y. Huang, Ultrasonic signal classification for beef quality grading through neural networks, Proceedings of the 1991 Symposium on Automated Agriculture for the 21st century, ASAE, St. Joseph, Michigan, 1991, p. 116.

72. Y. Liu, Automatic detection of the outline of longissimus dorsi muscle from cross-sectional ultrasonic images, Food Processing Automation III—Proceedings of the FPAC Conference, ASAE, St. Joseph, Michigan, 1994, p. 107.

73. J. D. McCauley, B. R. Thane, and A. D. Whittaker, Fat Estimation in beef ultrasound images using texture and adaptive logic networks, *Trans. ASAE 37*(3):997 (1994).

74. R. T. Elster and J. W. Goodrum, Detection of cracks in eggs using machine vision, *Trans. ASAE 34*(1):307 (1991).

75. J. W. Goodrum and R. T. Elster, Machine vision for crack detection in rotating eggs, *Trans. ASAE 35*(4):1323 (1992).

76. V. C. Patel, R. W. McClendon, and J. W. Goodrum, Detection of blood spots and dirt stains in eggs using computer vision and neural networks, Proceedings of the International Conference on Agricultural Engineering, AgEng '94, Milano, Italy, 1994, p. 833.

77. K. Das and M. D. Evans, Detecting fertility of hatching eggs using machine vision I: Histogram characterization method, *Trans. ASAE 35*(4):1335 (1992).

78. K. Das and M. D. Evans, Detecting fertility of hatching eggs using machine vision II: Neural network classifiers, *Trans. ASAE 35*(6):2035 (1992).

79. P. Marchal, V. Louveau, and P. Marty-Mahe, Detecting egg weight loss during incubation with a vision machine, Food Processing Automation III—Proceedings of the FPAC Conference, ASAE, St. Joseph, Michigan, 1994, p. 527.

80. M. Monnin, Machine vision gauging in a bakery, Food Processing Automation III—Proceedings of the FPAC Conference, ASAE, St. Joseph, Michigan, 1994, p. 62.

81. K. Unklesbay, N. Unklesbay, J. Keller, and J. Grandcolas, Computerized image analysis of surface browning of pizza shells, *J. Food Sci. 48*(4):1119 (1983).

82. L. Heyne, N. Unklesbay, K. Unklesbay, and J. Keller, Computerized image analysis of protein quality of simulated pizza crusts, *J. Can. Inst. Food Sci. Technol. 18*(2):168 (1985).

83. N. K. Okamura, M. J. Delwiche, and J. F. Thompson, Raisin grading by machine vision, *Trans. ASAE 36*(2):485 (1993).

84. M. Kassler, "ROBOSORTER": A system for simultaneous sorting of food products, Food Processing Automation III—Proceedings of the FPAC Conference, ASAE, St. Joseph, Michigan, 1994, p. 413.

85. A. Smolarz, E. Van Hecke, and J. M. Bouvier, Computerized image analysis and texture of extruded biscuits, *J. Texture Studies 20*:223 (1989).

86. P. M. K. Alahakoon, K. A. Sudduth, and F. Hsieh, Image analysis of the internal structure of corn extrudate, ASAE Paper No. 91-6540, ASAE, St. Joseph, Michigan, 1991, p. 16.

87. A. M. Barrent and M. Peleg, Cell size distributions of puffed corn extrudates, *J. Food Sci. 57*(1):146 (1992).

88. J. Tan, X. Gao, and F. Hsieh, Measuring extruded-food quality with machine vision, ASAE Paper No. 92-3586, ASAE, St. Joseph, Michigan, 1992, p. 11.

89. X. Gao and J. Tan, Analysis of extruded food texture by image processing, ASAE Paper No. 93-6568, ASAE, St. Joseph, Michigan, 1993, p. 20.

90. X. Gao, J. Tan, and H. Haymann, Evaluation of extruded-food sensory properties by image processing, ASAE Paper No. 94-3023, ASAE, St. Joseph, Michigan, 1994, p. 16.

91. L. Werth, Machine vision for agri-applications, *Agric. Eng. 68*:22 (1987).

92. J. N. Wagner, Inspecting the impossible, *Food Eng. 55*(6):79 (1983).

93. G. R. Gagliardi, D. Sullivan, and N. F. Smith, Computer aided video inspection, *Food Technol. 38*(4):53 (1984).

94. S. McClelland, A sweet sense of success for PA technology, *Sensor Review 7*(4):179 (1987).

95. T. A. Doney, Automatic glue placement inspection, Food Processing Automation II—Proceedings of the FPAC Conference, ASAE, St. Joseph, Michigan, 1992, p. 92.
96. A. R. Novini, Fundamentals of machine vision inspection in glass and metal container industries, Food Processing Automation II—Proceedings of the FPAC Conference, ASAE, St. Joseph, Michigan, 1992, p. 98.
97. C. F. Jones III and J. C. Griner, An image processing system to inspect foil-packaged food, Food Processing Automation III—Proceedings of the FPAC Conference, ASAE, St. Joseph, Michigan, 1994, p. 55.
98. S. D'Agostino, Generic machine vision system for food processing inspection, Food Processing Automation II—Proceedings of the FPAC Conference, ASAE, St. Joseph, Michigan, 1992, p. 130.

8

Computer-Based Fermentation Process Control

Georges Corrieu, Bruno Perret,
Daniel Picque, and Eric Latrille
National Insitute for Agronomical Research,
Thiverval-Grignon, France

I. INTRODUCTION

Biological transformations by microorganisms, grouped under the term *fermentation*, have an important, specific place in food industries. This specificity results from the use of microorganisms that will carry out a set of biological transformations characteristic of each fermented food product.

The progression of a fermentation process first presupposes the development of the population of microorganisms introduced (inoculum), which often reaches high levels (10^9 to 10^{13} cells per liter or per kg). In a subsequent step, these microorganisms carry out biotransformations that are very different from growth.

Microbiologically, several important factors are to be taken into consideration. First, the fermentations carried out are rarely done so by pure cultures. Complex mixtures of strains are used for organoleptic and technological reasons, and also to comply with regulations. These mixed cultures give rise to equilibria and population dynamics that are particularly difficult to measure and control. They assure biological functions that pure cultures or simple mixtures do not: lactic fermentation used to produce fresh cheeses requires a combination of four strains characterized by their acidifying, texturing, aromatic, and proteolytic properties.

Second, the selection of strains with the best performance is a constant preoccupation of the industry. Technological performance will directly depend on the choices made by considering both the characteristics of starting materials (culture media) and the typical nature of the finished products. This is a complex combination among factors such as quality, preparative treatments of raw materials, microorganisms selected, culture conditions (implying process control), and subsequent treatments determining the quality of a fermented product. Quality, however, can be defined only in comparison with a product, including concepts as broad as microbial and hygienic quality, organoleptic quality, nutritional quality, mechanical properties, etc.

Third, foods are fragile and often difficult to preserve. Fermentation processes generally tend to increase their lifetime and improve storage conditions. Transformations from milk to cheese, from grape juice to wine, of a mixture of ground meat into sausage are several examples. Risks of contamination associated with fermentations are considerable, however, resulting from numerous microorganisms, either pathogenic or not. Their simple presence or development constitutes a risk of product deterioration or, even more serious, a health risk for consumers. As a result, operations such as cleaning in place, disinfection, and sterilization are necessary to limit these risks and ensure a correct hygiene level in factories.

Finally, fermentation processes are slow, resulting from the low generation time of the microorganisms, from the quantitative importance of biotransformations (volumes treated, concentration of substrates and metabolites), and from reaction limitations (slow kinetics determined by complex inhibition mechanisms related to surrounding conditions).

As a consequence of these considerations the fermentation processes in food industries are batch processes. Continuous cultures are possible but rarely used, both to limit contamination and to obtain products with elevated and reproducible organoleptic quality. This industrial practice has considerable consequences for process methods and control procedures. Before discussing these in more detail, an overview of food fermentation processes currently used will place our discussion in context. With the exception of exotic processes, used primarily in developing countries, two types of fermentation are predominant:

1. *Lactic fermentations*, carried out with mixed cultures of mesophilic or thermophilic lactic acid bacteria, concern many products, the most important of which are fermented milk, cheeses, fermented vegetables, and some meat products. In the case of some products, such as sourdough French bread and wines, lactic fermentation is combined with other fermentations.
2. *Alcoholic fermentations*, carried out with specially selected yeasts, result in alcoholic beverages (beer, wine, cider, spirits).

In addition, starter production industries (lactic starters, yeasts for bakery products, beer making, and winemaking) should be mentioned. They supply food industries with large volumes of selected and controlled starters at competitive prices.

Finally, a number of fermented products are presented as individual units (bread, cheese, sausage). Their preparation is associated with complex microbial transformations that cannot be carried out in a fermentor. The concept of a "food bioreactor" [1] for cheese can be applied to them.

This introduction, concentrating on the microbiological aspects of fermentation processes, is meant to place the field in context. More detailed information can be obtained [2]. Our discussion will now be limited to the control of fermentation processes. The first part (Section II) describes the problem aspects, procedures, tools, and current research trends, and the second part (Section III) contains practical illustrations.

II. GENERAL APPROACH TO FOOD FERMENTATION PROCESSES CONTROL

A. Introduction

The main problem in the control of fermentation processes results from the lack of sensors, detectors, and measurement devices. When measurements are available, automatic control and sequential automation pose no particular difficulties, because high-performance tools exist. In the field of modeling, however, very different approaches are possible. This results in different models that are well suited to simulation or optimization studies, but whose real-time application is not always possible.

These aspects are developed in this section. The conclusion includes considerations on principal future orientations, currently the object of intensive research. There nevertheless exists a tremendous difference between the know-how of researchers and practices in an industrial setting. Theoretical and experimental work has led to the development of methods and algorithms that are often perfected and validated on the pilot scale.

Their industrial applications are capable of leading to considerable advances, but this often does not occur for a variety of reasons: retardation in industrial innovation, difficulty in evaluating the economic impact of the innovation, preference for empirical use of know-how without making significant investments, insufficient control of scale-up problems, etc.

Thus laboratory and pilot-scale research is carried out using instrument-equipped and -automated bioreactors. Industrial units, on the other hand, operate with the simplest instrumentation possible to limit problems of calibration, maintenance, etc. Although the past decade has seen intense development of automation procedures applied to industrial units justified by considerable gains in productivity and reliability, process optimization often remains a long-term objective.

B. Sensors and On-Line Measurements

Several reviews have dealt with measurement systems used in bioreactors [3–5]. Their classification is traditionally based on the type of variables (physical, chemical, biological) or on the measurement principles. We will not deal with on-line measurements that are considered to be reliable (Table 1); rather, we will concentrate on the more detailed identification

TABLE 1 Physical and Chemical Measured Variables for Fermentation Processes

Type of Variables	Variables	Principle
Physical	temperature	Pt100 resistance, thermocouple
	pressure	strain gauge
	rotation speed	inductive sensor, optical encoder
	weight	load cell
	level, volume	conductance or capacitance sensors
	gas flux	turbine meter, magnetic flow meter, thermal mass flow meter
	liquid flux	metering pump, electromagnetic meter, load cell
Chemical	pH	potentiometric sensor
	dissolved O_2	amperometric sensor (polarographic, galvanic)
	exit O_2 concentration	paramagnetic analyzer, mass spectrometry
	exit CO_2 concentration	infrared analyzer, mass spectrometry

of process variables that are not accessible in real time, try to understand why, and above all describe current research under way to resolve these problems in the field of sensor research.

Three parameters essential for following and controlling processes are not always available in real time: concentrations of microorganisms (and their biological activity), of substrates, and of metabolites. The difficulties in accessing these parameters on-line results primarily from the nature of biological media and from limitations in the use of biotechnological processes. Table 2 summarizes the principal specifications required of an on-line sensor for bioprocesses.

1. Measuring Biomass

Most off-line methods for measuring biomass [6] are unsuitable for on-line measurements. The most recent directions in the development of sensors are based on the optical and dielectric properties of cell suspensions or on the development of software sensors (this will be discussed later).

Optical Measurements Optical methods are based on the transmission, reflection, and scattering of light. Probes have been developed and tested in cultures of microbial or animal cells [7–9]. The wavelengths used are in the visible and near infrared, and laser diodes and optical fibers are generally used. Each type of sensor must be calibrated for each type of fermentation process. The range of linearity between the phenomenon measured and cell concentration is generally restricted. Polynomial regressions must be used for higher concentrations, introducing increased risk of error.

Components of the medium (suspended particles, gas bubbles) and of the process (mixing) may interfere with optical measurements. Signal-

TABLE 2   The Most Important Specifications for the On-Line Sensors for Fermentation Processes

Constraints	Comments
Sterilizable	Effect of temperature and pressure
Preserved the sterility of the reactor	
To be stable for long time	To avoid too frequent standardizations
Worked in complex medium	Select if sensors not disturbed by the flux, the bubbles, etc.
To be able to measure low and high concentrations in the same fermentation	Important changes in the measured concentrations

processing or mechanical-degassing systems have been developed to over-come these interferences. Several sensors are available, but their uses are yet limited.

Electrical Measurements An on-line instrument for measuring bio-mass based on a dielectric measurement is available on the market (Bug-meter, ABER Instruments Ltd., Aberdeen, Great Britain). Its operating principle is based on the inability of electrical charges to pass through cell membranes at low radio frequencies, which allows the membranes to act as insulators between intracellular and extracellular fluids [10]. The result-ing dielectric dispersion is proportional to the radius of the cell and to cellular volume. The sensor is equipped with four electrodes that are cleaned by electric pulses. Several publications have described the appli-cations of this instrument for on-line measurements of yeast and plant cell cultures [10–12]. Optimal frequencies, sensitivity, and the range of measurable concentrations vary considerably from one strain to another.

The effect of bubbles and mixing can be reduced by probe design. The use of electrical measurements remains limited because of a lack of sensitivity to cell concentrations lower than 5 g/L and interference created by the conductivity of the medium itself.

2. Measurements of Substrate and Product Concentrations

The on-line measurement of substrates and products in culture media also suffers from a lack of suitable sensors. Two paths have been explored to resolve this problem, one being measurements of samples taken from the medium, using conventional analytical instruments. The other, more am-bitious, route is the attempt to develop remote measurements without sampling and even without contacting the product.

Analyses of Samples Analytical laboratories are equipped with high-performance instruments for the measurement of biological molecules. The most widely used techniques are gas chromatography, high-perfor-mance liquid chromatography, and flow injection analysis. The idea in-volves coupling these systems to bioreactors. The main problem to resolve becomes the analyzer–reactor interface, i.e., the development of the sam-pling system [13]. Two cases are possible, depending on whether a gaseous or liquid sample is to be obtained.

In the first case, the sampler is composed of a flat or tubular hydro-phobic membrane swept by a carrier gas and placed in the fermentation medium or the gas phase [14–16]. Volatile molecules diffuse through the membrane and are transported by the carrier gas to the detector, which in most cases is a gas chromatograph. This system can analyze only volatile compounds and is used on the laboratory and pilot scales for the detection of molecules such as alcohols, aldehydes, organic acids, and esters that are synthesized during fermentations.

Several parameters affect the performance of these samplers [13], including the composition of the medium, pH, pressure, and temperature. These factors change thermodynamic equilibria between the liquid and vapor phases, leading to differing detector responses. There is no commercial instrument of these vapor-phase samplers today, but in some cases it is possible to use the outlet bioreactor gas.

The measurement of volatile compounds is rarely sufficient for controlling a process, explaining why the analysis of liquid samples, the second case, is more relevant. The sampling module is generally based on filtration membranes (organic or inorganic) that can be placed in an external recirculation loop [17, 18] or within the bioreactor itself [19–21]. The external system has the advantages of being able to accommodate large membrane surfaces and allowing the membrane to be changed during the process. On the other hand, there are considerable risks of reactor contamination and of damaging fragile cells. The principal problem is membrane fouling, which depends on the type of membrane used (surface, porosity, composition, etc.) and on the culture media. Filtration performance is well suited to simple media, but it decreases rapidly in complex media. Operating times can be lengthened by the use of countercurrent cleaning [20] or by rotating the membrane [21].

There are many associated analysis methods, including liquid and gas chromatography, biosensors, and flow injection analysis. Sample volumes required are low (several milliliters), and considerable dilutions are necessary because of the large concentration ranges of medium components. Several instruments are available on the market based on loops or an in situ filtration probe. Their development has nevertheless remained limited, partly because of the need for robust on-line instruments capable of operating in an industrial environment and requiring minimal maintenance.

Measurements Without Sampling Faced with the difficulty of implementing reliable sampling–analysis systems, researchers have turned to studying on-line measurements that do not require sample removal. The measurement is conducted directly inside the bioreactor, even without direct contact with the product. Spectroscopic techniques have been the most widely investigated, in particular those employing infrared vibrations. They are used in some food industries for determining moisture, fat, and protein contents in grains and milk. In the course of alcoholic and lactic fermentations, their potentialities have been shown for assaying the principal organic molecules in culture media: sugars, alcohols, and organic acid [22, 23]. They have the advantage of leading to the quantitative determination of several compounds using a single spectral analysis by attenuated total reflection or transmitted light. In parallel, the development of optical fibers with low signal attenuation in the middle infrared has led to the first remote measurements [24, 25]. Current efforts are aimed at developing fibers and probes that can work in the 900 cm^{-1} to 1300 cm^{-1} spectral zone, a zone that furnishes considerable information on molecules involved in bioprocesses [26]. Combined with the development of reliable chemical analysis methods, these investigations indicate that new on-line measurement instruments may soon make their appearance.

C. Automatic Control and Monitoring of Bioreactors

Process monitoring and control presupposes a consistent collection of process data, as well as the distribution of output signals toward control devices. The circulation of data from varied origins, as well as their processing, are thus at the heart of this type of system. Fermentations do not escape this rule. As early as 1975 this data collection could be organized for automatic processing by newly emerging computer systems, as opposed to the widespread use of recorders at that time. This new perspective was directly related to the advent of special components (analog to digital converters [ADC] and digital to analog converters [DAC]) that created a link between the analog and digital worlds.

1. Hardware

Electronic Environment of Fermenters

Sensors. All sensors contain an electronic device (transmitter) that produces an electrical signal from the parameter to be measured. The signal produced can be used by a computer system after passing through an ADC. The electrical signal from the transmitter is generally linearized and standardized. The most current standards, having arisen from analog uses, are 4/20 mA and 0/10 V. This results in a linear correspondence with the measurement range, e.g., 0/10 V corresponds to 0/150°C.

Actuators. The actuators used depend on the process in question. They are generally heating resistances, motors, solenoids, or motorized valves. Computer systems have only weak output signals to control these organs, and so relay systems (analog or optical transformations) are necessary to obtain the required power. Digital signals should be isolated to the greatest extent possible from these strong currents, which interfere with weak signals.

The Computer Environment   The computer environment of processes is broken down into two hierarchical levels.

Remote computer systems. This level is in physical proximity to the process. It carries out the analog/digital conversion of the signals from all the sensors present and an initial centralization of the data (measurements). This function should be carried out as close to the sensors as possible because analog signals are difficult to transport over long distances. The power of remote computer systems varies as a function of the planned uses. They may be stand-alone systems and locally assure all process monitoring and control functions with no connection to a central system. Conversely, they may be composed of a set of intelligent sensors/ actuators capable of self-diagnosis but which make no decision with respect to the environment. In this case they are distributed in the network and controlled by a central system that makes all decisions concerning process operation.

These remote systems are currently being developed at the expense of intelligent sensors/actuators. Because of a high-level integration of components and of analog/digital converters stored on the same board as the digital components, these organs have a minimal local computer system. If one breaks down, it does not affect the remaining instruments. The second feature favoring this type of solution is the recent development of so-called field networks, in particular because of the multiplication of measurement and control organs, whose density is constantly increasing.

Central computer systems. These systems must have an overview of system status at each moment of observation, explaining the need for high data-processing power and storage capacity. Data-processing power depends on the functional modes of the application (real time, multitask, modeling, optimal control, etc.). Depending on the applications, from 10 MIPS (millions of instructions per second) and 1 MFLOP (millions of floating point operations per second) to 100 MIPS and 40 MFLOP or more (real-time image processing) may be necessary. High storage capacity is required for archiving a steadily expanding volume of data: a minimum of 0.5 Gb is indispensable, with the possibility of saving on diskettes or digital cartridges, e.g., 2 Gb DAT (DAta Tapes). Compatible microcomputers have proven themselves in this field and are better accepted because they are inexpensive and user-friendly.

2. Software

In all cases, the application software uses the low-level subroutines of the operating systems. The essential functions of this software are described below. The differentiation between the two hierarchical levels discussed above will be made only if it is obvious and a consensus has been reached.

Calibration Following analog to digital conversion, calibration enables measurements to be transformed into physical and chemical parameters. The operation involves determining two coefficients (slope and Y intercept) because there is a linear correspondence between the electrical signal and the measurand. These coefficients can be calculated automatically (two-point calibration) or entered manually.

Configuration The better software packages always have a configuration function: its role is to adapt the software to the operating environment, contributing to its polyvalence. The inputs and outputs characteristic of the process are defined at this level.

Inputs are classified by taking into account computer procedures required for storing the data they represent. This classification considerably facilitates their development and confers robustness and flexibility on the software. Regardless of the class of data, the configuration must give a name and a measurement unit for each channel created, which is therefore customized. Actuators can be associated with each measurement channel regardless of its type. Three types will be discussed:

1. *Direct measurements.* They are characterized by a physical address corresponding to either a single sensor or a set of sensors (remote computer system). The data returned is the result of an analog/digital conversion, a frequency measurement, a time of action, or a binary value (alarm, on/off). In all cases, it is a number.
2. *Indirect measurements.* They are generated at the same frequency as direct measurements by the software that determines them. Calculation formulas or models can be preprogrammed or entered directly in the software that interprets them.
3. *Off-line measurements.* These are measurements entered nonautomatically by the operator at low frequencies, often a posteriori. They are used to compensate for insufficiencies of sensors or models, or for validating models.

Outputs are always direct and require the same definitions as inputs (address, order number, access protocol, name, and type). They are always associated with a direct or indirect input (control or alarm).

Data Logging This function enables the operator to store data concerning all or part of the process. The name of the file in which they are saved and the storage period are entered at this level. The opposite of this function enables logging to be stopped in time. Exporting data is necessary for their archiving or reprocessing.

Graphical Presentation of the Results Any combination of curves corresponding to measured or calculated parameters must be representable. Individual adjustment of the X and Y axes is desirable.

Alarms, Control, and Optimizing Alarm management is indispensable in every automatic system called on to run without permanent monitoring. The minimum control function on a measurement channel involves applying a setpoint the value of which is constant as a function of time. Most software manages at least one setpoint profile as a function of time and sometimes another measurement. To go further, one must determine the command algorithms to apply to the process for it to adopt a given behavior. A model must have first been defined in order to establish these algorithms. An optimal process control strategy can then be designed. This is done at the central level of computer hierarchy because of the often complex and iterative calculations that are necessary.

Descriptors It is often of interest to have available data taken from the changes of a process parameter. A number of descriptors, such as maximum rate, can be defined. These data do not correspond to measurements, but they represent another source of data useful to the understanding or control of a process.

Using Databases As acquisition systems are becoming increasingly widespread, the volume of data stored is expanding. They represent a very rich information potential, but users often do not know how to use them profitably. Descriptive or classification software has appeared to remedy this situation. It can run on machines other than the supervisor but can also generate information destined for the host computer, e.g., the preparation of reference libraries.

D. Process Modeling

Modeling fermentation processes is becoming increasingly interesting in food industries. The installation of computerized data acquisition systems, however, creates a huge daily mass of data that is often unused. In addition, the determination of relevant bioprocess variables (concentrations of substrate, metabolites, biomass and physiological state, specific rates of growth and production, etc.) cannot be done with sensors. Modeling the fermentation process creates a link between available data and these nonmeasurable variables. Generally, the procedure involves the establishment of a dynamic model of the bioprocess to access nonmeasurable variables by the use of observers or estimators, to optimize a given criterion (final concentrations, minimum fermentation time, etc.), to diagnose problems, or to control the process (control by internal model). Generally bioprocesses are nonlinear, posing more modeling difficulties than linear processes.

In spite of a large number of publications dealing with modeling of fermentation processes [27–29], one criticism is the absence of industrial applications [30]. One of the reasons for this is that methods used in the laboratory become economically unprofitable because of excessively complicated modeling procedures. This complexity arises because research aims more at increasing knowledge than at considering benefit–cost ratios. It is nonetheless clear that excessive dissociation between these objectives is undesirable.

We generally distinguish "structured" (or knowledge) models and "black box" models. The former require the development of models linking state variables and having physical or biological significance, while the latter are based on the empirical identification (based on experiments) of relationships between input and output variables. This distinction is often more symbolic than real: complex knowledge models cannot completely describe microbial physiology and metabolism. This is why a certain nuance is applied to the two concepts, which should be considered as a definition of different levels of complexity within a continuum. In addition, it is recognized that an increase in complexity, when we pass from a black box model to a knowledge model, is accompanied by a loss of robustness and identifiability.

We will first describe a general representation of knowledge models in the form of a state space model, based on the reaction scheme of a biotechnological process described by Bastin and Dochain [31]. Their applications for supplying software sensors will be discussed. We will then discuss black box models, of which one of the best representatives are neural networks that enable nonlinear state space models to be obtained. A synthesis of these two approaches will be sketched and designated by the terms *hybrid models* or *knowledge neural networks*.

1. Knowledge Models of a Bioprocess

The Functional Model The functional diagram of a process representing its mode of operation (batch, fed batch, continuous, recycle, etc.) allows a general expression to be established based on mass balances for each of the components.

In the case of a very general model whose parameters vary with time, the balance is expressed as

$$\frac{dX}{dt} = Yr(X) - DX + F - G \tag{1}$$

Here, X is the vector of the concentrations of the each of the components. It has as many terms as there are reactants considered. $Yr(X)$ describes the kinetics of the *biochemical and microbiological reactions* (developed below). It expresses the transformations of components according to a fermentation reaction scheme. The other terms describe the transport dynamics of the components through the bioreactor: DX expresses the dilution of each component due to volume variations within the reactor. D is the dilution rate (h^{-1}). F is the vector of input flows of the different components (g/L). G is the vector of output flows of the different components in the gaseous form. Note that the vectors X, F, and G all have the same dimension.

In the case of a batch process, D is zero. In the case of a fed batch, D varies with time; it is constant in a continuous process.

The Reaction Scheme   The term of biochemical and microbiological reactions, $Yr(X)$, is the product of a yield matrix Y and a reaction rate vector $r(X)$. If the N components involve M biochemical reactions, the dimension of the matrix Y will be $N \times M$. The vector $r(X)$ (dimension $= M \times 1$) is composed of the reaction kinetics of each of the reactions, which generally depend on the concentration of components. Several types of biochemical reactions exist:

Simple reaction: $\xi_1 + \xi_2 \rightarrow \xi_3 + \xi_4$
Catalytic reaction: $\xi_1 + \xi_2 \rightarrow \xi_2 + \xi_3$
Autocatalytic reaction: $\xi_1 \rightarrow \xi_2 + \xi_3$

The matrix Y is composed of yield coefficients. Each column represents a reaction, and so yields can be expressed in comparison with one of the compounds participating in the reaction. Components not participating in the reaction receive a zero coefficient.

The reaction scheme is not an exhaustive metabolic description; it involves only those components with significant concentration variations. In addition, the reaction scheme is not necessarily stoichiometric and so does not always agree with the mass balance. It is primarily a reflection of relationships existing among the different components we are interested in and is of interest only to the engineer. Thus the consumption or production of a water molecule is generally not considered in these reactions. For reasons of simplicity and effectiveness, the yield coefficients are initially considered to be constant. Regardless of the type of process operation, this yield matrix is identified by a multiple linear regression, provided that measurements of all variables are available (concentrations of compounds and flow rates). The choice among the different reaction schemes possible can be made by cross validation by defining one set of data used for identification and a different set of test data [32].

Thus the yield matrix Y is generally well known, but the same is not true for reaction kinetics $r(X)$. Depending on the objectives of the model, the vector $r(X)$ is considered as either varying with time or as a biochemical kinetic function (Michaelis–Menten, Monod, Haldane, etc.). The latter case enables a complete dynamic model of the process to be obtained. This is the most difficult objective to define for modeling a bioprocess.

Applications: Software Sensors   In a software sensor, the physical or physicochemical sensor is associated with an estimator. This software element uses the measurement delivered by the sensor to furnish a real-time estimation of state variables and process parameters [33]. This estimation is made possible by a certain a priori knowledge of the process and the measurements. Depending on the degree of knowledge available, different types of software sensors can be defined to carry out the indirect measurement, the partial estimation of state and of parameters, or the complete estimation of state.

Indirect measurement. The minimum knowledge required is a measurement model that mathematically reflects the relationship between the measurement delivered by a sensor and the state of the process. This is the case, for example, of the indirect measurement of biomass concentration by a measurement of absorbance, of the measurement of the sugar concentration by a measurement of the density of the fermentation medium, or the indirect measurement of the lactic acid concentration by a measurement of electrical conductivity during a lactic acid fermentation.

Furthermore, if the process mass balance equation is available and if the yield matrix is known, one can indirectly measure certain state variables based on the measurement of a portion of the state variables and measurements of input and output flows. The mathematical method used involves inverting a matrix, which is the case in the indirect measurement of the ethanol concentration from the measurement of the volume of CO_2 released during an alcoholic fermentation.

Partial estimation of state space and of parameters. If the process mass balance equation is available and if the yield matrix is known but kinetics unknown, one can estimate some state variables and some reaction kinetics from the measurement of a portion of the state variables and from measurements of input and output flows. The mathematical method is an adaptive observer [31, 33] that compensates for the estimation error as a function of the difference between measurements and their predictions by the model. Gain coefficients are chosen so that estimation errors tend toward zero. This is the case in the estimation of the specific growth rate determined from the indirect measurement of the biomass concentration during a lactic acid fermentation [34]. In this case, a nonlinear measurement model enables biomass to be measured indirectly from the measurement of the mass of sodium hydroxide added to control pH.

The interest of this method is that there is no need for models of kinetics, because they are only considered as varying with time. Nevertheless, it is not always possible to estimate all state variables and all kinetics, but only those that are "observable" from available measurements.

Complete estimation of state space. If the process mass balance equation, the yield matrix, and kinetic models are known (complete process dynamics), a complete state space estimation is possible based on the measurement of a portion of the state variables and measurements of input and output flows. The mathematical method used is an extended observer [31, 33] that compensates for the estimation error of observable variables as a function of the difference between measurements and their predictions by the model. Nonobservable state variables are simply simulated. Here again, gain coefficients are chosen such that the estimation errors tend toward zero. The disadvantage of this method is its sensitivity to errors in modeling reaction kinetics that do not always generate the best estimation of state [33].

The dynamic model used nevertheless enables the process to be simulated, the progression to be predicted (based on initial conditions or by using measurements available in real time), or the operation to be optimized. Research is under way to improve the complete estimation of state using numerical methods of vanishing horizon estimators [35].

2. Black Box Models of a Bioprocess

For the past decade, the application of neural networks has covered fields as varied as they are numerous. Resulting from artificial intelligence, formal (or artificial) neural networks have partially left this domain and joined statistics and nonlinear mathematics. Feed-forward neural networks have the two basic properties of being a Bayesien classifier and a universal, parsimonious approximator of multivariable and nonlinear functions. The latter property is used in modeling. Once the number of variables is at least three, neural networks have the advantage of requiring fewer parameters to identify than other conventional approximators (polynomials, Fourier series, etc.).

Depending on the type of model to be identified, static or dynamic, we can use either static neural networks [36–38] or recurrent networks that enable any dynamic system to be approximated [39]. In the case of a dynamic model, the neural model approximates a discontinuous state space model. The neural network input includes all state and command variables (operating variables) at time t, while the output contains the state variables at time $t + 1$.

The identification of a nonlinear model requires the definition of the broadest and most representative experimental plan possible to determine the influence of each variable. The experimental data are then distributed into two sets, one for identification (learning) and including about 70% of the data, and a test (or generalization) set containing the remaining data. This enables a cross validation of the model to be carried out. Network parameters are identified with classical algorithms for minimizing total square error. The most effective are the conjugated gradient and the quasi-Newton algorithms [40].

There are two modes involved in the method for identifying a dynamic system [41, 42]. The nonrecurrent mode is a directed or series–parallel learning. In this case, the network input contains variables measured at an instant t and the output contains variables measured at the following instant, $t + 1$. The set of time steps of each experient is considered, and the error between the calculated and measured values must be minimized. The recurrent mode is a semidirected or parallel mode during which the model operates for a certain time in parallel to the process with no measurement. This time defines a chosen prediction horizon. The error to minimize is that observed at the end of this time horizon.

The dynamic models obtained are then used either to conduct a complete reconstruction of state space from a partial state space measurement (indirect dynamic measurement or possibility of coupling with extended observer algorithms), or to carry out future predictions of changes in variables for a predictive or optimal control.

The general configuration of the neural model in the case of fermentation processes involves inputting concentrations of compounds (concentrations of substrate, product, and biomass), the command variable (feed flow), and operating conditions (values of pH and temperature, for example). The network outputs are only compound concentration values predicted in the prior time step. This representation enables fed batch [43] or batch [42] bioprocesses to be simulated.

3. Hybrid Models or Knowledge Neural Networks

Several situations may arise when attempting to model a dynamic process: (1) No mathematical knowledge sufficiently precise to be usable is available. In this case "black box" modeling described in Section II.D.2. is used. (2) A physical and biochemical analysis enables the state space model to be determined, which can be expressed as a set of nonlinear state equations, but which is not sufficiently precise to be usable (see Section II.D.1).

By using a discrete-time formulation, however, the model can be used in the form of a neural network including known functions with parameters having a physical or biological sense [44]. This model now becomes more precise, benefiting from the learning capacities of neural networks described in Section II.D.2. One of the major interests in this type of knowledge neural network is that it generally results from the decomposition of the overall problem into smaller subproblems. This decomposition is found in the structure of the neural network, thereby becoming "legible." This legibility in particular leads to the verification that each subnetwork carries out the function assigned to it, and detects problems of nonidentifiability when they exist. Thus learning can supply the values of physical or biochemical parameters that are poorly known or that deviate from theoretical values.

This procedure has been successfully used (in simulation) for the prediction and optimization of a fed batch microbial growth process [45].

E. Main Trends in Research and Development

The diversity of research shares the objective of increasing knowledge of bioprocesses and their progression. There are five principal trends:

1. On-Line Determination of Key Biological Variables

In most cases, modeling methods of differing complexities are used for the real-time determination of the concentrations of residual substrates and accumulated metabolites. In parallel, attempts are made to determine the concentrations of microbial biomasses. This is often the most delicate determination, first because measurement methods used as reference are imprecise, especially in complex culture media (milk, grape juice, must, ground meat, etc.), and second because the concept of biological activity reflecting the physiological status of microorganisms is not consistently correlated with their concentration.

2. On-Line Determination of Reaction Kinetics

Reaction kinetics (microbial growth, substrate consumption, production of metabolites), whether or not specific, is always a relevant representation of the progression of a bioprocess. Their often original information content can open the way to innovations in the area of process control. Their determination in real time is thus an essential objective.

According to the individual case, kinetics can be established from biological variables determined on-line (Section II.E.1) or from the measurement of other parameters representative of the bioprocess, for example, the rate of fermentation in winemaking reflected by the rate of release of CO_2 (Section III.A.1).

3. Concepts of Kinetic-Temporal Behavior and Reference Trajectory

Food fermentation processes are batch processes, and the on-line measurement of changing noncontrolled variables enables the corresponding kinetics to be established. It then becomes possible to analyze the changes of these kinetics as a function of culture time by taking into account complete curves or, more simply, particular descriptors of the curves. It is thus possible in controlled operating conditions to define an "ideal" (or optimal) trajectory. When the process is put into use subsequently, it is verified in real time that this trajectory is really followed. Any significant difference is signaled to the operator, who decides whether or not to intervene.

These concepts have recently been applied to certain lactic fermentations, such as the production of stirred yogurts and low-fat fresh cheese (Section III.B).

4. Determination of Critical Thresholds

Some fermented food products are subjected to very strict standards or qualitative criteria. One example is the ethanol content of an alcoholic beverage or the final pH of a yogurt. The determination and control of these parameters during the fermentation process have a strategic nature. Depending on the process under consideration, this parameter is determined in real time or is anticipated, and an appropriate control algorithm is written. This approach has been successfully used in the production of alcoholic beverages (Section III.A.2)

5. Predictive Modeling and Estimation of End of Fermentation Times

The interest in using a predictive modeling method is to obtain process progression data in an *anticipated manner*. These data may pertain to a biological variable (concentration, pH, gas release, etc.), a kinetic variable (fermentation rate), or a temporal parameter (particular instants of a batch culture, essentially the end of the process). Methods for establishing these models may use recurrent neural networks such as described in Section II.D.

III. PRACTICAL EXAMPLES OF FOOD FERMENTATION PROCESS CONTROL

A. Alcoholic Fermentations

1. Winemaking (Enology)

Considerable work has been devoted to this seasonal industry since 1985. The traditional approach based on daily measurements of density and temperature is progressively being replaced by continuous process monitoring, involving the automatic control of fermentation temperatures. This obviates the risk of an interruption of fermentation resulting from excessive temperature and has considerably improved the quality and regularity of wines. White wines are thus generally prepared at low temperatures (13 to 17°C in the Champagne region, lower than 22°C in other regions). Maximal working temperatures for red wines may reach 30 to 32°C, resulting in much more rapid kinetics than for white wines.

The most relevant method for determining these kinetics involves the on-line measurement of CO_2 release or density changes [46, 47]. In addition to calculating the rate of fermentation, these measurements enable the real-time determination of ethanol and residual sugar concentrations in the must. Linear relationships, whether or not they include the initial sugar concentration in the grape juice, could be established between physical measurements and biological variables [48, 49]. This represents considerable progress, illustrated by Figs. 1 and 2. This work has been extended in two directions:

(a)

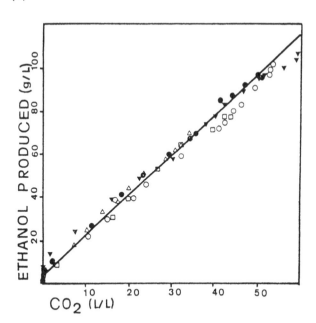

FIGURE 1   Relationships between the volume of CO_2 released and the concentrations of ethanol (a) and residual sugars (b) during wine production. Five trials with different initial sugar concentration: △: 160 g/L; □: 180 g/L; ●: 200 g/L; ○: 220 g/L; ▼: 240 g/L.

(b)

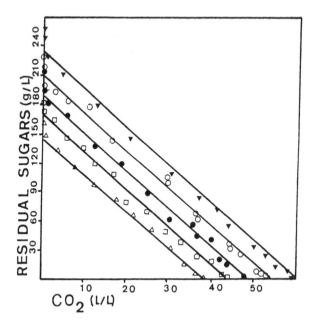

FIGURE 1 *(continued)*

1. The definition of piloting laws specific to the preparation of white and red wines. In the former case, fermentation temperature and then the rate of fermentation are automatically controlled in succession [49]. In the case of red wines, automatic control involves acting on must temperature so that its increase is progressive as a function of time of reaction advancement [47]. Figure 3 shows an example of each of these laws.
2. Models have been established for predicting the changes in fermentation kinetics as a function of certain operating conditions (temperature, initial must richness in sugars, rate of fermentation, initial acceleration of fermentation, etc.) [50, 51].

In conclusion, the control of alcoholic fermentation in enology today is centered on temperature. Centralized computer systems installed in wine cellars that can reach 60 to 90 tanks are functional. The use of the rate of fermentation for automatic process control remains marginal because of the seasonal nature of the industry, but interesting developments have been accomplished for certain specialized wines (Section III.A.2).

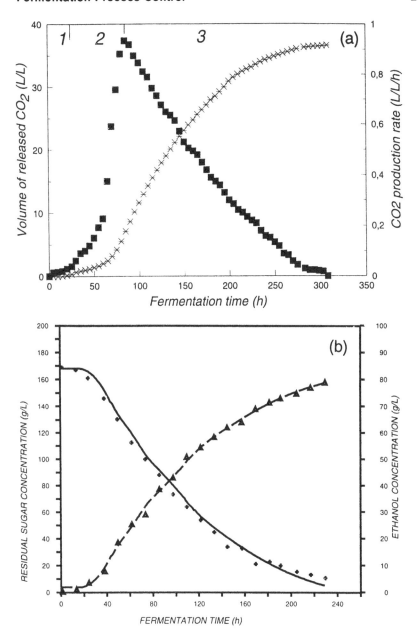

FIGURE 2 (a) Example of two experimental curves obtained in real time during wine production: volume of CO_2 released (✳); CO_2 production rate (■). (b) Residual sugar and ethanol concentrations evolution during wine production. +, sugar; ▲: ethanol; off-line analysis; —, --: on-line determination.

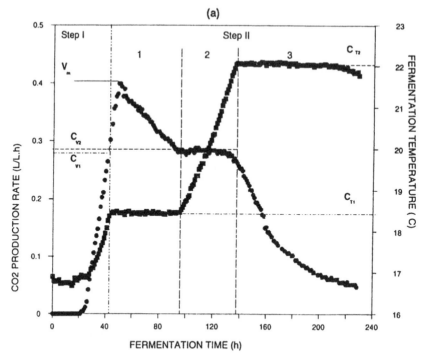

FIGURE 3 (a) Monitoring rule applied to white wine production. ■: evolution temperature; ●: CO_2 production rate. Step I: Free evolution of temperature. Step II: (1) Isothermal control of temperature (C_{T1}); (2) evolution of temperature during fermentation rate control (V_{C2}); (3) isothermal control at the maximum allowed temperature (C_{T2}). (b) Monitoring rule applied to red wine production. Evolution of the CO_2 production rate and the temperature as a function of fermentation progress. +: CO_2 production rate, ■: fermentation temperature.

2. Sparkling Wines and Muscats

Automatic Manufacture of "Pétillant de Raisin" "Pétillant de Raisin" is a sparkling beverage produced from partially fermented grape juice. According to French legislation, its alcohol content is limited to 3° GL.* To comply with this, alcoholic fermentation is stopped by rapidly chilling the must. The sparkling character of the wine is previously obtained by placing the fermentation tanks under pressure (4.5 bars).

To automate these two successive functions (pressurizing and chilling) as a function of the alcohol content of the must, a linear relationship

_____
* °GL = Gay-Lussac degrees, roughly equivalent to percent V/V at 20°C.

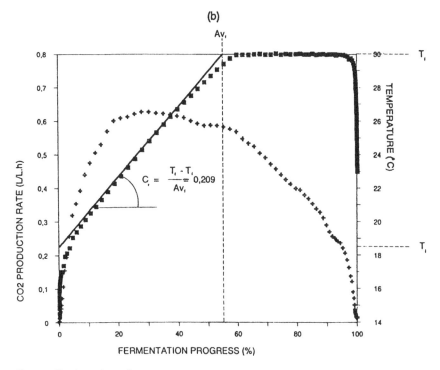

FIGURE 3 *(continued)*

has been established between CO_2 released and ethanol produced. When pressurization is started, no CO_2 has yet been released, and the ethanol content is calculated either from the pressure increase or by extrapolating data obtained in the phase of CO_2 release [52]. Satisfying results are obtained in both cases. The second, simpler solution, has been applied at the industrial scale. This is shown in Fig. 4, showing the changes in the principal data measured or calculated in real time. It can be seen that the maximal rate of fermentation is reached before the tank is pressurized. Chilling time is about 10 hours, but the yeasts cease activity once the temperature reaches 15°C.

Automation of Muscat Production   Muscat is a typical French wine partially fermented. Alcoholic fermentation is stopped by adding absolute ethanol, which inactives the yeast. This addition must be realized to obtain an accurate control of the final sugar (125 g/L) and ethanol (15°GL) concentrations of the Muscat. Depending on the production area, small variations of these concentrations are allowed.

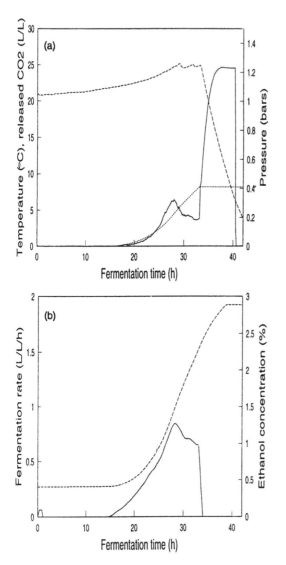

FIGURE 4 (a) Evolution of pressure, temperature, and released CO_2 during low ethanol sparkling wine production. —: pressure; ----: CO_2 released; --: temperature. (b) Evolution of calculated fermentation rate and ethanol concentration during low ethanol sparkling wine production. —: fermentation rate; ---: ethanol concentration.

To control this process, on-line density measurements were performed on several tanks [53]. The density decrease is controlled by fermentation temperature. The addition of ethanol (time and volume) is controlled by an algorithm specially developed to calculate the final concentrations of sugar and ethanol. Its main functions are the determination of the density at which addition must be performed and the volume of absolute ethanol to add [54]. This algorithm, including a predictive model, was implemented in the computer control system devoted to eight fermentation tanks equipped with one on-line density measurement device (Microbar-Imeca-France). The process has been working at the industrial scale since 1993.

3. Brewery

The brewery industry produces different varieties of beers in large outdoor tanks (200 to 1000 m^3). At the present time, two main parameters are automatically controlled. Fermentation temperature measured at different levels (two or three) in the tanks is controlled by the admission of chilled water into different jackets (two to four levels). The pressure of the tanks is also controlled, acting on the flow of released CO_2. Figure 5 is a typical example of variations of these controlled parameters as a function of fermentation time. Low temperatures (10 to 18°C) are used during alcoholic fermentation, following typical profiles linked to the type of beer and to the know-how of each brewery. At the end of the fermentation step, the temperature is decreased to near 0°C for the maturation step.

The carbon dioxide pressure is generally low (1 atm) at the beginning of alcoholic fermentation and is increased step by step (two or three steps)

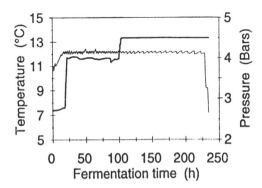

FIGURE 5 Typical evolution of controlled temperature (—) and pressure (–) during beer production.

during the process. It can reach 1.5 to 1.8 atm. These controls are usually done by automatic controllers that may or may not be included in computer-based systems. They do not present significant difficulties, but their innovative level is poor.

Considerable recent research has attempted to improve beer-making process monitoring taking quality into account. In this way, three different approaches were tested, one of them based on the development of structured models concerning kinetics and flavor [55–57]. The main disadvantage of these models is the large number of parameters to identify. As a consequence, their identification is very difficult. The second approach is based on new on-line measurements [58–62]. Third, some approaches have tried to establish relationships between fermentation control and beer quality. Temperature profiles to control fermentation foaming were defined [63]. The effect of CO_2 pressure on yeast sensitivity and the formation of volatile compounds was studied [64, 65]. The effect of fermentation conditions on volatile concentration profiles was also studied [66]. Finally, several authors used neural network modeling for predicting beer fermentation behavior. They focused their interest on estimating and predicting density and ethanol, sulfur dioxide, and diacetyl concentrations [67–70].

Most of this work is interesting and highly innovative, but until the present time, there has been no industrial application.

B. Fermented Milk and Dairy Products

1. Fermented Milk

During the production of fermented milk, the initial pH is close to 6.4 and then decreases to values as low as 4.5. Each type of fermented milk available on the market has its own characteristic final pH, which is a feature of the dairy product and is determined by the manufacturer. To achieve the desired final pH value, the operator monitors pH; and when the specific value is reached, the fermentation process is quickly stopped, generally by cooling. In a plant where many fermentations are performed simultaneously, the prediction, in real time, of the final time of each fermentation is crucial for scheduling the use of heat exchangers and fermentors. Moreover, the definition of feature points and indirect measurements of key variables enhance fermentation process monitoring and enable deviations in culture progress to be detected. We describe three methods applicable in the case of stirred yogurt production, combining instrumentation and models to obtain feature points of kinetic curves, indirect measurements, and predictive models based on neural network methods. See Chapter 15 for more discussion on this.

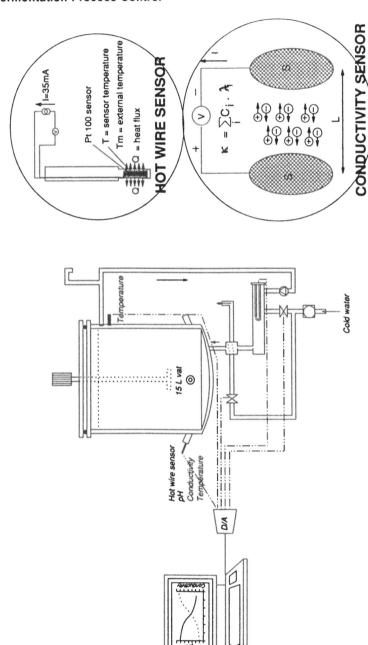

FIGURE 6 Instrumentation for monitoring yogurt production

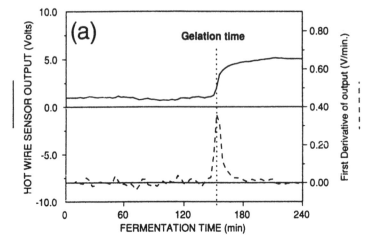

FIGURE 7 Evolution curves during yogurt production at 44°C (skim milk 13% w/ v). (a) Evolution of hot wire sensor signal and its first derivative versus time. Detection of gelation time. (b) Evolution of pH signal and its first derivative versus time. Detection of maximum acidification rate time (tpHm). (c) Evolution of electrical conductivity signal and its first derivative versus time. Detection of three feature points (tG1, tG2, and tG3).

Instrumentation and Determination of Feature Points   During yogurt making, lactic acid bacteria (*Streptococcus thermophilus* and *Lactobacillus bulgaricus*) produce lactic acid and small amounts of ammonia, decreasing the pH and increasing electrical conductivity [71]. When the pH reaches 5.3, the milk gels, which can be detected by a hot wire sensor [72]. Figure 6 shows some facilities for studying and monitoring yogurt production in a fermentor. The temperature and electrical conductivity probes and the hot wire sensor are linked to a computer system for real-time signal acquisition and data processing. The first derivatives of the signals were calculated to define feature points characterizing curve changes (Fig. 7). For example, the point of inflection on the curve of the hot wire sensor defines the gelation time (Fig. 7a). That observed on the pH curve changes indicates the maximum acidification rate (Fig. 7b). There are three points of inflection on the conductivity curve (Fig. 7c). The first (tG1) is related to the maximum rate of ammonia production (urea catabolism of *S. thermophilus*). The second (tG2) is related to the end of ammonia production, and the third (tG3) reflects the maximum rate of production of total lactic acid. The time at which these feature points occur and the values of the rate of change at this time are character-

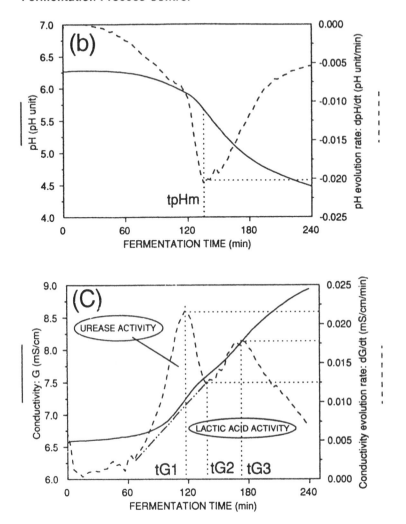

FIGURE 7   *(continued)*

istic of the bacterial strain used and of the operating conditions (medium composition, temperature, etc.). They were thus used to compare the effects of the type of starter, the temperature, and the culture medium. The potential application of the on-line determination of these feature points is related to diagnostic purposes. Because they furnish summarized information on process changes, they can be compared with reference values to indicate normal behavior or a deviation from standard cultures.

Indirect Measurements Stirred yogurt is produced in very large vats where pH measurements are difficult to obtain. The pH probes must be frequently calibrated and have a short lifetime due to fouling by dairy products. Thus it is possible to replace pH probes by electrical conductivity probes that are more reliable and less expensive in the long term. However, no theoretical linear relationship is known between these two variables because of the influence of the composition of the medium. Moreover, this relationship is sensitive to fermentation temperature, which activates the growth of *S. thermophilus* and the production of lactic acid [71]. To identify the nonlinear relationship between pH and electrical conductivity, we used a neural network model with two inputs (conductivity and temperature) and one output (pH) [73]. Experimental values of pH and conductivity were recorded for different temperatures and different types of starter. They were separated into a learning set and a test set to select the most appropriate number of hidden nodes. For new experiments, the model enables an indirect measurement of pH change to be performed using only two on-line measurements (electrical conductivity and temperature). The method is available for different types of starters

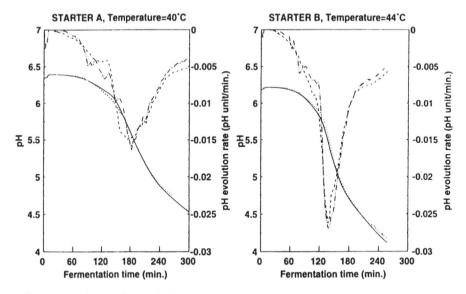

FIGURE 8 Comparison of direct and indirect measurement of pH during yogurt production. Indirect measurement is determined from electrical conductivity. —: pH measurement; ·····: indirect pH measurement; ———: *d*pH/*dt* determination; —·····—: indirect *d*pH/*dt* determination.

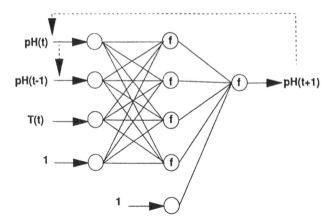

FIGURE 9 Recurrent neural network model for pH prediction during yogurt production. t: fermentation time; T: fermentation temperature.

and is sufficiently accurate to obtain an indirect measurement of the rate of pH change. Figure 8 shows the results obtained for two different starters (A and B) and for two temperatures (40 and 44°C).

Instrumentation and Predictions of the Final Process Time Dynamic models can predict future pH changes during culture for early detection of failure. Considering the final pH to be reached (for example, pH = 4.6), it is possible to determine the final fermentation time. For this purpose, a dynamic model was simulated by a recurrent neural network with on-line measurements of pH and temperature (sampling period of 3 min) in mixed cultures of *S. thermophilus* and *L. bulgaricus* [74]. Because temperature was an important parameter for explaining different dynamic behaviors of the process, it was explicitly taken into account. The general neural network structure, expressed in the form of a one-step-ahead predictor, is presented in Fig. 9. This structure can be viewed as a second-order model. A semidirected algorithm was used to minimize the error observed with a prediction horizon of 2 h.

Figure 10 shows that the error (difference between real and calculated values of final fermentation time) was greater at the beginning of fermentation and prediction greatly improved toward the end. To better judge the capability of making predictions over a long time horizon, the mean relative error (for 13 experiments) of the final fermentation time prediction (when pH measurement reaches 4.6) was 8%. This means that for a fermentation that will reach a pH of 4.6 in 2 h, the error in the predicted final fermentation will be less than 9 min. This performance is

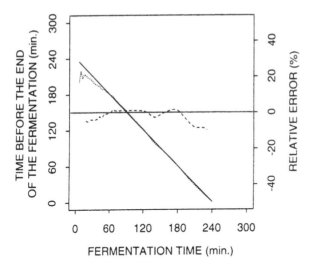

FIGURE 10 Yogurt production monitoring. Prediction of the remaining time before the end of the fermentation and relative error of this prediction as a function
of fermentation time. —: experimental value; ------.: predicted value; –––: relative
error.

quite sufficient for an industrial application. Dynamical models involving
electrical conductivity can also be used to obtain pH prediction with similar accuracy [73].

2. Other Dairy Products

Yogurt production is based on milk acid coagulation. Two other methods
of milk coagulation are widely used: coagulation by rennet, and mixed
coagulation (rennet and acid), in particular for the production of fresh
cheese and cottage cheese.

   The instrumentation and concepts described previously can be applied to these two processes. In this case, the aims are to determine coagulation time and above all the end-of-process time, defined by a level of
gel firmness or a reference pH to be reached. Although the know-how of
the cheese maker remains the reference in cheese making, several instrumentation approaches are possible.

   Numerous attempts to develop rheological measurement instruments have not resulted in industrial applications for on-line process moni-

toring [75]. The complexity and fragility of the instruments are undoubt-edly the reasons for these failures.

Current work is oriented toward the use of probes for measuring thermal or optical parameters. In the case of rennet coagulation, hot wire and optical probes allow detection of the beginning of the micelle aggrega-tion. Beyond, just the optical signal increases at the same time of the curd firmness (Fig. 11).

During mixed coagulation, the curd will be demineralized because of acidification of the medium by the lactic and acetic acid production. Around pH 5.2, micelle demineralization is almost achieved, and aggrega-tion of submicelles or disunited caseines begins. At this time, the evolution rate of the optical signal increases again (Fig. 12). The electric conductivity changes during the entire process, permitting a general monitoring and the determination of the two steps of coagulation. Moreover, a linear correlation is established between the conductivity and the lactate + ace-tate concentration [76].

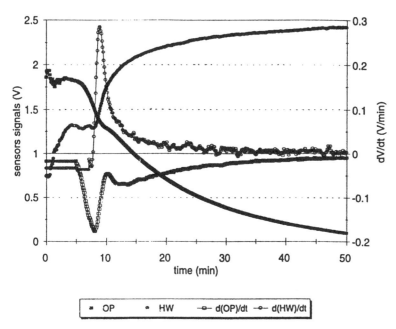

FIGURE 11 Instrumentation of the rennet coagulation process. Evolution of the signals and first derivatives of signals of optical (OP) and hot wire (HW) sensors during the rennet coagulation.

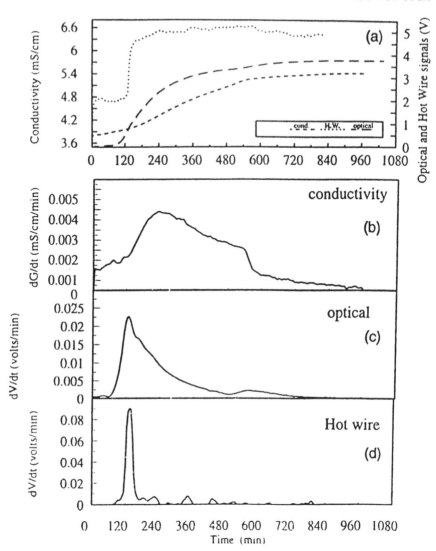

FIGURE 12 Instrumentation of the mixed coagulation process. (a) Evolution of the signals of conductivity, optical, and hot wire sensors versus fermentation time. (b) Conductivity, (c) optical, and (d) hot wire evolution rates as functions of fermentation time.

The determination of the end time of the fresh cheese process is not improved by the use of these sensors in comparison with the traditional method based on the settled final time. In fact, above pH 4.7, pH does not vary much, and consequently, changes are very small.

For cheese production, the determination of the cutting time is based on a prediction model taking into account some state variables and incorporating feature points from the signal sensors [77]. The standard errors of prediction are between 1 and 2.4 min. But an objective method of predicting cutting time should have a repeatability of 1 min [75]. Thus the present method must be improved, for example, by using other models as well as neural network–based ones.

C. Other Fermentation Processes

1. Production of Baker's Yeast

The production of baker's yeast (*Saccharomyces cerevisiae*) is an aerobic pH-controlled fed batch process. Recently, time-limited (5–7 days) continuous culturing was processed. In the two cases, the aim of substrate feeding is the control of the glucose concentration at very low levels (0.25 to 0.3 g/L) to avoid catabolite repression. This concentration cannot be measured on the industrial scale. Experiments using a biosensor were efficient on the pilot plant scale [78] but cannot be envisaged in industrial bioreactors. An alternative approach involves measuring ethanol exhausted into the effluent gas. A gas-sensitive semiconductor is sometimes used for this purpose [79]. The control of substrate flow rate as a function of ethanol emitted, however, has not yet been developed.

The concentration of dissolved oxygen is one of the main factors acting on the productivity of this process. It must be higher than 8–10% of saturation of the culture medium by air. Agitation and aeration conditions directly determine the dissolved oxygen concentration, but they are design parameters of the bioreactors. In industrial plants, they cannot be modified for the control of dissolved oxygen, the measurement of which has not yet been done.

The pH is another key parameter of this process. In many industrial plants, however, pH control (addition of an alkaline solution) is not done because the pH probes are not considered to be reliable enough. The process currently includes feeding profiles of the alkali. The profile is generally time dependent and may be slightly adapted to the operating conditions.

In conclusion, two feeding profiles (substrate, alkali) are processed. Their definition is based on previously acquired know-how. In any case, automatic control of feeding is realized. In this way, many improvements are still possible.

2. Vinegar Production

Vinegar is produced by the oxidation of ethanol to acetic acid using aerobic acetic acid bacteria (Acetobacter or Gluconobacter). Modern acetators involve submerged cultures, strongly stirred and aerated to transfer the oxygen needed for the biological oxidation of ethanol. Any oxygen starvation (air supply interruption) must be avoided: a short time (several seconds) of oxygen deprivation of the bacteria stops the rate of acetification irreversibly. It seems that oxygen starvation produces a breakdown of the acid resistance of the bacteria [80].

The process is run semicontinuously: 30 to 40% of the volume of the culture is harvested every time, and the submerged culture is then made up to volume with fresh medium. Thus the final concentration of acetic acid can reach 140 g/L. Temperature is automatically controlled at 30°C using recycled cooling water. The measurement of dissolved oxygen, necessary during the starting phase, is not generalized. The main problem for process control is the on-line determination of the residual ethanol concentration. The harvesting time must correspond to a very low ethanol concentration (2–3 g/L). Taking into account ethanol consumption (approximately 1.5 g/L·h), harvesting times have to be determined with an accuracy of about 30 min [81]. This emphasizes the importance of the on-line determination of ethanol concentration. Several methods have been tested, and some industrial equipment is available. This is the case of the Alkograph [82] commercialized by Frings GmbH (Bonn, Germany). Based on an ebullioscopic method, it allows continuous or sequential measurements. The main problems preventing its generalized use are linked to sampling, fouling of pipes, and to interference by other volatile components (ethylacetate, ethanal, etc.).

On the other hand, the use of a biosensor (immobilization of the yeast *Trichosporum brassicae*, Biotech analyzer, Toyo Jozo Co., Tokyo) or of a semiconductor sensor (tin oxide semiconductor element containing palladium, Figaro Engineering Inc., Osaka) was studied in Japan. But for selectivity, stability, repeatability, and cost, the equipment composing the sensors is unacceptable [83].

More recently, an important development has involved using gas chromatography to analyze the ethanol concentration of the outlet gas of the acetators [81]. Relationships between ethanol concentrations in the gas and liquid phases, and the concentrations of other volatiles (acetaldehyde, ethylacetate) were established. A model of gas exchanges in the acetator was also proposed. French industrial plants have been equipped with this analytical device, associated with data processing for on-line monitoring of the semicontinuous vinegar production process [81]. About 15% of French vinegar production is controlled in this way.

REFERENCES

1. J. Lenoir, G. Lamberet, J. L. Schmidt, and C. Tourneur, La maitrise du bioreacteur fromage, *Biofutur 41*:23 (1985).
2. B. J. Wood, *Microbiology of Fermented Foods*, vols. 1 and 2, Elsiever Applied Science, London, 1985.
3. R. I. Fox, Computer and microprocessors in industrial fermentation, *Topics in Enzyme and Fermentation Biotechnology* (A. Wiseman, ed.), Ellis Horwood, 1984, p. 125.
4. K. Schugerl, On-line analysis of broth, *Biotechnology, Measuring, Modeling and Control*, (H.-J. Rehm and G. Reed, eds.), VCH, New York, 1991, p. 149.
5. G. Locher, B. Sonnleitner, and A. Fiechter, On-line measurement in biotechnology: Techniques, *J. Biotechnol. 25*:23 (1992).
6. C. M. Harris and D. B. Kell, The estimation of microbiol biomass, *Biosensors 1*:17 (1985).
7. T. Yamané, Application of an on-line turbidimeter for the automation of fed-batch cultures, *Biotechnol. Prog. 9*:81 (1993).
8. B. H. Junker, J. Reddy, K. Gbewonyo, and R. Greasham, On-line and in-situ monitoring technology for cell density measurement in microbial and animal cell cultures, *Bioprocess Eng. 10*:195 (1994).
9. P. Wu, S. S. Ozturk, J. D. Blackie, J. C. Thrift, C. Figueroa, and D. Navesh, Evaluation and application of optical cell density probes in mammalian cell bioreactors, *Biotechnol. Bioeng. 45*:495 (1995).
10. G. Markx, C. Davey, and D. Kell, The permittistat: A novel type of turbidostat, *J. General Microbiology 137*:735 (1991).
11. D. B. Kell, G. Markx, C. Davey, and R. Todd, Real-time monitoring of cellular biomass: Methods and applications, *Trends in Anal. Chem. 9*:190 (1990).
12. R. Fehrenbach, M. Comberbach, and J. O. Petre, On-line bioprocess monitoring by capacitance measurement, *J. Biotechnol. 23*:303 (1992).
13. Y. Marc, Couplage de bioréacteurs et d'analyseurs, *Capteur et mesure en biotechnologie*, (J. Boudrant, G. Corrieu, and P. Coulet, eds.), Tec. and Doc., Lavoisier, Paris, 1994, p. 243.
14. T. Yamané, M. Matsuda, and E. Sada, Application of porous teflon tubing method to automatic fed-batch culture of microorganisms. I. Mass transfer through porous teflon tubing, *Biotechnol. Bioeng. 23*:2493 (1981).
15. T. Yamané, M. Matsuda, and E. Sada, Application of porous teflon tubing method to automatic fed-batch culture of microorganisms. II. Automatic constant-value control of fed-batch substrate (ethanol) concentration in semi-batch culture of yeast, *Biotechnol. Bioeng. 23*:2509 (1981).

16. M. N. Pons, P. Ducouret, and J. Bordet, Etude de transfert de matière dans un capteur à membrane tubulaire hydrophobe: Cas des alcools, *Entropie 123*: 21 (1985).

17. W. Knol, M. Minekus, A. G. F. Angelini, and J. Bol, In-line monitoring and process control in beer fermentation and other biotechnological processes, *Monatsschrift für Brauwissenschaft 7*:281 (1988).

18. Y. Marc, F. Blanchard, E. Ronat, and J. M. Engasser, Contrôle de la fermentation de la bière à l'aide d'un système de prélévement d'échantillons liquides et d'estimateurs par la méthode de Kalman, First French Process Engineering Congress (Lavoisier, ed.), Paris 1987, p. 174.

19. W. Kuhlman, H. D. Meyer, K. H. Bellgardt, and K. Schugerl, On-line analysis of yeast growth and alcohol production, *J. Biotechnol. 1*:171 (1984).

20. Y. Marc, F. Blanchard, E. Ronat, and J. N. Rabaud, Automatic sampling device for a biological fluid, French patent 89 06647 CNRS, 1989.

21. D. Picque and G. Corrieu, Performances of aseptic sampling devices for on-line monitoring of fermentation processes, *Biotechnol. Bioeng. 40*:919 (1992).

22. P. Fairbrother, W. O. George, and J. M. Williams, Whey fermentation: On-line analysis of lactose and lactic acid by FTIR spectroscopy, *Appl. Microbiol. Biotechnol. 35*:301 (1991).

23. D. Picque, D. Lefier, R. Grappin, and G. Corrieu, Monitoring of fermentation by infrared spectroscopy: Alcoholic and lactic fermentations, *Anal. Chim. Acta 279*:67 (1993).

24. E. K. Kemsley, R. H. Wilson, and P. S. Belton, Potentiel of fourier transform infrared spectroscopy and fiber optics for process control, *J. Agric. Chem. 40*:435 (1992).

25. E. K. Kemsley, R. H. Wilson, G. Poulter, and L. L. Day, Quantitative analysis of sugar solutions using a novel sapphire ATR accessory, *Appl. Spectroscopy 47*(10):1951 (1993).

26. X. H. Zhang, H. I. Ma, C. Blanchetière, K. Le Foulgoc, J. Lucas, J. C. Farcy, D. Picque, and G. Corrieu, Tellurium halide IR fibers for remote spectroscopy, Proceedings of SPIE Congress, Los Angeles, January 1994, vol. 21, p. 31.

27. N. M. Fish, Modelling bioprocesses, *Modelling and Control of Fermentation Processes* (J. R. Leigh, ed.), Peregrinus, London, 1987.

28. C. Kleinstreuer, Analysis of biological reactors, *Advanced Biochemical Engineering* (H. Bungay and G. Belfort, eds.), John Wiley and Sons, New York, 1987.

29. N. F. Thornhill and P. N. C. Royce, Modelling fermentors for control, *Measurement and Control in Bioprocessing* (K. Carr-Brion, ed.), Elsevier Science, London, 1991.

30. P. N. Royce, A discussion of recent developments in fermentation monitoring and control from a practical perspective, *Crit. Rev. Biotechnol. 13*:117 (1993).

31. G. Bastin and D. Dochain, On-line estimation and adaptive control of bioreactors, *Process Measurement and Control, 1*, Elsevier, Amsterdam, 1990.

32. V. Chotteau, G. Bastin, I. M. Y. Mareels, and L. Fabry, Reaction mechanisms and cell density estimators for animal cell cultures, Preprints of the Sixth Int. Conference on Computer Applications in Biotechnology, Garmisch-Partenkirchen, Germany, 1995, p. 66.

33. A. Chéruy and J. M. Flaus, Des mesures indirectes à l'estimation en ligne, *Capteurs et mesures en biotechnologie* (J. Boudrant, G. Corrieu, and P. Coulet, eds.), Tech. and Doc., Lavoisier, Paris, 1994, p. 444.

34. G. Acuna, E. Latrille, C. Béal, G. Corrieu, and A. Chéruy, On-line estimation of biological variables during pH controlled lactic acid fermentations, *Biotechnol. Bioeng. 44*:1168 (1994).

35. L. Boillereaux and J. M. Flaus, Adaptive receding horizon state estimation for non linear processes, ACASP'95, IFAC Congress, Budapest, Hungary, 1995.

36. G. Cybenko, Approximation by superpositions of a sigmoidal function, *Math. Contr. Sign. Syst. 2*:303 (1989).

37. K. I. Funahashi, On the approximate realization of continuous mappings by neural networks, *Neural Networks 2*:183 (1989).

38. K. Hornik, M. Stinchcombe, and H. White, Multilayer feedforward networks are universal approximators, *Neural Networks 2*:359 (1989).

39. K. I. Funahashi and Y. Nakamura, Approximation of dynamical systems by continuous time recurrent neural networks, *Neural Networks 6*:801 (1993).

40. C. M. Bishop, Neural networks and their applications, *Rev. Sci. Instrum. 65*:1803 (1994).

41. O. Nerrand, P. Roussel-Ragot, L. Personnaz, and G. Dreyfus, Neural networks and non-linear adaptive filtering: Unifying concepts and new algorithms, *Neural Computation 5*:165 (1993).

42. E. Latrille, P. Teissier, G. Acuna, B. Perret, and G. Corrieu, Estimation en ligne de la croissance microbienne dans des procédés de fermentation discontinus à l'aide de réseaux de neurones recurrents, *Récents progrès en génie des procédés 9* (J. P. Corriou, ed.), Lavoisier, Paris, 1995, p. 87.

43. J. Thibault and V. Van Breusegem, Modelling and on-line optimization of a fedbatch fermentation using neural networks, *Bioautomation 12*:79 (1992).

44. I. Rivals, P. Personnaz, G. Dreyfus, and J. L. Ploix, Modélisation, classification et commande par réseaux de neurones: Principes fondamentaux, méthodologie de conception et illustrations industrielles, *Récents progrès en génie des procédés 9* (J. P. Corriou, ed.), Lavoisier, Paris, 1995, p. 1.

45. D. C. Psichogios and L. H. Ungar, A hybrid neural network-first principles approach to process modeling, *AIChE J. 38*:1499 (1992).

46. P. Barre, J. Chabas, G. Corrieu, A. Davenel, N. E. El Haloui, P. Grenier, J. M. Navarro, D. Picque, J. M. Sablayrolles, F. Sevila, and C. Vannobel, On-line prediction and control process for alcoholic fermentations, and device of its realization, Brevet Inra/Cemagref 86 157 19 et 87 4025430 (1986).

47. N. E. El Haloui, D. Picque, and G. Corrieu, Mesures physiques permettant le suivi biologique de la fermentation alcoolique en oenologie, *Sci. des Aliments 7*:241 (1987).

48. N. E. El Haloui, D. Picque, and G. Corrieu, Alcoholic fermentation in winemaking: On-line measurement of density and carbon dioxide evolution, *J. Food Eng. 8*:17 (1988).

49. G. Corrieu, N. E. El Haloui, Y. Cleran, and D. Picque, Control of alcoholic fermentation in wine production, Fifth International Congress on Engineering and Food (W. E. L. Spiess and H. Schubert, eds.), Elsiever Applied Science, 1989, p. 593.

50. N. E. El Haloui, G. Corrieu, Y. Cleran, and A. Cheruy, Method for on-line prediction of kinetics of alcoholic fermentation in wine making, *J. Ferment. Bioeng. 68*:131 (1989).

51. Y. Cleran, J. Thibault, A. Cheruy, and G. Corrieu, Comparison of prediction performances between models obtained by the group method of data handling and the neural networks for the alcoholic fermentation rate in oenology, *J. Ferment. Bioeng. 71*:356 (1991).

52. D. Picque, B. Perret, E. Latrille, and G. Corrieu, On-line ethanol estimation and prediction: Application to the production of low alcoholic wines, *Process Biochemistry 26*:173 (1991).

53. D. Picque, B. Perret, E. Latrille, and G. Corrieu, Adaptation de mesures physico-chimiques, *Capteurs et mesures en biotechnologie* (J. Boudrant, G. Corrieu, and P. Coulet, eds.), Tec. and Doc., Lavoisier, Paris, 1994, p. 197.

54. B. Perret, E. Latrille, F. Merican, and G. Corrieu, Suivi et contrôle de la fermentation de vins doux naturels (1996, Reune Française d'oenologie, submitted).

55. J. M. Engasser, Y. Marc, M. Moll, and B. Duteurtre, Kinetic modeling of beer fermentation, EBC Proceedings 18° Congrès, Copenhagen, 1981, p. 579.

56. D. A. Gee and W. R. Ramirez, A flavour model for beer fermentation, *J. Inst. Brew. 100*:321 (1994).

57. A. I. Garcia, L. A. Garcia, and M. Diaz, Modelling of diacetyl production during beer fermentation, *J. Inst. Brew. 100*:179 (1994).

58. W. Knol, M. Minekus, S. A. Angelino, and J. Bol, In-line monitoring and process control in beer fermentation and other biotechnological process, *Monatsschrift für Brauwissenschaft 7*:281 (1988).

59. M. Lenoel and C. Mathis, Mesure en ligne et contrôle du procédé de fabrication de la bière. Une approche systémique, *Cerevisia 15*(2):127 (1990).

60. I. S. Daoud and B. A. Searle, On-line monitoring of brewery fermentation by measurement of CO_2 evolution rate, *J. Inst. Brew. 96*:297 (1990).

61. P. Stassi, G. P. Goetzke, and J. F. Fehring, Evaluation of an insertion thermal mass flowmeter to monitor CO_2 evolution rate in plant scale fermentations, *MBAA Technical Quarterly 28*:84 (1991).

62. J. Miller, H. Patino, M. Babb, and W. Michener, Comparison of exotherm and carbon dioxide measurements in brewing fermentations, *MBAA Technical Quarterly 31*:95 (1994).

63. P. Stassi, J. F. Rice, T. J. Kieckhefer, and J. H. Munroe, Control of fermentation foaming using temperature profiling, *MBAA Technical Quarterly 26*:113 (1989).

64. L. Kruger, A. T. W. Pickerell, and B. Axcell, The sensitivity of different brewing yeast strains to carbon dioxide inhibition: Fermentation and production of flavour-active volatile compounds, *J. Inst. Brew. 98*:133 (1992).
65. R. S. Renger, S. H. Van Hateren, and K. Luyben, The formation of esters and higher alcohols during brewery fermentation; the effect of carbon dioxide pressure, *J. Inst. Brew. 98*:509 (1992).
66. J. M. Sendra, V. Todo, F. Pinaga, L. Izquierdo, and J. V. Carbonell, Evaluation of the effect of yeast strain and fermentation conditions on the volatile concentration profiles of pilot plant lager beers, *Monatsschrift für Brauwissenschaft 10:115* (1994).
67. T. D'Amore, G. Celotto, G. D. Austin, and G. G. Stewart, Neural network modeling: Applications to brewing fermentations, EBC 23 Congress, 1993, p. 221.
68. M. J. Syu, G. T. Tsao, G. D. Austin, A. G. Celotto, and T. D'Amore, Neural networks modelling for predicting brewing fermentations, *J. Amer. Soc. Brewing Chemists*: pgs. 52, 1, 15 (1994).
69. G. Gvazdaitis, S. Beil, U. Kreibaum, R. Simutis, I. Havlik, M. Dors, and F. Schneider, Temperature control in fermenters: Application of neural nets and feedback control in breweries, *J. Inst. Brew. 100*:99 (1994).
70. U. Schrader, U. Huge, T. Enders, and V. Denk, On-line measurement of physical and chemical process parameters to achieved improved control, feedback control and time optimization of fermentation process using neural networks and fuzzy logic, Third Congress on Automatic Control of Food and Biological Processes (J. J. Bimbenet, E. Dumoulin, and G. Trystram, eds.), Elsiever Applied Science, 1994, p. 115.
71. E. Latrille, D. Picque, B. Perret, and G. Corrieu, Characterizing acidification kinetics by measuring pH and electrical conductivity in batch thermophilic lactic fermentations, *J. of Fermentation and Bioeng. 74*(1):32 (1992).
72. T. Hori, Objective measurement of the process of curd formation during rennet treatment of milks by the hot wire method, *J. Food Sci. 50*:911 (1985).
73. E. Latrille, G. Acuna, and G. Corrieu, Application des réseaux de neurones pour la modélisation de bioprocédés discontinus, *Automatique productique informatique industrielles*, in press (1996).
74. E. Latrille, G. Corrieu, and J. Thibault, Neural network models for final process time determination in fermented milk production, *Computers & Chemical Eng. 18*:1171 (1994).
75. A. C. M. van Hooydonk and G. van den Berg, Control and determination of the curd-setting during cheesemaking, *Bull. IDF 225*:2 (1988).
76. D. Picque, E. Latrille, B. Perret, F. Remeuf, and G. Corrieu, Evaluation of sensors to monitor mixed coagulation, Third Congress on Automatic Control of Food and Biological Processes (J. J. Bimbenet, E. Dumoulin, and G. Trystram, eds.), Elsiever Applied Science, 1994, p. 89.
77. F. A. Payne, C. L. Hicks, S. Mandangopal, and S. A. Shearer, Fiber optic sensor for predicting the cutting time of coagulating milk for cheese production, *Trans. ASAE 36*:841 (1993).

78. D. Picque and G. Corrieu, On line monitoring of fermentation process using a new sterilizable sampling device, Fifth Int. Congress on Eng. and Food (W. E. L. Spiess and H. Schubert, eds.), Elsiever Applied Science, Amsterdam, 1989, p. 771.

79. D. Picque and G. Corrieu, Problèmes liés a l'utilisation des semi-conducteurs sensibles aux gaz pour la détermination de l'éthanol, *Ind. Alim. Agri. 10*:899 (1988).

80. H. Ebner, Vinegar, *Prescott and Dunn's Industrial Microbiology* (G. Reed, ed.), AVI, Westport, Connecticut, 1982, p. 802.

81. J. Y. Burlot, C. Ghommidh, and J. M. Navarro, Pilotage de la fermentation acétique en vinaigrerie, *Récents progrès en génie des procédés* (G. Antonini and Ben Aim, eds.), Lavoisier, Paris, 1991, p. 157.

82. H. Ebner and A. Enenkel, Der Frings Alkograph, *"GIT" Fachzeitschrift Lab. 13*:651 (1969).

83. M. Akihiko, Technical trends in the on-line measurement and control of alcohol concentrations in bioreactors for vinegar production, *Instrumentation & Control Eng. in Japan 76*:18 (1989).

9

Process Control for Thermal Processing

A. Ryniecki

Institute of Food Technology, University of Agriculture, Poznan, Poland

D. S. Jayas

University of Manitoba, Winnipeg, Manitoba, Canada

I. INTRODUCTION

The most important factors in food preservation are quality, safety, and low cost of processing. Thermal processing refers to heating, holding, and cooling of a food product to eliminate the potential of foodborne illness. Pasteurization and sterilization are two examples of thermal processes. In pasteurization, product is heated to a temperature and for a length of time such that many pathogenic organisms are destroyed and thermoduric organisms (e.g., lactic acid bacteria) survive. The pasteurized products are not shelf-stable and require refrigeration for their short storage life (1–2 weeks). In sterilization, product is heated to a temperature and for a length of time that the product becomes free of living microorganisms. The sterilized product is shelf-stable and has long (>6 months) storage life at room temperatures. Traditionally, sterilization refers to in-container sterilization, in which food product is filled into a container, the container is sealed and heated using steam to a specified temperature, and then it is cooled to room temperature for distribution. In-container sterilization is used as an example, to illustrate control of a thermal process. Other methods of achieving sterilization currently under investigation by food researchers are ohmic heating, aseptic processing, and irradiation. In controlling thermal processes, the objective is to meet the desired level of bacterial inactivation for the process, irrespective of any variation in the input and reference conditions and with a minimum of overprocessing.

Thermobacteriology defines an optimal level of bacterial inactivation (cumulative lethality) for every product [1]. If the cumulative lethal effect delivered by a control system is under the desired optimal value, it may seriously endanger public safety; if it is over, energy is wasted and nutrient properties of the product are overreduced because of overheating.

II. FEED-FORWARD CONTROL OF BACTERIAL INACTIVATION

A. Controlled Variable

The most important factor for bacterial inactivation during thermal processing of canned food is sterilization value or cumulative lethality, F. There are several methods for calculating thermal process lethality [1, 2], but usually cumulative lethality is calculated using the general method of Ball and Olson [3] based on the product temperature at the slowest heating and cooling spot in a container:

$$F(t_n) = \int_{t_0}^{t_n} L(t) \, dt = \int_{t_0}^{t_n} 10^{(T_{sp}(t) - T_{ref})/z} \, dt \tag{1}$$

where

$F(t_n)$ = sterilization value or lethality accumulated for the time from t_0 to t_n, min; the symbol $F_0(t_n)$ is used when the reference temperature T_{ref} is equal to 121.1°C

t_0 = time at the beginning of the bacterial inactivation, s

t_n = moment of evaluation of cumulative lethality, s

The temperature T_{sp} is at the slowest heating or cooling point in the can. The T_{sp} is usually measured using Ecklund (Cape Coral, Florida) and the Ellab (Ro-dovre, Denmark) type-T thermocouple systems [4]. Specially designed thermocouples, connectors, and cables are used for measurement of heat distribution in retorts and cans. Correction factors have been published [5] to correct errors resulting from heat conducted into conduction-heated food products by the thermocouples and fittings. No correction is normally needed for convection-heated products or for thermocouples more than 5 cm long used to measure temperature in all other products [6]. Resistance temperature sensors, e.g., thermoresistors (made of conductors) and thermistors (made of semiconductors), are also used in the canning industry [7].

In the general method it is assumed that if the desired value of F is satisfied at the slowest heating and cooling point in a can, then it is also satisfied at all other points in the can. The $T_{sp}(t)$ depends on the retort temperature $T_R(t)$, the dimensions of the can, and the thermal properties of the product. The sterilization value F is commonly used as an indirect or direct controlled variable in control systems of thermal processing.

During thermal processing, canned foods are heated to the sterilization temperature, held for a specified period, and then cooled to near-ambient temperature for storage, distribution, and consumption of the product. The inactivation of bacteria is achieved during both the heating and the cooling periods. Thus the final lethal effect of thermal processing, F^f, is the following sum:

$$F^f(t) = F^h + F^c \tag{2}$$

where F^c, F^h are lethalities accumulated during the cooling and heating phases of the process, respectively.

B. Control Idea

The aim of controlling batch thermal processes is to achieve equality between the final value of the "controlled variable" F and its desired value, i.e.

$$F^f = F^d$$

where F^d is the desired value of cumulative lethality for the product. The time of heating is commonly used as the "manipulated variable" that influences the final level of bacterial inactivation the most.

It is possible to determine the heating time (t_h) for known $T_{shp}(t)$, $T_{scp}(t)$, the cooling time, and the given values of T_{ref}, z, and F^d [e.g., from Eq. (1)]. Traditionally, the approximate value of t_h is determined for a given heating medium temperature T_R, can size, product, and the F^d value. In the Ball formula method [3], the approximate value of the t_h is determined on the basis of the difference between the retort temperature and the product temperature at the slowest heating spot. The retort is switched from the heating to the cooling phase when time of heating exceeds the t_h, i.e., $t \geq t_h$. However, controlling a thermal process using the setpoint t_h, calculated prior to the process or determined using the Ball formula method, cannot respond to the process deviations, e.g., unexpected variations in the heating medium temperature T_R. Therefore Datta et al. [8] proposed a method of correcting for process deviations in which the heating time t_h is determined on-line during processing. They proposed to check every few seconds during heating if the lethality $F^h(t)$ accumulated so far (in the time from the beginning of heating to the moment t), together with the predicted contribution from cooling, F^c, exceeds the desired final bacterial inactivation. If it does, i.e.,

$$\text{if } F^h(t) + (\text{predicted } F^c) \geq F^d, \qquad \text{then } t \geq t_h \qquad (3)$$

and then the retort should be switched from the heating to the cooling phase. The predicted cumulative lethality F^c was calculated with the assumption that retort is switched to the cooling phase at moment t. Their system was capable of achieving the desired final cumulative lethality even in the case of unexpected deviations in the heating-medium temperature. The heating turn-off decisions in their controller were based on the simulation of the cooling phase in addition to calculating can center temperature and cumulative lethality. The foregoing system was later confirmed [9, 10].

There is no way of controlling cumulative lethality F^c once cooling is under way. The contribution of the F^c in the F^f could be as much as 40% or greater, especially in large cans in a conduction-heating situation. Because the F^c cannot be controlled and is not a constant, it has to be predicted on-line during the thermal process in individual time increments before a control decision is made. There is no other way to correct for process disturbances. Therefore on-line calculations of cumulative lethality F^c have to be done. From the above it is clear that the model-based control systems are essential in thermal processing. Other reasons for intelligent computer-based controls are

1. Automatic record keeping
2. Possibility of monitoring activities in retort room (including equipment malfunctions) from the office area
3. Tighter control of retort temperature and pressure in comparison with conventional systems
4. Flexibility

A schematic diagram of a typical computer-based control system for retort-based thermal processing is presented in Fig. 1. The retort is armed with sensors and transducers to measure temperature, pressure, and water level. It is equipped with various valves and switches to control the flow of steam, water, and air into and out of the retort. The communication between the retort and the computer controller is achieved via various converters and amplifiers. The computer-based controller consists of a

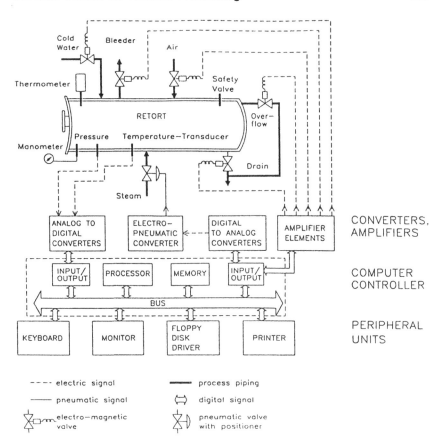

FIGURE 1 Schematic diagram of a computer-based retort control.

PC equipped with typical peripherals (keyboard, monitor, and floppy and hard disk drives). It can also be made of a microprocessor controller consisting of a processor, memory, and input/output units, and equipped only with a simple keyboard and display.

A computer-based controller should be programmed to perform various control actions. Most important are the feed-forward control of bacterial inactivation (e.g., as presented in Figs. 2 and 3), and the feedback control of retort temperature (presented also in this chapter). The accuracy of the feed-forward control depends strongly on the accuracy of prediction of the controlled variable F and requires accurate mathematical models of the product temperature changes during processing.

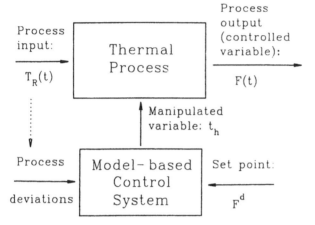

Figure 2   Block diagram of the feed-forward control of bacterial inactivation in thermal processing (for definitions of symbols, see Notation).

1.   Input data: - necessary to calculate $T_{shp}(t)$ and $T_{scp}(t)$
                     - z, T_{ref}, and F^d
                     - T^{end}

2.   Start heating: - put steam on and vent
                     - start the controller of retort temperature
                     - let the character variable CYCLE$ = "HEATING"

3.   Read and store retort temperature $T_R(t)$

4.   Calculate and store the temperature in the slowest heating and cooling spot in a can $(T_{shp}(t)$ or $T_{scp}(t))$.

5.   Calculate and store lethality $L(t)$ and cumulative lethality $F(t)$

6.   If the CYCLE$ = "HEATING", predict cumulative lethality of the cooling cycle F^c with the assumption that cooling is initiated at moment t.

7.   If the sum $F(t)$ + predicted $F^c \geq F^d$, proceed with cooling; let the variable CYCLE$ = "COOLING".

8.   If the temperature $T_{scp}(t) > T^{end}$, jump to point 3, otherwise finish cooling.

Figure 3   Algorithm of the feed-forward control of bacterial inactivation during thermal processing.

III. MATHEMATICAL MODELS OF THE PRODUCT TEMPERATURE CHANGES

A. Fourier Model

It is the heat transfer phenomenon that governs temperature profiles achieved within a food container during sterilization or pasteurization process. Fourier's equation of heat transfer is usually used to model heat penetration in conduction-heated canned foods:

$$\frac{\partial T}{\partial t} = \alpha \nabla^2 T \tag{4}$$

where

T = product temperature at any point in container at any time, °C
α = thermal diffusivity of the product, m²/s
∇^2 = Laplace operator

In rectangular coordinates:

$$\nabla^2 T = \frac{\partial^2 T}{\partial x^2} + \frac{\partial^2 T}{\partial y^2} + \frac{\partial^2 T}{\partial z^2}$$

In cylindrical coordinates:

$$\nabla^2 T = \frac{\partial^2 T}{\partial r^2} + \frac{\partial T}{r \, \partial r} + \frac{\partial^2 T}{r^2 \, \partial \theta^2} + \frac{\partial^2 T}{\partial z^2}$$

where

x, y = horizontal positions in the container, m
z = vertical position in the container, m
θ = angular position in cylinder, rad
r = radial position in cylinder, m

Many researchers [11–14] have used the numerical finite difference method, and others [15–17] have used the finite element method to solve Fourier's equation to predict the product temperature at the slowest heating and cooling points in a can [$T_{shp}(t)$ and $T_{scp}(t)$].

For the finite difference method as well as for the finite element method, prior knowledge of the thermal diffusivity of the conduction-heated foods is essential. It is possible to determine thermal diffusivity of a homogeneous food product [12, 18]. In a heterogeneous mixed food, however, variations in the proportion of ingredients of such a food product from one batch of canning to another batch can cause significant deviations in thermal properties of the food from the prior estimated values. When a prior estimated value of a thermal diffusivity of such a mixed food is used in a computer control of the retort thermal process, it may cause under- or oversterilization. Therefore, Ryniecki and Jayas [10, 19] proposed a step-response model of $T_{shp}(t)$ and $T_{scp}(t)$ temperature changes, and a method for automatic determination of the model parameters for canned heterogeneous mixed food.

B. Step-Response Model

A block diagram of the heating or cooling process of canned food subjected to sterilization is shown in Fig. 4. The transfer operator or model of the process characterizes the dynamic and physical properties of the process. If the transfer operator is known, the response of the process to a known input can be predicted. Ryniecki and Jayas [10] assumed the most general mth-order exponential lag transfer function (including a dead-time delay) for the transfer operator:

$$G(s) = \frac{Ke^{-s\tau_1}}{(1 + s\tau_2)^m} \tag{5}$$

where $G(s)$ is the ratio of the Laplace transforms of lag's response and input, assuming zero initial conditions.

When the step function is used as an input, the method of prediction of the response is called a step-analysis method [20], and a model of the response is called a step-response model. Assuming that the changes of retort temperature $T_R(t)$ at the beginning of heating and at the beginning of cooling can be simplified to the positive and negative step functions respectively, one can write

$$
\begin{aligned}
T_R(t) &= T_{Ri} && \text{for } t < 0 \\
T_R(t) &= T_{Rh} && \text{for } 0 \leq t < t_h \\
T_R(t) &= T_{Rc} && \text{for } t_h \leq t < t_c
\end{aligned}
\tag{6}
$$

where T_{Ri}, T_{Rh}, and T_{Rc} are the initial, heating, and cooling retort tempera-

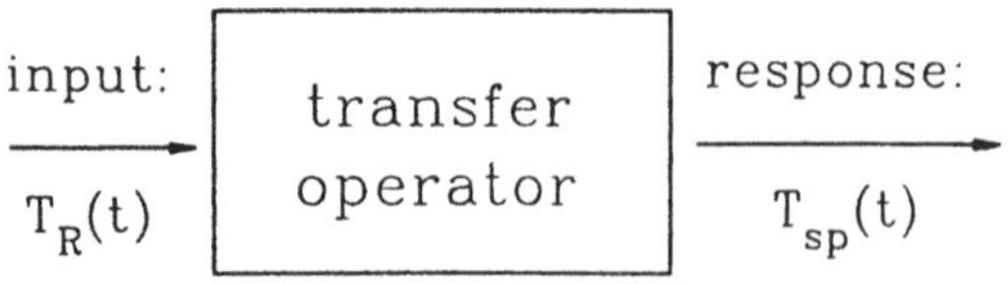

input:
$T_R(t)$ → transfer operator → response: $T_{sp}(t)$

FIGURE 4   Block diagram representation of heat transfer between the retort heating/cooling medium and canned food during sterilization or pasteurization (for definitions of symbols, see Notation).

tures, respectively. At time t_h, the heating phase of retort ends, and at time t_c, cooling ends.

The mth order transfer function without dead-time delay in Laplace domain has the form

$$G(s) = \frac{K}{(1 + s\tau_2)^m} \tag{7}$$

A general solution of such a transfer function (with a unit step function as input) in the time domain has been proposed by Skoczowski [21]:

$$T_{sp}(t) = K\left(1 - e^{-t/\tau_2} \sum_{i=0}^{m-1} \frac{t^i}{i!\tau_2^i}\right) \tag{8}$$

His solution and method of finding parameters the m and τ_2 were applied for modeling temperature changes at the slowest heating point in a can, $T_{shp}(t)$. The initial condition for the solution was $T_{shp}(t \leq 0) = T_{ih}$. Taking into account this initial condition, the positive input step function (Eq. 6), Skoczowski's solution (Eq. 8), and inverse transform of the dead-time delay, one can write the step solution for the response temperature T_{shp} in the time domain:

$$T_{shp}(t) = T_{ih} + [T_{Rh}(t) - T_{ih}]\left[1 - e^{(\tau_{1h} - t)/\tau_{2h}} \sum_{i=0}^{m_h - 1} \frac{(t - \tau_{1h})^i}{i!\tau_{2h}^i}\right] \tag{9}$$

Equation (9) is the mth-order model of the food temperature response at the center of a can $[T_{shp}(t)]$ to the step change of retort temperature $T_R(t)$ during heating.

In a similar manner, using the negative input step function for retort temperature and applying the initial condition for cooling $[T_{scp}(t \leq t_h) = T_{ic}]$, one can write the mth-order model of the temperature changes during cooling for $t \geq t_h$:

$$T_{scp}(t) = T_{Rc}(t) + [T_{ic} - T_{Rc}(t)]e^{(\tau_{1c} + t_h - t)/\tau_{2c}} \sum_{i=0}^{m_c - 1} \frac{(t - t_h - \tau_{1c})^i}{i!\tau_{2c}^i} \tag{10}$$

Roots [22] explained the relationships between parameters of such step-response models and the heat transfer coefficients of thermal processes.

C. Automatic Determination of Parameters for the Step-Response Model

Automatic determination of the dead-time delay, τ_{1h}, at the beginning of heating is relatively simple. It is the period of time after the start of heating in which the temperature T_{shp} at the center of a can does not change from its initial value (Fig. 5).

Calculation of the heating model constants m and τ_{2h} is based on automatic searching for the heat penetration curve inflection point. In typical heat penetration curves, the inflection point (Fig. 5) occurs for temperatures lower than those which occur at the beginning of the calculation of cumulative lethality F. This provides enough time to calculate model parameters.

The following algorithm for determination of model constants m_h and τ_{2h} was developed by Riniecki and Jayas [10, 19].

1. Find the inflection point of the heat penetration curve $T_{shp}^{mv}(t)$. It is the point of maximum value of $\Delta T_{shp}^{mv}/\Delta t$, assuming that Δt approaches zero (point I on Fig. 5).

   a. After each time interval Δt (about 0.15% of the process duration), measure the temperature T_{shp}^{mv} and calculate $\Delta T_{shp}^{mv}/\Delta t$.

   b. Compare the new calculated value of $\Delta T_{shp}^{mv}/\Delta t$ with the value from the previous time interval.

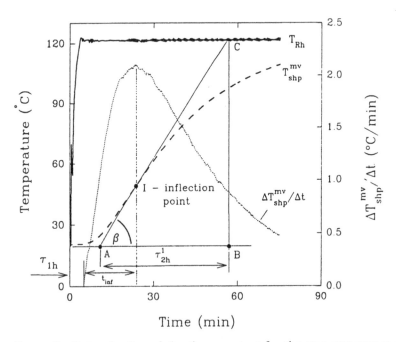

FIGURE 5 Determination of the time constant for the step-response model of temperature in the slowest heating spot in a can; $\Delta T_{shp}^{mv}/\Delta t$ is the speed of the temperature rising; for other symbols, see text and Notation.

    c. Continue comparison until the moment of inflection t_{inf} (maximum $\Delta T_{shp}^{mv}/\Delta t$) is found. The temperature $T_{shp}^{mv}(t_{inf})$ and value of $\Delta T_{shp}^{mv}/\Delta t$ at this moment, which is defined as $tg\beta$, are stored.

2. Calculate the time constant of the first-order model, τ_{2h}^1 (triangle ABC, Fig. 5):

$$\tau_{2h}^1 = \frac{T_{Rh} - T_{ih}}{tg\beta} \tag{11}$$

3. Calculate model constants m_{hr} and m_h as [21]

$$m_{hr} = 1 + 2\pi \left(\frac{t_{inf}}{\tau_{2h}^1}\right)^2 \tag{12}$$

The constant m_h is the integer approximation of m_{hr}.

4. Calculate the time constant of the mth-order model, τ_{2h}, as [21]

$$\tau_{2h} = \frac{t_{inf}}{m_{hr} - 1} \tag{13}$$

Ryniecki and Jayas [10] calculated the cooling model parameters on the basis of heating model parameters: $\tau_{1c} = r_1\tau_{1h}$, $\tau_{2c} = r_2\tau_{2h}$, $m_{cr} = r_m m_{hr}$, and m_c is the integer approximation of m_{cr}. The assumption was made that ratios r_1, r_2, and r_m depend only on the thermal properties of the cooling and heating mediums and not on the sterilized food product. These ratios were found by running several experiments (prior to the main experiments) and fitting the experimental data to the model.

In the computerized determination of parameters for the step-response model, made prior to the main sterilization, it is possible to correct values of parameters by fitting the model to the measured temperature data. Ryniecki and Jayas [10, 19] used the optimum gradient method (described by Groover [23]) with an objective function in the form of the sum of the squares of the differences between the measured and predicted values of the product temperature.

IV. PREDICTION OF CUMULATIVE LETHALITY

Heating turn-off decisions are based on the calculation of cumulative lethality. The model of temperature changes at the slowest heating point in a can (e.g., Eq. 9) together with the model of thermal inactivation of food spoilage–causing organisms (e.g., Eq. 1) can be used to calculate cumulative lethality of the heating phase, F^h, for the time from t_0 to t_h. It is possible to use the measured temperature at the geometrical center of a can to calculate F^h [24, 25], but in commercial practice this method is prohibitively expensive (taking into consideration production efficiency) and is viewed as impractical [26].

At time t_h, the retort is switched from the heating to the cooling phase. During the dead-time delay (the time from t_h to $t_h + \tau_{1c}$), temperature at the geometrical center of a can is still rising because of the temperature gradient in the can. The step-response model of changes in temperature $T_{scp}(t)$ does not include this phenomenon because Eq. (10) was developed for the uniform initial temperature T_{ic} [$T_{ic} = T_{scp}(t \le t_h)$]. So, when using a step-response model of the product temperature changes, the cumulative lethality of the cooling phase F^c should be predicted using the temperature $T_{shp}(t)$ (Eq. 9) for the time from t_h to $t_h + \tau_{1c}$ and the temperature $T_{scp}(t)$ (Eq. 10) for the rest of the cooling:

$$F^c = \int_{t_h}^{t_h + \tau_{1c}} 10^{(T_{shp}(t) - T_{ref})/z} \, dt + \int_{t_h + \tau_{1c}}^{t_c} 10^{(T_{shp}(t) - T_{ref})/z} \, dt \tag{14}$$

The trapezoidal rule for numerical integration is most useful for on-line computer evaluation of cumulative lethality. The integral of the function of lethality is replaced in this method by the sum of finite areas ΔF_i under the curve of lethality [27]. The formula has the following form:

$$F(t_n) = \int_{t_0}^{t_n} L(t) \, dt \approx \sum_{i=1}^{n} \Delta F_i = \sum_{i=1}^{n} \frac{L(t_{i-1}) + L(t_i)}{2} \, \Delta t \tag{15}$$

where

I = counter of time intervals Δt
n = number of time intervals in the time from t_0 to t_n
ΔF_i = elementary area under the curve of lethality, min
Δt = time interval $(t_i - t_{i-1})$, s

The accuracy of the trapezoidal method depends on the time interval Δt; the smaller Δt is, the higher the accuracy. Time intervals of 6–12 s for the process duration of 1200–7200 s give sufficiently accurate evaluation of cumulative lethality.

V. FEEDBACK CONTROL OF RETORT TEMPERATURE

The aim of retort temperature control is to provide a designed temperature–time profile of T_R independent of any disturbances. Typically, T_R is kept constant or, for maximum retention of nutrient properties in some applications, should vary according to a designed function $T_R(t)$ [28, 29]. A steam flow rate is usually used as a manipulated variable that influences heat input to the retort and the retort temperature. Traditionally, retort

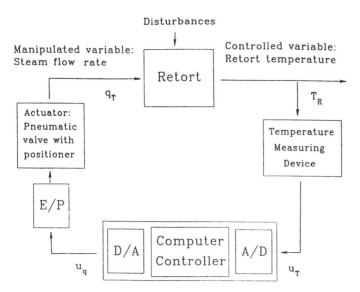

FIGURE 6 The block diagram of the feedback control of retort temperature; A/D and D/A: analog to digital and digital to analog converters, respectively; E/P: electropneumatic converter (for definitions of other symbols, see text).

temperature controllers (analog pneumatic or electric type) have used a continuous PID (proportional–integral–derivative) control algorithm. In a computer-based retort control one can replace the analog PID by a discrete or digital PID control action.

Figures 1 and 6 show a retort equipped with a temperature measuring device (based on thermocouple sensors) and a pneumatic valve with positioner through which a continuously varying heat input is supplied. The computer serves as a controller. The control algorithm for a computer-based controller with the discrete PID control action is as follows:

1. Set initial values of output signal, u_q, counter of time increments, I, and error signal, e.
2. Measure the value of the retort temperature T_R (controlled variable) and transfer the temperature signal u_T from the sensor to the computer memory via the analog to digital converter. Change the value of the counter of time increments: $I = I + 1$.
3. Compare the value of the measured retort temperature signal, $u_T(t_i)$, with the desired value of the retort temperature, $u_T^d(t_i)$ (stored in the computer memory) and form an error signal, $e(t_i)$ = $u_T^d(t_i) - u_T(t_i)$.

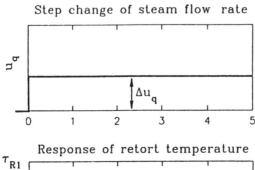

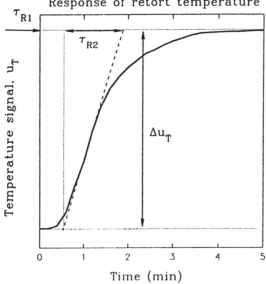

FIGURE 7 The open loop response of the retort temperature signal u_T to a step change of the steam flow rate, proportional to the computer output signal u_q; the dead-time delay τ_{R1} and the time constant τ_{R2} of the response are determined using the Zigler–Nichols method [30].

4. Calculate the value of an output signal of the computer controller, $u_q(t_i)$, as the following discrete PID function (in the incremental form) of current error [30]:

$$u_q(t_i) = u_q(t_{i-2}) + c \left\{ e(t_i) - e(t_{i-1}) + \frac{\Delta t}{\tau_1} e(t_i) \right.$$

$$\left. + \frac{\tau_D}{\Delta t} [e(t_i) - 2e(t_{i-1}) + e(t_{i-2})] \right\}$$

(16)

where

$u_q(t_i)$ = current value of output signal (the steam flow rate should be dependent on that signal)

c, τ_I, τ_D = parameters of the PID controller (they should be fitted to the dynamics of retort temperature)

5. Adjust the level of the output signal and transfer through the interface and converters to the steam valve. The steam flow rate is held constant over each interval Δt.

6. Wait to the end of the period Δt and go back to step 2.

Parameters c, τ_I, and τ_D of the PID controller can be determined using the Ziegler–Nichols method [30]:

$$c = \frac{1.2\tau_{R2}}{K_R \tau_{R1}}, \qquad \tau_I = 2\tau_{R1}, \qquad \tau_D = 0.5\tau_{R1} \tag{17}$$

where

K_R = gain of retort heating; $K_R = \Delta u_T / \Delta u_q$ (Fig. 7)

τ_{R1}, τ_{R2} = parameters of the response of retort temperature to the step change of computer output signal u_q (as presented in Fig. 7)

VI. ASEPTIC PROCESSING

Aseptic processing involves sterilization of the food product and package separately and filling of the sterilized product in a sterile environment. The technique is used in the food industry for liquid foods such as milk and juices, but it is not used for liquid foods with particulates. Researchers are developing this technique further for such applications [31–33]. The control of the process still requires that the food product be properly heated to sterilization temperature and held at that temperature for a specified period without under- or overprocessing. The equipment used, however, is different from that in traditional in-container sterilization systems. Aseptic processing technology lends itself to continuous processing of food products.

NOTATION

F	Accumulated amount of sterilization (cumulative lethality) (min)
F_0	Lethal effect equivalent to number of minutes at temperature 121.1°C $\equiv 250°F$ (min)
F^c	F value achieved during cooling cycle (min)
F^d	Desired level of bacterial sterilization (min)
F^f	Final lethal effect of thermal processing (min)

F^h	F value achieved during heating cycle (min)
$G(s)$	The ratio of Laplace transforms of the response of the thermal process and input, assuming zero initial conditions
K	Gain of a thermal process (for heating it is $T_{rh} - T_{ih}$, and for cooling $T_{ic} - T_{rc}$) (°C)
L	Lethality
m	The lag order of a thermal process
m_c	The lag order of the cooling process (integer value)
m_{cr}	The lag order of the cooling process (real value)
m_h	The lag order of the heating process (integer value)
m_{hr}	The lag order of the heating process (real value)
r_1	Ratio of the dead-time delay in the lag model (cooling/heating), τ_{1c}/τ_{1h}
r_2	Ratio of time constant in the lag model (cooling/heating), τ_{2c}/τ_{2h}
r_m	Ratio of the lag order in the lag model (cooling/heating), m_{cr}/m_{hr}
s	Complex variable in a Laplace transform
t	Time (s)
t_c	Time at the end of cooling (s)
t_h	Time at the end of the heating cycle of retort (s)
t_{inf}	Time at which heat penetration curve $T_{shp}(t)$ has inflection (s)
t_n	Moment of evaluation of cumulative lethality (s)
T_{ic}	Initial food product temperature for cooling (°C)
T_{ih}	Initial food product temperature for heating (°C)
T_{ref}	Reference temperature (typical value used for sterilization is 121.1°C $\equiv$ 250°F, and for pasteurization 66–70°C $\equiv$ 150–158°F) (°C)
T_R	Retort temperature (°C)
T_{Rc}	Retort temperature during cooling (°C)
T_{Rh}	Retort temperature during heating (°C)
T_{Ri}	Initial retort temperature (°C)
T_{scp}	Temperature at the slowest cooling point in a can—predicted value (°C)
T_{scp}^{mv}	Temperature of the slowest cooling point in a can—measured value (°C)
T_{shp}	Temperature at the slowest heating point in a can—predicted value (°C)
T_{shp}^{mv}	Temperature of the slowest heating point in a can—measured value (°C)
T_{sp}	Temperature of the slowest heating and cooling point in a can—predicted value (°C)
T_{sp}^{mv}	Temperature of the slowest heating and cooling point in a can—measured value (°C)
T^{end}	Desired temperature of the slowest cooling point in a can—a constant value used in control algorithm to finish the cooling cycle (°C)
z	Constant rate of bacterial inactivation for a given type of bacteria and T_{ref}; it is the temperature difference required for the thermal death-time curve to traverse one logarithmic cycle (°C)
α	Thermal diffusivity of the product (m²/s)
τ_1	Dead-time delay of the mth-order lag model (s)

τ_{1c}	Dead-time delay of the mth-order lag model of cooling (s)
τ_{1h}	Dead-time delay of the mth-order lag model of heating (s)
τ_2	Time constant of the mth-order lag model (s)
τ_2^1	Time constant of the first-order lag model (s)
τ_{2c}	Time constant of the mth-order lag model of cooling (s)
τ_{2h}	Time constant of the mth-order lag model of heating (s)

REFERENCES

1. C. R. Stumbo, *Thermobacteriology in Food Processing*, Academic Press, New York, 1973.
2. T. Smith and M. A. Tung, Comparison of formula methods for calculating thermal process lethality, *J. Food Sci. 47*(2):626 (1982).
3. C. O. Ball and F. C. W. Olson, *Sterilization in Food Technology: Theory, Practice and Calculations*, McGraw-Hill, New York, 1957.
4. G. R. Bee and D. K. Park, Heat penetration measurement for thermal process design, *Food Technol. 32*(6):56 (1978).
5. O. F. Ecklund, Correction factors for heat penetration thermocouples, *Food Technol. 10*(1):43 (1956).
6. *Ecklund Bulletin. Custom Thermocouples*, P.O. Box 279, Cape Coral, Florida 33910-0279, 1989.
7. B. P. Lappo and M. J. W. Povey, A microprocessor control system for thermal sterilization operations, *J. Food Engg. 5*:31 (1986).
8. A. K. Datta, A. A. Teixeira, and J. E. Manson, Computer-based retort control logic for on-line correction of process deviations, *J. Food Sci. 51*:480 (1986).
9. T. A. Gill, J. W. Thompson, G. LeBlanc, and R. Lawrence, Computerized control strategies for a steam retort, *J. Food Engg. 10*:135 (1989).
10. A. Ryniecki and D. S. Jayas, Automatic determination of model parameters for computer control of canned food sterilization, *J. Food Engg. 19*:75 (1993).
11. A. A. Teixeira, J. R. Dixon, J. W. Zahradnik, and G. E. Zinsmeister, Computer determination of spore survival distributions in thermally processed conduction-heated foods, *Food Technol. 23*(3):78 (1969).
12. A. A. Teixeira, J. R. Dixon, J. W. Zahradnik, and G. E. Zinsmeister, Computer optimization of nutrient retention in the thermal processing of conduction-heated foods, *Food Technol. 23*(6):137 (1969).
13. J. E. Manson, C. R. Stumbo, and J. W. Zahradnik, Evaluation of thermal processes for conduction-heating foods in pear-shaped containers, *J. Food Sci. 39*:276 (1974).
14. A. A. Teixeira and J. E. Mason, Computer control of batch retort operations with on-line correction of process deviations, *Food Technol. 36*(4):85 (1982).
15. D. Naveh, I. J. Kopelman, and I. J. Pflug, The finite element method in thermal processing of foods, *J. Food Sci. 48*:1086 (1983).
16. D. Naveh, I. J. Kopelman, L. Zechman, and I. J. Pflug, Transient cooling of conductive heating products during sterilization: Temperature histories. *J. Food Proc. Preservation 7*:259 (1983).

17. D. Naveh, I. J. Pflug, and I. J. Kopelman, Transient cooling of conductive heating products during sterilization: sterilization values, *J. Food Proc. Preservation 7*:275 (1983).

18. M. A. Tung, G. F. Morello, and H. S. Ramaswamy, Food properties, heat transfer conditions and sterilization considerations in retort process, *Food Properties and Computer-Aided Engineering of Food Processing Systems* (R. P. Singh and A. G. Medina, eds.), Kluwer Academic Publishers, Norwell, Massachusetts, 1989, p. 49.

19. A. Ryniecki and D. S. Jayas, A computer program for control of canned food sterilization, ASAE Paper No. 93-3063, St. Joseph, Michigan, 1993.

20. N. Anderson, Step analysis method of finding time constant, *Instruments and Control Systems 36*(11):130 (1963).

21. S. Skoczowski, Some remarks on the approximation of plants with self-regulation [in German], *Regelungdtechnik 31*:231 (1983).

22. W. K. Roots, *Fundamentals of Temperature Control*, Academic Press, New York, 1969.

23. M. P. Groover, *Automation, Production Systems and Computer-Aided Manufacturing*, Prentice-Hall, Englewood Cliffs, New Jersey, 1987.

24. S. Navankasattusas and D. B. Lund, Monitoring and controlling thermal processes by on-line measurement of accomplished lethality, *Food Technol. 32*(3):79 (1978).

25. S. J. Mulvaney and S. H. Rizvi, A microcomputer controller for retorts, *Trans. ASAE 27*(6):1964 (1984).

26. A. A. Teixeira and C. F. Shoemaker, *Computerized Food Processing Operations*, Van Nostrand Reinhold, New York, 1989.

27. J. Wojciechowski and A. Ryniecki, Computer control of sterilization of canned meat products [in German], *Fleischwirtschaft 69*(2):268 (1989).

28. A. A. Teixeira, G. E. Zinsmeister, and J. W. Zahradnik, Computer simulation of variable retort control and container geometry as a possible means of improving thiamine retention in thermally processed foods, *J. Food Sci. 40*:656 (1975).

29. I. Saguy and M. Karel, Optimal retort temperature in optimizing thiamine retention in conduction-type heating of canned foods, *J. Food Sci. 44*(5):1485 (1979).

30. J. R. Leigh, *Applied Digital Control: Theory, Design and Implementation*, Prentice-Hall International, Englewood Cliffs, New Jersey, 1985.

31. V. M. Balasubramaniam and S. K. Sastry, Liquid-to-particle heat transfer in continuous tube flow: Comparison between experimental techniques, *Int. J. Food Sci. Technol.* (in press) (1995).

32. H. S. Ramaswamy, K. A. Abdelrahim, B. K. Simpson, and J. P. Smith, Residence time distribution (RTD) in aseptic processing of particulate foods: A review, *Food Res. Int. 28*(3):291 (1995).

33. G. Tewari and D. S. Jayas, Heat transfer during thermal processing of liquid foods with or without particulates: A review. *Agric. Eng. J.* (submitted for publication) (1995).

10

Automatic Control of Drying Processes

Francis Courtois

High School of Food Science and Technology, National
Institute for Agronomical Research, Massy, France

I. INTRODUCTION

Drying can be seen as an operation where the objective is to reach some final food quality using for this purpose some optimally spent thermal energy. From this point of view, three secondary objectives appear.

First, dryers are some kind of reactors; thus, multiple quality criteria are generally considered. At least we need to decrease the product moisture content under a certain limit to ensure a specified preservation ability. But secondary qualities are often contradictory to this (e.g., paper texture, color and vitamins of food). The quality objective is clearly a complex combination difficult for the production head to pursue.

Second, the cost of this operation depends largely on the way energy is supplied (in space and time) to the product. For quality purposes, the maximum energy is generally given at the beginning of the drying, when the product is at its highest moisture content (e.g., cereal dryers).

Third, although we would like to keep the output product moisture content constant, we must admit that input moisture content is generally highly variable. Thus the need for automatic control is clear.

Induced benefits of an automatic control apparatus can vary from US dollars 67,000 to 272,000 (for rice and soybeans, respectively) per

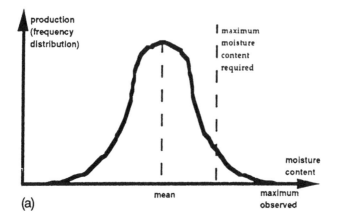

(a)

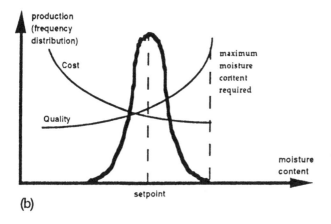

(b)

FIGURE 1   (a) Manual control implies the use of overdrying to meet the specifications. (b) Automatic control reduces the standard deviation and thus the overall cost of drying while ensuring better quality.

dryer and per year [1]. An extra benefit can also come from the increased quality of the product: it is less overdried and more homogeneous. A value of US $37 per bushel of corn is an extra cost due to an overdrying of 1% [2].

We can give three more reasons to use some form of automatic control on dryers:

First, as can be seen in Fig. 1, the cost of drying increases exponentially with decreasing moisture content. On the other hand, target moisture content must be considered more as an upper limit than a real setpoint. Thus, depending on the standard deviation around the mean, the operator generally needs to overdry to ensure that all the product is below the desired final moisture content, which leads to extra cost.

Second, the relationship between operating variables and final moisture content of the product is not trivial. In fact it is highly nonlinear. To explain this, it is generally admitted that, whatever the product, the drying kinetic can be modeled as

$$X(t) = X(t = 0)e^{-kt} \qquad (1)$$

where t is time (s), X the product moisture content (d.b.), and k is a complicated (nonlinear) function of the operating variables. Thus reducing the influence of the disturbances is not trivial work for the operator.

Third, depending on the combination of product and process, the response time can be very short (e.g., seconds in paper drying, spray drying) or very long (e.g., hours in cereal drying). In these cases, the operator's work becomes very difficult, involving continuous activity or memory.

For all of these reasons, people working on dryers are always interested in implementing some type of automatic control apparatus. But the cost of sensors and the sophistication of nonlinear control algorithms have limited the number of applications.

One should remark that only three communications dealt with process control in the recent Drying '94 symposium. One of the reasons may be that more than 90% of control theory is limited to the linear systems.

II. METHODOLOGY: A SURVEY OF THE BIBLIOGRAPHY

As previously mentioned, the number of papers concerning "automatic control of dryers" is very low. Moreover, only a few of them discuss methods that can be easily generalized. Some of these methods have been tested only by simulation or on small-scale pilot plants. Moreover, the drying range is often very narrow and disturbances are not as drastic as in other industries.

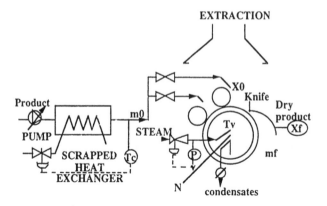

FIGURE 2 Schematic representation of a drum dryer. (From [3].)

A. Steady State Analysis

Before attempting to implement a control loop on a process, it is necessary
to study its steady state behavior. Among the things we need to decide
are:

> The range of operating conditions
> The choice, location, and characteristics of actuators
> The choice, location, and technology of the sensors

Using experimental design techniques, it is possible to obtain normographs showing interrelations between operating variables and quality criteria. Trystram and Vasseur [3] followed this methodology to find optimal
settings of a drum dryer (Fig. 2) depending on the control objectives.
Figure 3 shows how the rotation speed of the main cylinder influences

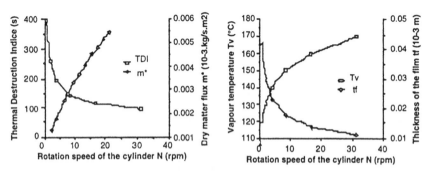

FIGURE 3 Graphical representation of the interrelations between the main drying
parameters of a drum dryer. (From [3].)

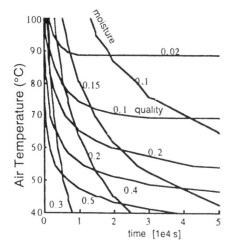

FIGURE 4 Graph showing the relations between air temperature, quality, and moisture content during thin-layer batch drying of corn. (From [4].)

the thermal destruction index, the thickness of the film, and the like. Using these graphs one can directly get the optimal setpoint. As an example, the thickness of the film has a strong influence on the rehydration capability of the product. To reduce this parameter, the drum speed should be increased, but at the same time the thermal destruction index would also be decreased.

This work would be considerably reduced if a model is available. A few days of simulation can be sufficient to get the required information.

Figure 4 shows how a "first-principle" model can be useful to construct normographs for choosing the correct air temperature setpoint to ensure desired final moisture content with product quality preserved.

B. Dynamics Analysis

Studying the dynamics of the process is moving one step forward. Knowing the settings that the controller should maintain is not enough. We need also some knowledge of the transient state behavior of the process, including:

What are the characteristics of disturbances?
What are the characteristics of manipulated variables?
Are the signals noisy?
Are there any couplings between controlled variables?
Where are the nonlinearities?

TABLE 1 Simulated Step Responses for Final Moisture Content (FMC) and Final Quality (FQ) for Variations of Initial Moisture Content (IMC), Mass Flow Rate (MF), Air Temperature (AT), and Air Flow (AF) in a Mixed-Flow Corn Dryer

Manipulation step		Controlled variable					
		FMC			FQ		
Manipulated variable	Change	Time delay (s)	Gain	Rise time (s)	Time delay (s)	Gain	Rise time (s)
AF stage #1	−10%	18,054	−0.59	7,876	18,142	0.223	9,735
	10%	18,408	−0.55	7,257	18,054	0.18	7,257
	20%	18,408	−0.54	7,788	18,054	0.17	8,761
AF stage #2	−10%	9,027	−0.37	7,080	9,027	0.32	7,788
	10%	9,381	−0.35	6,719	9,027	0.245	8,142
	20%	9,321	−0.34	7,051	9,115	0.21	7,434
MF	−10%	10,027	1.09	17,523	10,708	−0.36	17,036
	10%	8,841	1.14	13,939	8,602	−0.55	14,098
	20%	7,575	1.16	12,815	7,788	−0.7	12,744
AT stage #1	−10%	18,585	−0.9	6,980	18,585	0.304	7,080
	10%	18,585	−0.93	6,106	18,585	0.227	6,549
	20%	18,585	−0.94	6,372	18,496	0.195	6,018
AT stage #2	−10%	10,354	−0.83	5,753	9,646	0.74	6,549
	10%	10,354	−0.88	5,841	9,735	0.467	6,460
	20%	10,354	−0.91	6,018	9,646	0.37	6,726
IMC	−10%	24,603	1.5	1,947	24,160	−0.28	2,390
	10%	24,603	1.54	1,947	24,603	−0.37	2,047
	20%	24,603	1.54	1,416	24,603	−0.42	1,947

Gains are normalized to allow comparisons.

As an example, Table 1 shows the characteristics (delay, gain, and rise time) of the step responses of a corn dryer (Fig. 5) in final moisture content (FMC) and final quality (FQ). Different disturbances were tested: initial moisture content (IMC), mass flow rate (MF), air temperature (AT), and air flow (AF). Results, obtained through simulations [5], are given for one setpoint only. The tested mixed-flow dryer is a two-stage one. This study has shown that delay and rise time of the FMC response were highly variable depending on the height of the disturbance. The opposite remark could be made for the FQ: the gain is highly dependent on the disturbance. This all means that the system is strongly nonlinear.

Table 1 shows also the difficulties that may arise when trying to decouple moisture and quality in two separate control loops.

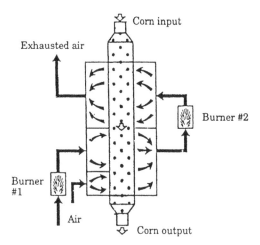

Corn input

Exhausted air

Burner #2

Burner #1

Air

Corn output

FIGURE 5 Simplified representation of a mixed-flow corn dryer.

C. Choice of the Controlled Variable(s)

The controlled variable is the sensor measure that the controller maintains as close as possible to the setpoint.

In most cases, the problem is reduced to a SISO (single-input, single-output) case. Mainly, this is because classical PID (proportional–integral–derivative) controllers can manage only these kinds of systems. Measuring the output moisture content of the product may appear necessary, but its cost may be significant depending on the technology used (resistive, capacitive, or infrared measurements, etc.). This can explain why many older applications use exhaust air (or steam) temperature measurement instead.

Output air temperature is interesting because it is generally measured before the product exists from the dryer; this allows faster feedback in the control loop. In spite of this, the relation between air temperature and product moisture content is not trivial, especially in the case of a countercurrent exchange. Therefore it should be used more as an indication than a direct sensor. In any case, it is an open loop strategy.

Sometimes the temperature of the air/product mixture is considered (e.g., spray drying). The preceding remarks apply.

When a product boils during drying, moisture content and temperature are physically related by the boiling curve coming from the desorption isobar measures. Rodriguez et al. [6] used an infrared remote temperature

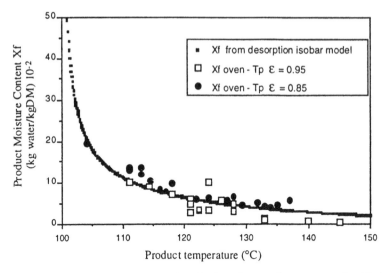

FIGURE 6 Desorption isobar of maltodextrin powder compared with industrial samples. (From [6].)

sensor combined with the desorption isobar model as a smart sensor for moisture content measurement (Fig. 6).

In the case of steam drying, the steam temperature seems to be a better indication of the product moisture content because this measure, knowing the pressure, indicates clearly the amount of transferred energy (remember the ideal gas law).

In a few cases, another quality criterion besides product moisture is measured or estimated. There are two main reasons: solids concentrations, flavors, and textures are generally difficult (or impossible) to access on-line, and the cost is often too high. There are several exceptions: weight of paper [7], wet-milling quality of corn [4, 8].

D. Choice of the Manipulated Variable(s)

The manipulated variable is the actuator that the controller uses to reach its objectives. Two cases are generally encountered: thermal flux (e.g., air or steam temperature, gas flow rate) or product flow rate (e.g., screw conveyor speed, discharge rate). The first controls the amount of energy received per amount of time (and the level of energy used). The second controls the sharing of the energy per amount of product (and the time of exposure). Whatever the case, the quality is equally influenced.

The choice of the product flow rate as the manipulated variable is generally cheaper (and simpler), but it should be avoided in case of a continuous flow production because it modifies locally the production capacity.

Sometimes, special objectives lead to the addition of a new actuator. Rodriguez et al. [9] have studied the use of an inductor to homogenize local moisture content gradients. It was not possible to reduce moisture gradients along the width of the drum cylinder using conventional actuators (Fig. 7).

E. Classical Control Approach

Classical here means "use of PID controllers." Generally, the final product moisture content is measured, compared with its setpoint, and then a new command is applied to the system. The general equation of a PID is very simple. The command is a linear combination of (see Chapter 4 for more details)

A proportional term: command is proportional to the current error.
An integral term: command still increases as long as error is not zero.
A derivative term: command is proportional to the change in error.

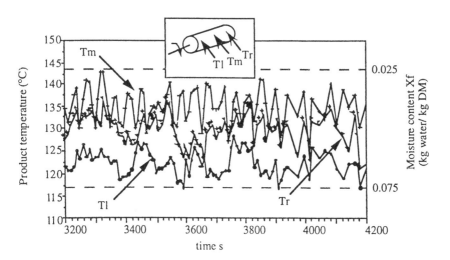

FIGURE 7 Temperature (i.e., moisture) gradients along the width of a drum dryer. (From [9].)

Many techniques exist to help the operator adjust PID parameters (e.g., Ziegler–Nichols). But the main difficulty comes from the nonlinearities of the system. A PID method may work under conditions of small nonlinearities. This means low disturbances (stable inputs) and rare changes of setpoint. Therefore it is generally used for inorganic or pre-transformed materials because their initial moisture contents are generally more stable than those of agricultural products, for example. The second possibility is to slow down the PID controller to increase its robustness [10]. Overall performance is thus decreased, but the global amount of work needed to implement an automatic control is considerably reduced. The third possibility is to adapt the PID parameters to the dryer setpoint either with a known law or with adaptive technics.

A MIMO (multiple-input, multiple-output) problem can generally be solved with a combination of several PID controllers. Rodriguez et al. [10] have used two PID to control the moisture content of the product and its gradient along the width of the drum (Fig. 8). Kiranoudis et al. [11] have used also two PID loops for a conveyor-belt dryer for grapes, but their results were obtained only by simulation.

F. Advanced Control

There are several definitions of an "advanced control strategy." One of them could be "any controller needing a computer to be implemented." It can also be seen as any controller other than a PID.

More sensible, one could speak about strategies implying the maximum knowledge of the process. In most cases modeling is at the center of the method. "Model" is a generic term emcompassing a variety of concepts and methods. Readers should refer to the chapters to make a survey of the different techniques of modeling. In any case, only dynamic models can be used for control purposes. These models often imply ordi-

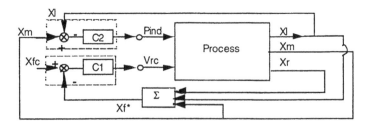

FIGURE 8 Use of two controllers (C1 and C2) to control the product moisture content (X) in three locations (left, middle, and right) using the rotation speed (Vrc) and an inductor power (Pind) as the manipulated variables. (From [10].)

nary or partial differential equations possibly coupled with algebraic ones. When linear systems are considered, equations can be represented as transfer functions (in continuous or discrete form) or, in more general manner, as a state representation.

The most important thing is that first-principle models can be used as a validation test of a control algorithm or in the heart of a model-based predictive control strategy. It is obvious that the latter implies intensive calculations and, thus, could not be developed until computer technology allowed it.

Several authors [5, 11–15] use a complex first-principle dynamic model as a simulator to validate their control algorithm to minimize costly experiments. In most cases, their models have been previously designed for steady state simulation for computer-aided design (CAD) purposes [16]. Trelea et al. [17] used a first-principle model to train a recurrent neural network as a moisture content and quality predictor to speed up the real-time calculations. Many [5, 13–15, 18] have used Eq. (1) implicitly to design a nonlinear adaptation of the classical PID. They noted that Eq. (1) could be easily linearized as

$$\log[X(t)] = \log[X(t = 0)] - kt \tag{2}$$

Thus, using the logarithm of the moisture content and the residence time respectively as the controlled and manipulated variables, they could decrease the gain variations (Table 1). Using an indexed sampling period led to shorter variations in the delay and rise time of the system to control [8]. Some authors combined feedback and feed-forward control as in Fig. 9 [5, 13]. Feed-forward control allows faster rejection of initial moisture

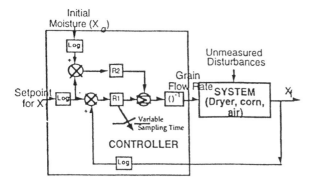

FIGURE 9 Nonlinear PID control of a corn dryer with combined feedback and feed-forward. (From [5, 13].)

disturbances, but global performance is smoothly increased due to the single actuator limitations. Because this-technique needs a second-moisture sensor, the choice can be debated.

Some [12, 19] have used classical linear or ARMAX (autoregressive moving average with external input) models to design an R-S-T controller using pole-placement techniques. To handle the strong nonlinearities, adaptive techniques were added to change one parameter of the model. Adaptive methods are only a partial solution of the problem because they treat the system as if it was unsteady but linear. Only nonlinear techniques can manage large and sudden disturbances or setpoint variations. Courtois et al. [5] has shown on a semi-industrial mixed-flow corn dryer that automatic control could start up a dryer full of wet product (see Section III.A.2).

An interesting approach was used [12] to treat the problem of crossflow drying of corn, known to be a distributed parameter system. The idea is to consider different layers of product inside the dryer, having different residence times. This method is interesting because this dryer considered has an important internal load and, thus, cannot be considered globally.

Advances in this area were made [13] using a shrewdly preprogrammed adaptive technique for a nonlinear controller. A first-principle model was used to simulate the change in the dynamics of the dryer at different drying temperature, and the parameter of the controller was recalculated in each case. A general relation between these parameters and the temperature setpoint of the dryer was found. This illustrates how useful a dynamic model of the process can be.

Two control strategies were tested [20] for a spray drying unit for milk: multiple PID controllers and internal model-based predictive control. The problem was multivariable, and thus the latter approach, while being less usual, led to improved performance. Trelea et al. [4] treated the complex problem of the batch drying of corn combining uncertainties on measures, quality constraints, and fixed final moisture content.

Some authors have used unconventional approaches, often called artificial intelligence. Zhang et al. [21] used fuzzy techniques to help control the final breakage susceptibility of corn. Najim [22] used a self-learning algorithm based on probability calculations for a phosphate dryer. The generalization of these approaches to industrial applications remains to be shown. Interested readers can refer to Chapters 5 and 6.

Many authors agree that model-based techniques perform better. While nonlinear system theory remains more a research area than a well-known toolbox, it seems that good industrial results are occurring. These come from a conjunction of two phenomena: increase of computer performance and validation of accurate first-principle models.

III. COMPARING DIFFERENT CONTROL STRATEGIES

We now consider the example of corn drying presented in Fig. 5. A dynamic model of the process is available and has been validated industrially for steady state simulation [16] and for unsteady state simulations [23, 24]. To illustrate the wide choice of control strategies, we present a few methods tested either by simulations or by experiments.

A. Methods Without the Need for a Model

These methods are essentially based on the PID controller. Although they do not explicitly require a model to work, it is useful for adjusting the control parameters.

1. Classical PID Feedback

While being overtaken now by more recent R-S-T controllers, PID controllers are still widely used in industry. Its popularity comes from its facilities for manual adjustment and its legendary robustness. The structure of the feedback loop is very simple (Fig. 10).

This may explain why it is common to compare current algorithms to old-fashioned PID. Figure 11 presents results from a classical PID feedback loop when a small increase of initial moisture content arises. Stability is low, and speed had to be reduced drastically (initial rise time was less than 20,000 s). Whatever the settings of the controller, its stability margins are so low that it cannot handle the strong nonlinearities.

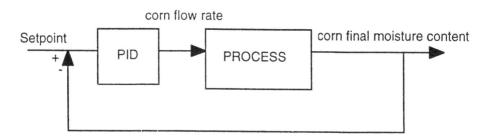

FIGURE 10   PID basic feedback loop.

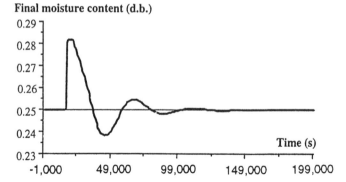

FIGURE 11 Simulated final moisture content response (PID controller in a feedback loop) to a step variation of initial moisture content (+10%).

2. Nonlinear PID Feedback

As discussed previously, the nonlinear system could be easily (pseudo-)linearized with log and inverse transformations and a preprogrammed adaptive sampling frequency [8] (Fig. 12). Then tuning the PID becomes easy. Performances are interesting (speed and stability), and the robustness is sufficient to start up the dryer, full of wet grain, with only the help of the controller.

Figure 13 shows that the automatic control behaved as the simulation predicted. While the inlet moisture content varied widely, the outlet moisture content converged quickly to the desired setpoint, through a wide drying range. Such performance is impossible to obtain with classical linear controllers.

3. Nonlinear PID Feedback plus Feed-Forward

The principle of the feedback loop is to correct the effects of disturbances *as soon* as they affect the controlled variable. From another point of view one can ask, why we should wait for the effects to correct the cause?

Knowing that the main disturbance is variation of the inlet moisture content, it is obvious that measurement of the inlet moisture would allow better control performance. Any change modifies (feed-forward action) the command issued from the feedback loop (Fig. 9).

Figure 14 shows the moisture response during a negative step variation (at time 0) in the inlet moisture content of the product. The simulated curve increases (feed-forward action) until the disturbance effects appear at the outlet. This kind of response has the advantage of a better balance

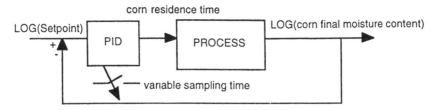

corn residence time

FIGURE 12 Nonlinear PID with variable sampling time.

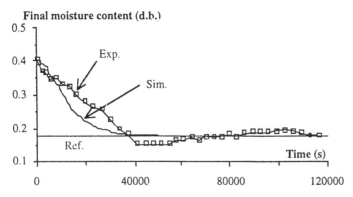

FIGURE 13 Experimental and simulated final moisture content during dryer start-up procedure (nonlinear PI controller in a feedback loop).

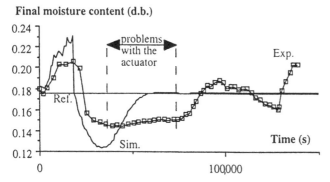

FIGURE 14 Experimental and simulated corn moisture content responses to a step variation (−21%) of initial moisture content (nonlinear PI controller in a feedback + feed-forward structure).

between under- and overdrying. With a single feedback, we would have a strong overdrying phenomenon.

We note that in Fig. 14 the experimental curve has a constant section due to problems with the actuator.

The tuning of the feed-forward effect is not simple, and the additional cost not negligible depending on the sensor technology. Gains in performance here are poor because the command acts on the entire dryer content. Thus controlling the inlet can conflict with the outlet control.

B. Methods Needing a Model

We recognize that there are several kinds of models. For this reason, we will overview three methodologies using different models (linear state, first-principle, and neural network models).

1. Nonlinear Linear Quadratic Gaussian Feedback

Here, we consider the linearization technique previously shown combined with a linear quadratic Gaussian (LQG) control [25]. The resulting controller is thus globally nonlinear.

The LQG technique has some advantages:

It treats MIMO (multivariable) as well as SISO systems.
It is known to be a robust technique.
If some states are unknown, it can estimate them (loop transfer recovery [LTR] techniques).

But LQG needs a precise dynamic model within the bandwidth of the system, and some tuning aspects are not intuitive (e.g., addition of an integral term, settings of ponderation matrix).

In the present case, the SISO problem was treated with a state model derived from the identification of a continuous transfer function. The same

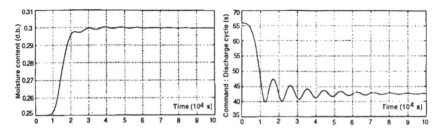

FIGURE 15 Simulated corn moisture content response to a step variation in the setpoint. Command is the discharge rate at the bottom of the mixed-flow dryer (nonlinear LQG with cancellation of zeros).

linearization technique was used as seen above. A Kalman filter was used to estimate the unknown states generated by the conversion from the Laplace domain to the state space domain (LTR algorithm).

The results shown in Fig. 15 confirm that LQG (with the linearization) gives better performance, particularly in the rise time. This comes from the high-order transfer function, which was identified: zeros are precisely modeled, permitting their cancellation by the controller.

In actual experiments, performance would not have been so good because the high-frequency dynamics are more difficult to identify than in simulation. In experiments, we have shown that moisture trajectory is less stable than predicted in Fig. 15.

This remark can be generalized for the drying of foods because this type of biomaterial is highly variable. It is evident that obtaining a very precise model, even in the high frequencies, is impossible.

2. Nonlinear Model-Based Predictive Control

Using a simpler transfer function (no zeros) and the linearization technique, the model was implemented as a predictor in a model-based predictive control [26]. A nonlinear optimization method (e.g., simplex) is implemented in the controller and evaluates the optimality (cost function) of a future command trajectory (Fig. 16).

The difference between prediction and reality is assumed to be a measure of the model bias and distortions. A low-pass filter is used to predict future values.

The command is assumed to be a linear combination of a step, a ramp, and a parabola. The desired trajectory for converging to the setpoint trajectory is chosen to be an exponential function. The cost function is a classical balance of squared errors and squared command variations.

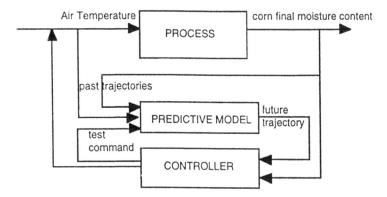

FIGURE 16   Model-based predictive control scheme.

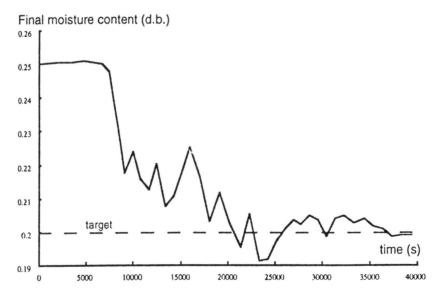

Final moisture content (d.b.)

Figure 17 Simulated corn moisture content response to a step variation in the setpoint (nonlinear model-based predictive optimal control).

The optimization problem is then to obtain the optimal combination of step, ramp, and parabola commands to minimize the cost function, knowing the past and simulating the future.

Because of the lack of linearity assumptions, more calculations are needed, particularly at the beginning. From a different point of view, the tuning of the algorithm is very intuitive. The essential step is to select the rise time to get back to the setpoint.

Figure 17 shows results obtained by simulation using the first-principle model (reference model). Whatever the settings, the controller cannot speed up the response because of the unmodeled zeros, which strongly affect the dynamics. Performances are comparable with those of the nonlinear PID and LQG (when not canceling the zeros).

This approach is interesting because it can use any kind of model, including first-principle ones. This may be one of the best ways to incorporate knowledge into the control strategy.

3. Nonlinear Model-Based Multivariable Predictive Control

A study in progress [4] concerns the optimization of batch drying. The study focuses on the fixed-bed drying of a thin layer of corn. The wet-milling quality [27, 28] of the corn is considered as well as its moisture content (Fig. 18).

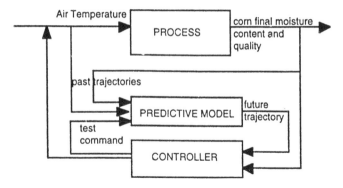

FIGURE 18 Bivariable nonlinear model-based predictive control scheme.

Two recurrent neural networks has been identified to speed up the on-line prediction of moisture content and wet-milling quality of corn [17]. These models are coupled with optimization techniques to find out the best air temperature for achieving the correct final moisture content at the desired drying time, and for ensuring that the quality stays within required boundaries. Uncertainties in the state variables are taken into account.

Figure 19 shows the results of an experiment conducted to observe the efficiency of the controller during a simulated disturbance (heating

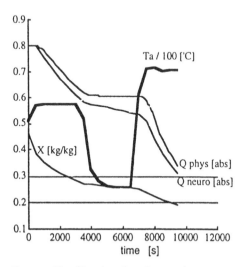

FIGURE 19 Simulated and experimental corn moisture content and wet-milling quality during a thin-layer batch drying with a temporary failure of the air heating device. Predicted (neural network) quality is compared with reference model simulation.

resistors were decoupled for one hour). The algorithm has succeeded: final moisture content is under the desired value, and the quality is maintained above its limit.

The interest in the method is its ability to be generalized. The lack of assumptions concerning the model, the general formulation, and very good performance are important advantages to consider.

IV. CONCLUSIONS

Because of its wide variety of products, drying technologies, production objectives, etc., the drying process presents some specific problems for automatic control. Nonlinearity is probably the main problem. We have shown that several interesting strategies are already used or in validation study. A linearization technique has proven, in several applications, to be efficient and simple. It can be used in most cases to increase the robustness of the controllers.

Model-based predictive control appears to be one of the most promising techniques to treat nonlinear and multivariable problems as are found in drying. The model type is no longer a limitation for this method, so that it is interesting to combine it with a first-principle model.

The next areas to study to increase controller performance are (1) to consider the measurement uncertainties, and (2) letting the problem stay an MIMO problem (do not simplify to an SISO one).

REFERENCES

1. P. L. Douglas and G. R. Sullivan, Automatic moisture control of process dryers in the agriculture and food industries, *Drying '94* (V. Rudolph and R. B. Keey, eds.), 1994, p. 327.
2. R. G. Moreira and F. W. Bakker-Arkema, Grain dryer controls: A review, *Cereal Chemistry 69*(4):390 (1992).
3. G. Trystram and J. Vasseur, Modélisation et simulation d'un procédé de séchage sur cylindre, *Entropie 152*:43 (1989).
4. I. C. Trelea, F. Courtois, and G. Trystram, Non-linear optimal control of a cereal dryer, Proceedings of SACCS'95, 5th International Symposium on Automatic Control and Computer Science, Iasi, Romania 1995.
5. F. Courtois, J. L. Nouafo, and G. Trystram, Control strategies for corn mixed-flow dryers, *Drying Technol. 13*(1/2):147 (1995).
6. G. Rodriguez, J. Vasseur, and G. Trystram, Automatization of a drum dryer for the food industry, *Automatic Control of Food and Biological Processes*, Developments in Food Science, vol. 36 (J. J. Bimbenet, E. Dumoulin, and G. Trystram, eds.), Elsevier, 1994, p. 359.

7. T. Cegrell and T. Hedqvist, Successful adaptive control of paper machines, *Automatica 11*:53 (1975).

8. I. C. Trelea, F. Courtois, and G. Trystram, Dynamics analysis and control strategy for mixed flow corn dryer, submitted to the *Journal of Process Control*.

9. G. Rodriguez, J. Vasseur, and F. Courtois, Design and control of drum dryers for the food industry. 1: Set-up of a moisture sensor and inductive heater, *J. Food Engg.*, to appear.

10. G. Rodriguez, J. Vasseur, and F. Courtois, Design and control of drum dryers for the food industry. 2: Automatic control, *J. Food Engg.*, to appear.

11. C. T. Kiranoudis, G. V. Bafas, Z. B. Maroulis, and D. Marinos-Kouris, MIMO control of conveyor-belt drying chambers, *Drying Technol. 13*(1/2): 73 (1995).

12. R. G. Moreira and F. W. Bakker-Arkema, Digital control of crossflow grain dryers, Paper No. 89-6535, ASAE, St. Joseph, Michigan, 1989.

13. D. M. Bruce and N. J. B. McFarlane, Control of mixed-flow grain dryers: An improved feedback-plus-feedforward algorithm, *J. Agricultural Engg. Res. 56*:225 (1993).

14. R. D. Whitfield, Control of a Mixed-Flow Drier: Part 1. Design of the Control Algorithm, AFRC, Silsoe, Divisional Note 1393, 1987.

15. R. D. Whitfield, Control of a Mixed-Flow Drier: Part 2. Testing the Control Algorithm, AFRC, Silsoe, Divisional Note 1394, 1987.

16. F. Courtois, Computer-aided design of corn dryers with quality prediction, *Drying Technol. 13*(5–7):1153 (1995).

17. I. C. Trelea, F. Courtois, and G. Trystram, Modélisation de la cinétique de séchage et de la dégradation de la qualité amidonnière du maïs par réseaux de neurones, *Automatique et maîtrise des procédés complexes*, Récents Progrés en Génie des Procédés, vol. 40, Lavoisier, 1995, p. 135.

18. N. J. B. McFarlane and D. M. Bruce, Control of mixed-flow grain-driers: Development of a feedback-plus-feedforward algorithm, *J. Agricultural Engg. Res. 49*:243 (1991).

19. T. G. Nybrant, Modelling and adaptive control of continuous grain driers, *J. Agricultural Engg. Res. 40*:165 (1988).

20. J. P. Desplans, P. Lebois, and M. Delouvrie, Commande prédictive multivariable par modèle interne appliquée au cas d'une tour de séchage de lait en poudre, Proceedings of Congrès de Applications des Techniques Gazières, Paris, 1990.

21. Q. Zhang, J. B. Litchfield, and J. Bentsman, Fuzzy predictive control for corn quality control during drying, Proceedings of Food Processing Automation, ASAE, St. Joseph, Michigan, 1990.

22. K. Najim, Modelling and learning control of rotary phosphate dryer, *Int. J. Systems Science 20*(9):1627 (1989).

23. F. Courtois, A. Lebert, J. C. Lasseran, and J. J. Bimbenet, Dynamic modelling and simulation of industrial corn dryers, *Computer Chem. Engg. 17*(suppl.):S209 (1993).

24. F. Courtois and G. Trystram, Study and control of the dynamics of drying processes, *Automatic Control of Food and Biological Processes*, Developments in Food Science, vol. 36 (J. J. Bimbenet, E. Dumoulin, and G. Trystram, eds.), Elsevier, 1994, p. 289.
25. G. Stein and M. Athans, The LQG/LTR procedure for multivariable feedback control design, *IEEE TAC AC 32*:105 (1987).
26. J. Richalet, Industrial applications of model based predictive control, *Automatica 29*(5):1251 (1993).
27. F. Courtois, A. Lebert, A. Duquenoy, J. C. Lasseran, and J. J. Bimbenet, Modelling of drying in order to improve processing quality of maize, *Drying Technol. 9*(4):927 (1991).
28. F. Courtois, A. Lebert, J. C. Lasseran, and J. J. Bimbenet, Corn drying: Modelling the quality degradation, *Developments in Food Engineering*, Part 1, Blackie Academic and Professional, 1994, p. 334.

11

Computerized Food Freezing/Chilling Operations

Hosahalli S. Ramaswamy and Shyam S. Sablani
McGill University, Ste. Anne-de-Bellevue, Quebec, Canada

I. INTRODUCTION

Food preservation by freezing dates back to prehistoric times when primitive man observed that at continued low temperature climates, perishable foods would keep in good condition as long as they were maintained in the frozen state. In the second half of the nineteenth century, the development of reliable mechanical refrigeration systems laid the foundation for the present-day cold storage system and food freezing operations. Since then, freezing has become one of the more common processes for preservation of foods. In the food freezing process, the product temperature is lowered to levels sufficiently below 0°C, causing retardation of microbial and enzymatic activities. In addition, the crystallization of the water to form ice within the product reduces the availability of free water, which otherwise supports deterioration reactions.

The quality of food preserved by freezing is generally considered to be superior to that from other methods of long-term preservation. Properly controlled freezing is effective in retaining the flavor, color, and nutritive value of food, while textural qualities are somewhat adversely affected in some instances. However, the quality of the frozen food is invariably influenced by the freezing process and storage conditions. It is also recognized that the quality of frozen food depends strongly on the size and number of ice crystals formed, which depend on the rate of freezing.

The temperature conditions during storage also influence the quality of frozen food. Temperature fluctuations during storage tend to result in the agglomeration of ice crystals, thereby reducing product quality. Process engineers dealing with food freezing are therefore faced with major tasks of (1) estimating the refrigeration requirement during product freezing as well as storage; and (2) designing the necessary equipment and processes to achieve rapid freezing. Various freezing equipment is available, each type designed to achieve efficient product freezing. One must be able to predict the freezing rate and/or freezing time and know the temperature history of the frozen food during storage to optimize the design of freezing equipment and to exercise control over the quality of the end product. The optimum freezing process also depends on the product characteristics.

Freezing time/rate is the most important factor associated with the selection of a freezing system that ensures optimum food quality. The problem of accurate prediction of freezing time is complex because of the dramatic influence of the freezing process on thermophysical properties. Over the years, many mathematical models have been proposed for predicting freezing times of foods. Most simple models are based on a number of assumptions and approximations. Hence the accuracy of any prediction model depends on how closely the experimental conditions match the model assumptions. In recent years the increased availability of computers has led to development of many numerical models to simulate the food freezing process. These models have been used to estimate freezing time as well as the necessary refrigeration load for the design of refrigeration equipment.

Present-day freezing equipment is designed not only to take into account the required rapid freezing of food samples, but to provide a wide range of options based on economic considerations, versatility of operation for a wide range of food products of different shapes and sizes at a rapid freezing rate, and adaptability to a selection of refrigerants including cryogenics.

It is generally recognized that although freezing permits maximal retention of nutritional quality, and frozen food traditionally is thought by consumers to be the best alternative to fresh foods, freezing nevertheless results in irreversible texture changes from the damage caused by the formation of ice crystals. Hence quality-conscious consumers are seeking other processes to retain better sensory characteristics than possible with frozen foods. Chilling processes with a rapid distribution cycle, the cook–chill processes currently popular in catering circles, and refrigerated sous-vide processes are examples of such processes.

This chapter focuses on the freezing systems with a background on the use of computerized approaches for freezing time forecasting, provides a survey of available modern equipment for freezing of foods, and finally gives a short discussion on chilling systems.

II. NATURE OF THE FREEZING PROCESS

Freezing processes can be divided into three distinct phases: a precooling period, where the product is cooled from the initial temperature to the beginning of freezing process; a phase change period in which all the latent heat is released, resulting in a change of phase from liquid to solid; and a tempering period in which the product temperature is lowered to the required final temperature. The factors contributing to the heat load in the precooling period are the thermal properties of the unfrozen material and temperature difference. The tempering period is governed by the properties of the frozen material and the corresponding temperature difference. Crystallization of water to ice characterizes the phase change period. The crystallization process consists of two distinct steps: nucleation and crystal growth. Nucleation is the association of water molecules into an orderly particle that can serve as an active site for the crystal growth. The crystal growth implies orderly addition of more water molecules to the nucleus. During freezing, both nucleation and crystal growth may occur simultaneously, but for the latter to occur there must be an initial nucleation. The relative occurrence of nucleation and crystal growth depends on the temperature and rate of freezing.

III. FROZEN FOOD PROPERTIES

Any model for the prediction of freezing time requires reliable values of thermophysical properties of food material. The properties of major interest are density, heat capacity, thermal conductivity, thermal diffusivity, and product enthalpy. Initial freezing point and surface heat transfer coefficient values are also required in the prediction of freezing time. Property evaluations related to food freezing can be divided into experimental investigations and prediction models [1]. Excellent reviews and tabulated data on these properties have been reported [2–6]. These sources will provide a good starting point; however, because of the large variation in origin and composition of same food, it may become necessary to evaluate food properties under test conditions. Thermal properties of food materials mainly depend on the composition, the properties of their constituents, and temperature. Empirical and mathematical relationships have been developed to predict the properties on this basis. Miles et al. [7] presented an extensive compilation of property correlations for various foods and also discussed the validity of the various models. Excellent descriptions of common measuring techniques for the measurement of thermophysical properties have been presented by Mohsenin [4], Ohlsson [8], and Singh [9]. More recently, Lind [10] reviewed the common measuring methods and the calculation models for thermal properties of foods.

IV. ESTIMATION OF FREEZING TIME

Freezing time represents the residence time that a food product is exposed to the freezing environment to achieve the desired level of freezing. However, there is lack of a consistent definition for freezing time in the published literature on freezing of foodstuffs. Because the temperature distribution within the product during the freezing process varies considerably, freezing time has to be defined with respect to a given location. Four major methods have been identified to describe the freezing rates: (1) time–temperature methods, (2) velocity of ice front, (3) appearance of the specimen, and (4) thermal methods [11]. All methods have some limitations. The effective freezing time as defined by the International Institute of Refrigeration (IIR) [12] is the more acceptable definition. According to IIR the effective freezing time is the total time required to lower the temperature of the product from its initial value to a given temperature at the thermal center. The final temperature of the product is an important parameter in the design of a freezing system.

Several methods to predict freezing times of food have been reviewed by Ramaswamy and Tung [13]. Procedures for determining freezing time can be separated into three groups: (1) experimental models involving evaluation of freezing time through actual measurement of food temperature during the freezing operation, (2) theoretical models obtained by solving heat transfer equations under appropriate boundary conditions, and (3) semitheoretical models based on a combination of the first two methods.

Freezing time determination based on purely experimental procedure is probably the most widely used method because of its simplicity and accuracy, and it is almost the perfect method for a particular situation under study [13]. It simply involves monitoring the temperature of the material to be frozen at its thermal center until target temperature is achieved. However, these methods are specific in nature, and the freezing time data cannot be generalized. Further, in these situations product quality optimization will not be easy and will involve many experiments. Also, in some cases, it may be very difficult to gather temperature data during experiments.

Theoretical models for the prediction of freezing times are reasonably accurate provided proper boundary conditions and thermophysical property values are used in the calculations. These models can again be classified into three groups. The first group comprises those methods where simplifying assumptions are made. They lead to the solution of partial differential equations to yield completely analytical and explicit formulas. Most of these simplified analytical models are based on modifications of Plank's equation.

The second group of methods uses a mathematical basis and a modification of the theoretical model based on experimental findings. These methods are more versatile than experimental methods alone and simpler than the numerical method; however, they are not without limitations. The third group is based on numerical finite difference and finite element solutions. This approach can handle most types of boundary conditions, shapes, and variation in the thermal properties of material to be frozen.

A. Numerical Methods

The freezing and thawing of food product is complex because of continuous changes in thermophysical properties during the process. The thermophysical properties also depend on various factors such as variety, maturity/age, growing practice, and composition of food material. The process of food freezing can be modeled by a partial differential equation of heat conduction in an appropriate coordinate system, with initial and boundary conditions. With the increasing availability of computing systems, these nonlinear equations can be solved in a short time using appropriate numerical methods. Temperature-dependent properties, nonregular geometries, and all types of boundary conditions can easily be accommodated by numerical methods.

The freezing/thawing process can be described mathematically by Fourier's heat conduction equation. Two different approaches have been reported to solve numerically the partial differential equation of heat conduction. In the first approach, Fourier's equation is solved using a conventional formulation of an apparent specific heat approach [14–17] with one dependent variable, temperature:

$$\rho C_{p,app}(T) \frac{\partial T}{\partial t} = \text{div}[k(T) \cdot \text{grad } T] \tag{1}$$

The thermal conductivity of food product, k, the product density, ρ, and the heat capacity of the product, C_p, are all considered to be temperature dependent. In the temperature zone where the phase change takes place, this temperature dependence of food product properties is strong. The analytical solution for such a nonlinear general equation is not available. The boundary conditions at the surface of the food product can be of the first kind (constant temperature), the second kind (constant heat flux), or the third kind (convection heat transfer). In numerical solution, all three of these boundary conditions can be taken in account:

$$T = T_s \qquad \text{Constant temperature type} \tag{2}$$

$$-k\frac{\partial T}{\partial x} = q_s \qquad \text{Constant flux type} \tag{3}$$

$$-k\frac{\partial T}{\partial x} = h(T_a - T_s) \qquad \text{Convective type} \tag{4}$$

In addition to this, a symmetry condition is used at the center of the geometry, and the condition of uniform temperature throughout the geometry is ensured as an initial condition:

$$\frac{\partial T}{\partial x} = 0 \qquad \text{Symmetry condition} \tag{5}$$

$$T = T_i \qquad \text{Initial condition} \tag{6}$$

A. Cleland and Earle carried out extensive analyses of freezing time computation using various finite difference schemes such as explicit, Crank–Nicholson, and Lees schemes. In their finite difference numerical solutions, they preferred a three-time-level scheme [18] because it has major advantages over other schemes. This scheme is conditionally stable and convergent, and it has the following form for the heat conduction equation [19]:

$$\frac{-2(\Delta x)^2}{3\,\Delta t} k^m_{i+1/2} T^{m+1}_{i+1}$$

$$+ \left[C^m_i + \frac{2(\Delta x)^2}{3\,\Delta t}(k^m_{i+1/2} + k^m_{i+1/2}) \right] T^{m+1}_i - \frac{2(\Delta x)^2}{3\,\Delta t} k^m_{i-1/2} T^{m+1}_{i-1}$$

$$= \frac{2(\Delta x)^2}{3\,\Delta t} [k^m_{i+1/2}(T^m_{i+1} - T^m_i + T^{m-1}_{i+1} - T^{m-1}_i)$$

$$- k^m_{i-1/2}(T^m_i - T^m_{i-1} + T^{m-1}_i - T^{m-1}_{i-1})] + C^m_i T^{m-1}_i \tag{7}$$

In the second approach, the heat conduction equation is converted from temperature to enthalpy. The enthalpy formulation uses two dependent variables: enthalpy is treated as the primary dependent variable, while temperature is treated as the secondary dependent variable. The advantage of using the enthalpy form is that the change in the relative enthalpy content of the product during freezing/thawing is continuous with temperature, which is a desirable feature in the numerical schemes based on enthalpy formulation:

$$\frac{\partial H}{\partial t} = \text{div}[k(H) \cdot \text{grad } T(H)] \tag{8}$$

In conventional formulation the apparent heat capacity, $C_{p,app}$, change appears as a large peak at the initial freezing point.

Initial and boundary conditions expressed in the apparent heat capacity formulation can also be used with the enthalpy formulation. However, the temperature dependence of enthalpy is required to relate primary and secondary dependent variables. Using an explicit finite difference scheme based on the enthalpy formulation, Mannapperuma and Singh [20, 21] obtained a solution for freezing/thawing time prediction for the case of one-dimensional geometries under a convective boundary condition. They developed the geometric factor concept based on the area and volume concept to use a similar scheme for all three coordinate systems. Mannapperuma and Singh [22] further extended the enthalpy formulation approach to two- and three-dimensional geometries and for constant temperature and constant heat flux type boundary conditions. Pham [23] also used the enthalpy formulation based three-time-level finite difference scheme to predict freezing time.

Two kinds of numerical methods are often used. Finite difference methods are generally preferred for food with uniform composition and of regular shape. This type of method has been used extensively. For irregular geometries and heterogeneous composition, finite element methods may have to be used. These methods, however, involve more computation than finite difference methods. Rebellatio et al. [24] and Purwadaria and Heldman [25] used the finite element method based on an apparent heat capacity formulation to simulate the food freezing of two-dimensional geometry under all three kinds of boundary conditions. Pham [26] reported the use of a lumped capacitance finite element method with an explicit and three-time-level enthalpy formulation. Bonacina and Comini [27], Hayakawa et al. [17], and D. Cleland [28] also used finite element methods with various types of boundary conditions. The finite element methods have been used by Mallikarjunan and Mittal [29, 30] to predict the chilling time of beef carcasses.

Salvadori and Mascheroni [31] have developed algebraic equations for the prediction of freezing/thawing time as a function of thermal properties of food material and process parameters. A finite difference method was used to calculate freezing/thawing times for a variety of conditions. Calculated freezing/thawing times were then correlated with initial and final product temperature, cooling medium temperature, initial freezing temperature, Fourier number, and Biot number, using a regression model. The regression equation predicted the freezing times with an error of 5% and thawing time with error 4.2%.

Mostly, temperature-dependent thermal properties have been taken into account while using numerical methods. Recently Wang and Kolbe [32] simulated the food freezing conditions for a plate freezer while considering temperature-dependent thermal properties (heat capacity, thermal conductivity, and density), and time-dependent boundary conditions (overall heat transfer coefficient and ambient temperature). Using a commercial PC-based finite element package—ANSYS in this simulation—the effect of model parameters on freezing time prediction was also studied. Freezing time was strongly affected by all the processing parameters, apparent specific heat effect being the most important and density effect least important among all the parameters. Earlier, Hsieh et al. [33] studied the influence of product properties on the prediction of freezing time. Foods with lower initial freezing points, higher initial water content, and higher initial product density were found to have longer freezing times.

Methods developed for the estimation of freezing time need data on thermal properties of food products and relevant heat transfer coefficients. Recently Mihori and Watanabe [34] proposed a two-stage mathematical model that does not require the knowledge of thermal data for the food being frozen. The procedure uses the time/temperature data collected from the early stages of cooling. The data collected during the cooling are separated into two groups: (1) before freezing (precooling stage) and (2) after freezing (tempering stage). The procedure requires the analytical solution of the heat conduction equation. The system parameters in the analytical solution, mainly initial freezing temperature, Biot number, and time constant, are determined using known temperatures at various locations with time during the initial cooling period. Later these system parameters are used to predict the time/temperature profile in the remainder of the freezing process. Mihori and Watanabe [35] demonstrated the applicability of this procedure to the on-line prediction of freezing time.

V. FREEZING EQUIPMENT

In food freezing processes the product temperature is lowered from its original value to the storage temperature. For this, product is exposed to the low-temperature medium in freezing equipment. The heat from the innermost part of the product must be removed by conduction to the surface. The surface heat is then removed by the refrigeration medium. There are several methods of classifying freezing equipment: direct contact or indirect contact system, batch type or in-line processing, or based on method of achieving heat transfer. The operating characteristics and salient features of various freezing equipment are discussed in this section.

A. Air Blast Freezers

This type of freezer uses cold air, with temperature varying from -18 to $-40°C$ and air velocities up to 20 m/s, for the heat transfer. Since air is the most common freezing medium, this method of heat transfer probably has the largest range of designs. These include stationary tunnel freezers, belt freezers, carton freezers, and fluidized-bed freezers. In its simplest form, the air freezing of food product is achieved by placing the packed or unpacked food product in rooms with large refrigeration capacity. In room freezing, however, heat transfer rates are relatively low because of lower air velocities, resulting in long freezing times. The slow rate of freezing also favors the formation of larger ice crystals, mostly located in intercellular spaces, thereby compressing and disrupting the tissue structure. Hence, upon thawing, the quality of the frozen cellular food products will be generally poor due to breakdown of tissue structure. The objective of most modern freezers is therefore to achieve rapid freezing, which is expected to result in the formation of many small ice crystals distributed through the intracellular structure of the food. Upon thawing these crystals are believed to melt within the cells, thereby preserving the tissue structure.

1. Tunnel Freezers

The classical stationary tunnel is an improved design over room freezing. It is an insulated chamber, equipped with a refrigeration unit and large fans for circulating air over and around the product in a controlled fashion. Trays with the food (packed or unpacked) are placed on racks within the tunnel. Improved heat transfer rates are obtained because of higher air velocities around the product. The manpower requirements are high for this type of batch system freezing. The in-line continuous system that uses a pushing mechanism helps to reduce the labor costs. In this modification, the racks are pushed on rails through a tunnel with regulated speed.

2. Belt Freezers

The continuous air blast freezing system reduces the problem of materials handling and also results in a significantly high throughput with smaller factory space. Early continuous freezing systems consisted of a wire mesh belt conveyor passing through a blast room. Nonuniformity of air flow distribution over the product often caused poor heat transfer and nonuniform freezing. More recent versions of belt freezers use vertical air flow for improved product and air contact. Multiple-layer belt design systems consist of single feed/single discharge with multiple pass or multiple infeeds and multiple discharge. Advanced design in belt freezers, e.g., air blast contact freezers from Sandvik Process Systems (Totowa, New Jersey) (Fig. 1), uses stainless steel belts, which combine the benefits of air blast freezing and the high conductivity of the steel belts. This results in reduced freezing times and low energy consumption. Another versatile design (PolyTray) from HSI company (Lancaster, Pennsylvania) in air blast belt freezers can handle a wide range of package sizes, shapes, and weights. Most modern designs of belt freezers use a spiral belt mounted on rails and air flowing vertically upward. The required residence time of the product to be frozen can be achieved by controlling the speed of the spiral belts. Both unpacked and packed products are frozen, and the freezer can accommodate a wider range of products. The spiral belt freezer

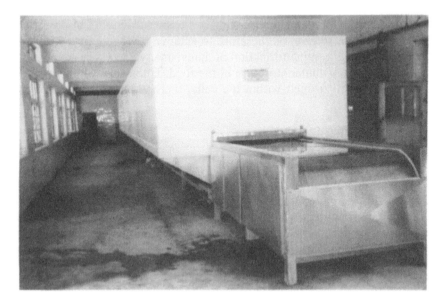

Figure 1 An air blast contact freezer that combines the benefits of air blast freezing and the high conductivity of steel belts (courtesy Sandvik Process Systems, Inc., Totowa, New Jersey).

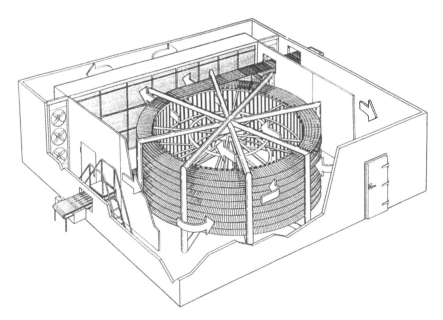

FIGURE 2 Cut-away section showing a spiral belt freezing system (courtesy Northfield Freezing System, Inc., Northfield, Minnesota).

requires less factory space because of its compact design. Spiral freezers are manufactured by several companies, such as Northfield Freezing Systems (Northfield, Minnesota) (Figs. 2 and 3), York Food System (Preston, Washington), and Liquid Carbonic Freezing Systems (Chicago). These freezers are simple in design, are operator-friendly with fewer moving parts, and have less complicated controls. Liquid Carbonic manufactures spiral freezers with both horizontal and vertical air flow design suitable for a variety of products. In the SpiralTRAK system (York Food System) (Figs. 4 and 5), the design of center drum drive eliminates chains, wagons, sprockets, and continuously lubricated parts are used for less maintenance and better hygiene. The freezer's improved double impingement air flow design provides fast two-sided freezing in the product, and it is reported to reduce by 15% to 30% freezing time over conventional air flow designs (company data). Two-sided freezing produces better color on sensitive products because of uniform freezing. In some commercial freezing systems (e.g., SpiralTRAK and SuperSTAK [Fig. 6], both from York Food System) the freezing operation is monitored and programmed with a programmable logic controller (PLC) based control panel with a liquid crystal display (LCD).

FIGURE 3 A spiral freezer (courtesy Northfield Freezing System, Inc., North-field, Minnesota).

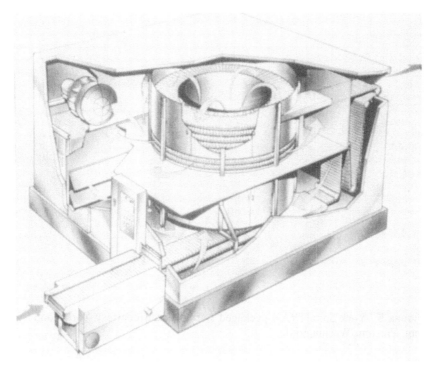

FIGURE 4 A line diagram of the York SpiralTRAK spiral freezer (courtesy York Food System, Preston, Washington).

3. Carton Freezers

The carton freezers are in-line continuous freezing systems used for the handling of products packaged in cartons. In this type of freezing system, large boxes are loaded onto carrier shelves, which move through the freezing section blasting with cold air.

4. Fluidized-Bed Freezers

Fluidizing-bed freezing is one of the most important methods of freezing in the food industry because of its use for individual quick freezing (IQF). The principle involved is to create a fluidized bed of food particles to be frozen by supplying the cold air in a vertical direction at above the minimum fluidization velocity. The minimum fluidization velocity of air is the velocity needed to lift the particles from the surface and keep them suspended in the low-temperature air. The air velocities employed in fluid-

FIGURE 5 York SpiralTRAK packaged spiral freezer (courtesy York Food System, Preston, Washington).

FIGURE 6 York SuperSTAK spiral freezer (courtesy York Food System, Preston, Washington).

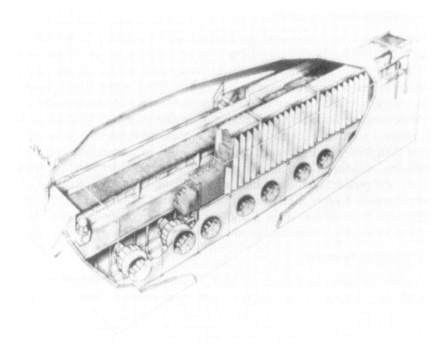

FIGURE 7 Cut-away section of the York SuperFLOW iqf belt freezer (courtesy York Food System, Preston, Washington).

ized-bed freezing are in the range of 2–10 m/s depending on the bed height and particulate shape and size. In a batch type system, the fluidized bed is fixed, and in continuous systems, the fluidized bed has a moving bed of particles. Fluidized-bed freezing facilitates better product handling and packing of particulate foods. The method is popular for the freezing of berry fruits, diced fruits, and vegetables. In fluidizing-bed freezing the convective heat transfer coefficients between air and particle are quite high, which results in rapid freezing of the product. The latest design of York's SuperFLOW iqf (York Food System, Preston, Washington) (Fig. 7) includes a two-belt, in-line freezer equipped with a unique fluidizing zone to handle delicate products. Product is conveyed in stainless steel wire mesh belt with variable speeds to allow for a wide range of freezing times for maximum product flexibility. Two-belt design allows two-stage freezing with an initial precooling/crust freezing followed by a final deep-freezing for optimum product quality.

5. Impingement Freezer

Impingement freezing is a modified version of conventional air blast freezing with freezing rates comparable with cryogenic freezing. In this type of freezer, cold air is introduced with very high velocity through many small orifices opening on product surfaces for fast freezing. This enhances the rate of convection heat transfer from the surface of a food product to the cooling medium (air) by breaking away the boundary layer of air on a product surface. Freezing time in impingement freezers is reduced to less than 25% of the freezing times achievable in conventional air blast freezing techniques. Impingement freezing reduces dehydration (yield loss), and product quality is comparable with liquid nitrogen freezing. An impingement freezer (JETFreeze) manufactured by York Food System (Preston, Washington) (Figs. 8 and 9) can accommodate capacities from 500 to 5500 kg/h of thin food products like raw patties or thin fish fillets.

6. Flexible Freezers

Flexible freezers can freeze and chill simultaneously a wide variety of products and package sizes with varying temperature profiles. HSI com-

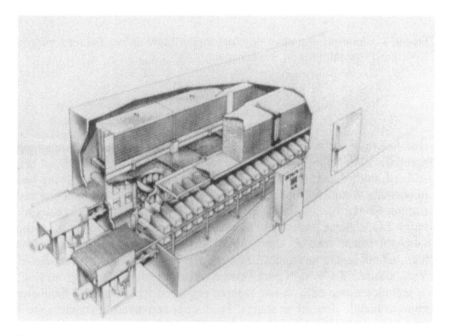

FIGURE 8   A line diagram of the York JETFreeze impingement freezer (courtesy York Food System, Preston, Washington).

FIGURE 9 York JETFreeze impingement freezer (courtesy York Food System, Preston, Washington).

pany (Lancaster, Pennsylvania) manufactures a variable-retention freezing/cooling system that is marketed under the name PolyFlex and Poly-Tray (Fig. 10). Intec, Inc. (Redmond, Washington) manufactures and services the Variable Retention Time (VRT) freezers (Fig. 11), developed by System Technology International Ltd. (Auckland, New Zealand). These freezers are controlled by an integrated computer system by which the retention time of box lots of product, temperature, and air profiles are controlled throughout the VRT and which can be displayed and altered on demand for new products. Inside the VRT freezer, end flow air distribution ensures even treatment of all products irrespective of their position within the VRT. This system provides in-line freezing flexibility for a variety of products such as ice cream, dairy, meat, fish, poultry, bakery, and other foods. The products to be cooled follow a specific flow cycle in the VRT. The packaged products are accumulated outside the VRT in shelf size batches according to their specification or retention time and moved into the VRT by in-feed conveyor and then transferred to shelves. Each shelf then is lowered or lifted to a programmed level and injected into the rack. Product goes through desired residence time in VRT before moving to exit conveyor through a mechanical transfer system.

FIGURE 10 PolyTray packaged product freezing system (courtesy HSI Company, Lancaster, Pennsylvania).

B. Contact Freezers

The most common type of contact freezing systems are plate freezers consisting of series of parallel plates. In contact plate freezers, the food product is held under pressure between two hollow flat metal plates in which a refrigerant flows to maintain the plates at temperatures adjustable from -20 to $-60°C$. The hollow refrigerated metal plates apply slight pressure to create good contact between food product and plate surface. This arrangement provides very good heat transfer, resulting in shorter freezing time. The space between plates can be adjusted depending on the thickness of the food product to be frozen. Typical food products to be frozen in plate freezers are in the shape of a slab with uniform thickness. Uneven thicknesses of the products can create void spaces, which can lead to poor heat transfer. In the case of packaged product, the package has to be filled and in uniform contact with the plates. The plate freezers

1. PLC UNIT & MONITOR.
2. PRODUCT ACCUMULATION PRIOR TO TRANSFER
   ON INFEED CONVEYOR.
3. PRODUCT SUPPLY INTO VRT FREEZER
   - INFEED CONVEYOR (VIA AIR LOCK).
4. LOAD ELEVATOR SYSTEM.
5. SHELVES LOADING & TRANSFER TO RACK.
6. RACK STORAGE OF SHELVES.
7. UNLOAD ELEVATOR SYSTEM.

8. PRODUCT OUT OF VRT FREEZER
   - OUTFEED CONVEYOR (VIA AIR LOCK).
9. FAN UNITS.
10. EVAPORATOR UNITS.
11. COLD AIR BLAST OFF EVAPORATORS.
12. INTERNAL CLADDING FOR AIR FLOW CONTROL.
13. AIR RECIRCULATION.
14. EXTERNAL INSULATED ENCLOSURE.

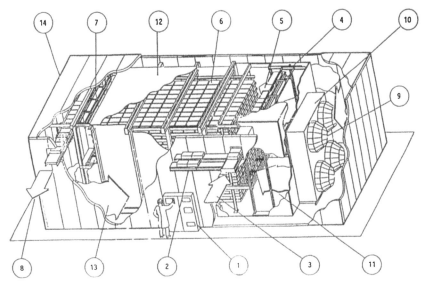

FIGURE 11   A schematic diagram of the Variable Retention Time freezing system (courtesy Intec Inc., Redmond, Washington).

are operated in both batch and in-line mode. Plate freezers are available with horizontal or vertical plate arrangements.

C. Cryogenic Freezers

Freezing by means of cryogenic liquids is the fastest growing area in the food freezing industry. In cryogenic freezing systems, liquified gases at very low temperature are used for the freezing of food product. Liquid nitrogen and liquid carbon dioxide are the most widely used refrigerants in cryogenic freezers. The low boiling points of cryogenic liquids ($-196°C$ for liquid nitrogen and $-98°C$ for liquid carbon dioxide) allow very fast heat transfer, resulting in short freezing times. Cryogenic freezing is also a convenient method of freezing for IQF products. Although the initial capital costs of cryogenic freezers are low, because of the high consumption of cryogenic liquids (1.0 kg of cryogenic liquid per 2.0 kg of the product) the operating costs are higher.

1. Cryogenic Tunnel Freezing

Straight tunnel cryogenic freezers have been used to freeze a variety of food products. In these freezers, product moves through the freezer on a continuous conveyor belt and liquid refrigerant is sprayed by a series of atomizing nozzles onto product surface close to product outlet from the tunnel. The refrigerant quickly evaporates on contact with the product, and vapor is passed countercurrent to the product flow by means of fans for the precooling of the product at the tunnel inlet. This also conserves cryogen and optimizes freezer performance. In some tunnels, product also goes through a zone of equilibrium after passing the spray zone to allow conduction of heat within the product. The process of cooling is controlled by varying the flow rate of liquid refrigerant to the freezing tunnel and the speed of the conveyor belt. In a recent design of tunnel freezers utilizing either liquid nitrogen or liquid carbon dioxide, from Liquid Carbonic (Chicago) (Fig. 12), an integrated computer system has been used to control all primary functions of the freezers. Ultra-Freeze cryogenic freezers from Air Liquide America Corporation (Houston, Texas) have a three-tier tunnel with a variable-speed fan and a conveyor drive; they also have dual cryogen capabilities.

   Advanced designs of cryogenic freezers use spiral belts to maximize freezing capacity in minimum floor space. The most recent spiral freezer design from Liquid Carbonic has a fan located at the top center of the spiral, which draws cold refrigerant vapor through the core and recircu-

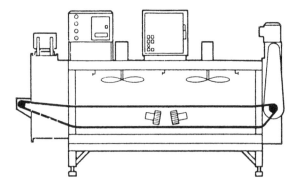

FIGURE 12 A schematic of a cryogenic in-line tunnel freezer (courtesy Liquid Carbonic, Chicago, Illinois).

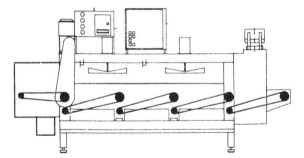

FIGURE 13 A schematic of a cryogenic flighted tunnel freezer (courtesy Liquid Carbonic, Chicago, Illinois).

lates it through the belt tiers, resulting in more uniform product temperature. These systems have product capacities of 1000 to 4000 kg/h.

Another design in tunnel freezing especially suitable for individually quick frozen or crust frozen types of food freezing applications is a *flighted tunnel freezer* (Liquid Carbonic) (Fig. 13). In this freezer, food pieces tumble on a slightly inclined-upward belt and are transferred to series of belts before coming out from the tunnel. This way the food particles being frozen are separated from each other and get maximum exposure to the cryogen. The capacity of this type of freezer reaches 2500 kg/h.

For smaller volume or batch type cryogenic freezing, cabinet freezers are available. This type of freezer consists of a number of trays resting on shelves that can be removed from the cabinet. The packaged or unpackaged food is kept on trays for freezing.

2. Immersion Freezing

In another commercial version of direct contact cryogenic freezers, the freezing is achieved by simply immersing the product in a liquid refrigerant. Food pieces are conveyed through a belt that drops into a bath of liquid refrigerant. As product is dropped into the refrigerant the crust freezes immediately, causing liquid refrigerant to boil. The residence time of product in the refrigerant is determined by the speed of the conveyor. In this type of freezing, the process is completed in two stages: the first stage involves an immersion treatment; and in the second stage the product goes through an in-line belt tunnel where product temperature is equilibrated at a slower rate. In Europe immersion freezers such as CRYODIP, manufactured by Air Products PLC (Middlesex, England), are widely used for the freezing of fruits and seafood [36].

FIGURE 14   A photograph of a Sandvik Rotoform indirect liquid freezer (courtesy Sandvik Process Systems, Inc., Totowa, New Jersey).

3. Indirect Liquid Freezing

Indirect liquid freezers are specifically designed to freeze liquids and pastes by taking advantage of cryogenic freezing. In this process, the lower surface of a stainless steel belt is cooled by recirculating liquid nitrogen, and liquid product or paste in the form of droplets is introduced on the top surface of the belt. Most recent designs of indirect liquid freezers combine the benefits of cryogenic and air blast freezing to enhance the quality of frozen products. Indirect liquid freezing systems comes in many designs, such as Cryo-Stream tunnel and Rotoform freezer manufactured by Sandvik Process Systems (Totowa, New Jersey) (Fig. 14). In such systems, freezing of pastilles from 1 to 40 mm in diameter can be handled.

VI. CHILLING OPERATIONS

Chilling refers to reducing the temperature of food product, at its slowest cooling point up to 4–5°C. The main reason for chilling is to minimize bacterial growth and thus extend shelf life. Most meat, immediately after the slaughtering, is preserved by chilling. The most widely used method for chilling is cold air blast chilling. In chilling of beef carcasses, the rate of chilling is a crucial factor: it affects the tenderness of beef [30]. The rate of temperature drop also affects bacterial growth. During chilling there is considerable loss of mass due to evaporation, which is of economic importance to the meat industry. Hence chilling of meat involves simultaneous heat and mass transfer.

Design engineers must know the temperature and mass profile of carcasses during chilling, as well as the effect of process conditions on the chilling rate of carcasses, to minimize evaporation losses and maximize the tenderness of beef carcasses. Bailey and Cox [37] developed nomographs for predicting center temperatures of a round muscle for different carcass masses, air velocities, and ambient temperatures. Temperature and mass profiles can also be estimated using appropriate heat and mass transfer models. These models require thermal properties and heat and mass transfer coefficients as input parameters. Daudin and Swain [38] evaluated heat and mass transfer coefficients and the water activity in chilling and storage of meat by considering the basic equation describing simultaneous heat and mass transfer. Mallikarjunan and Mittal [29] developed a model to simulate the simultaneous heat and mass transfer during chilling of beef carcasses using a finite element method. Beef carcasses was modeled as an irregular two-dimensional geometry consisting five sections with different composition. The model also included temperature-dependent thermal properties and computed temperature and mass histories that compared well with experimental values. Recently Mallikarjunan and Mittal [30] developed a set of algebraic equations to estimate chilling time, mass loss at the end of chilling, temperature history at the geometric center of the round muscle, and mass history during chilling. These regression equations were also validated using a finite element–based model and experimental results.

Beef carcasses are cooled in batch chills. In an industrial setup, for greater efficiency and better process control, continuous processes are always desirable. Drumm et al. [39] investigated the feasibility of a line (or continuous) process for chilling of beef carcass. In this study, beef sides were subjected to a particular chilling regime consisting of a sequence of "zones" while passing through a long serpentine tunnel. Air temperature and speed were precisely controlled in each zone. Depending on the desired condition at the end of chilling and the carcass type, the number of zones and residence time would vary. The temperature at the thermal center of an individual carcass was predicted using nomographs developed by Bailey and Cox [37], and these temperatures were reported to compare well with measured temperatures. Later, Drumm and McKenna [40] studied the factors influencing the chilling process, quality (tenderness), evaporation loss, and energy consumption. In model line chilling, better process control was achieved compared with commercial batch chilling. In different types of chilling regimes, significant differences in meat temperatures and shrinkage values were reported. Although no significant difference was found in toughness of meat, shrinkage was reduced by 33% (from 1.2 to 0.81%) with a concurrent 35% increase in power consumption. The success of the process will depend on economic viability. The major factors in deciding the optimum chilling regime for meat will be (1) the cost of electrical energy, (2) the cost of shrinkage, and (3) meat residence time to achieve desired thermal center temperature at the end of chilling [40].

Cook–chill systems are commonly used today in restaurant and institutional catering businesses, offering the advantages of superior-quality products and remarkably reduced preparation times prior to serving. The product is generally prepared in large commercial kitchens, packaged into bulk plastic bags, and chilled in specially designed refrigerated equipment, which gently tumbles the product to achieve rapid cooling. Immersion facilities are more common, chilled water being the common cooling medium. The cooled product is temporarily held at close to freezing temperatures and distributed under cold chain for final preparation, reheating, and serving at destination kitchens.

VII. SUMMARY

While it has been recognized that the quality of frozen food is affected by the size and number of ice crystals, quantitative understanding of ice crystallization processes is necessary to achieve high-quality frozen foods. Accurate data on frozen food properties and understanding of mathematical models describing heat transfer during freezing are needed for the optimized design of freezing equipment and for achieving optimum freezing processes. Freezing equipment manufacturers are continually improving and refining the many types of food freezers and updating them to meet changing market needs. In the last decade, process development has focused on improving the heat transfer between refrigerant and food product, thereby improving overall cooling capabilities, making efficient use of refrigerant to reduce freezing times, and increasing throughput for a given size of freezer. Equipment development has mainly concentrated on (1) accessibility, ease of cleaning, and maintenance, and (2) electronic control for efficient operations. In future, computers are expected to play a major role at every stage in the freezing operations, e.g., more complex models for the determination of frozen food properties, freezing times/ rates, and process control.

ACKNOWLEDGMENTS

The authors acknowledge the following companies for their help in supplying information and photographs of food freezing equipment:

1. Air Liquide America Corporation, 3535 West 12th Street, Houston, Texas 77008
2. HSI Company, Inc., P.O. Box 4785, Lancaster, Pennsylvania 17604
3. Intec, Inc., 3942-150th Avenue N.E., Redmond, Washington 98052
4. Liquid Carbonic, 135 South La Salle Street, Chicago, Illinois 60603-4282
5. Northfield Freezing Systems, 1719 Cannon Road, P.O. Box 98, Northfield, Minnesota 55057
6. Sandvik Process Systems, Inc., 21 Campus Road, Totowa, New Jersey 07512
7. York Food Systems, 8130-304th Avenue S.E., Preston, Washington 98050

NOTATION

C	Volumetric specific heat, $J/(m^3 \cdot {}^{\circ}C)$
C_p	Heat capacity, $kJ/(kg \cdot K)$
h	Heat transfer coefficient, $W/(m^2 \cdot K)$
H	Enthalpy, kJ/kg
k	Thermal conductivity, $W/(m \cdot K)$
q	Heat flux, W/m^2
T	Temperature, ${}^{\circ}C$
t	Time, s
x	Distance measured along x axis, m
ρ	Product density, kg/m^3

Subscripts/Superscripts

a	Ambient
app	Apparent
i	Initial condition, nodal point
i	ith component of the food
m	Time level
s	Surface

REFERENCES

1. S. D. Holdsworth, Physical and engineering aspects of food freezing, *Developments in Food Preservation*, vol. 43 (S. Thorne, ed.), Elsevier Applied Science, London, 1987, p. 153.
2. R. W. Dickerson Jr., Thermal properties of foods, *The Freezing Preservation of Foods* (K. K. Tressler, E. B. van Arsdel, and M. J. Copely, eds.), AVI, Westport, Connecticut, 1968, p. 26.
3. A. E. Kostaropoulos, W. E. L. Spiess, and W. Wolf, Reference values for thermal diffusivity of foodstuffs, *Lebensm.-Wiss. u. Technol. 8*(3):108 (1975).
4. N. N. Mohsenin, *Thermal Properties of Foods and Agricultural Material*, Gordon and Breach, New York, 1980.
5. S. L. Polley, O. P. Snyder, and P. A. Kotnour, Compilation of thermal properties of foods, *Food Technol. 34*(11):76 (1980).
6. M. A. Tung, G. F. Morello, and H. S. Ramaswamy, Food properties, heat transfer conditions and sterilization considerations in retort processes, *Food Properties and Computer-Aided Engineering of Food Processing Systems* (R. P. Singh and A. G. Medina, eds.), Kluwer Academic Publishers, Dordrecht, The Netherland, 1989.
7. C. A. Miles, G. van Beek, and C. H. Veerkamp, Calculation of thermophysical properties of foods, *Physical Properties of Foods* (R. Jowitt, F. Escher, B. Hallstrom, H. F. Th. Meffert, W. E. L. Spiess, and G. Vos, eds.), Applied Science Publishers, London, England, 1983.
8. T. Ohlsson, The measurement of thermal properties, *Physical Properties of Foods* (R. Jowitt, F. Escher, B. Hallstrom, H. F. Th. Meffert, W. E. L. Spiess, and G. Vos, eds.), Applied Science Publishers, London, England 1983.
9. R. P. Singh, Thermal diffusivity in thermal processing, *Food Technol. 36*(2): 87 (1982).
10. I. Lind, The measurement and prediction of thermal properties of food during freezing and thawing—a review with particular reference to meat and dough, *J. Food Eng. 13*:285 (1991).
11. O. R. Fennema, W. D. Powrie, and E. H. Marth, *Low Temperature Preservation of Foods and Living Matter*, Marcel Dekker, New York, 1973.
12. International Institute of Refrigeration (IIR), *Recommendations for the Processing and Handling of Frozen Foods*, IIR, Paris, 1972.
13. H. S. Ramaswamy and M. A. Tung, A review on predicting freezing times of foods, *J. Food Proc. Engg. 7*:169 (1984).
14. C. Bonacina, G. Comini, A. Fasano, and M. Primicerio, Numerical solution of phase-change problems, *Int. J. Heat Mass Transfer 16*:1825 (1973).
15. A. C. Cleland, *Heat transfer during freezing of foods and prediction of freezing times*, Ph.D. thesis, Massey University, New Zealand, 1973.
16. A. C. Cleland and R. L. Earle, A comparison of analytical and numerical methods for predicting freezing times of foods, *J. Food Sci. 42*:1390 (1977).
17. K. I. Hayakawa, C. Nonino, J. Succar, G. Comoni, and S. Del Giudice,

Two dimensional heat conduction in food undergoing freezing: Development of computerized model, *J. Food Sci. 48*(6):1849 (1983).

18. M. Lees, A linear three level difference scheme for quasilinear parabolic equations, *Mathematics of Computation 20*:516 (1966).

19. A. C. Cleland and R. L. Earle, The third kind of boundary condition in numerical freezing calculations, *Int. J. Heat Mass Transfer 20*(10):1029 (1977).

20. J. D. Mannapperuma and R. P. Singh, Prediction of freezing and thawing times of foods using a numerical method based on enthalpy formulation, *J. Food Sci. 53*(2):626 (1988).

21. J. D. Mannapperuma and R. P. Singh, Thawing of foods in humid air, *Int. J. Refrig. 11*(3):113 (1988).

22. J. D. Mannapperuma and R. P. Singh, A computer-aided method for the prediction of properties and freezing/thawing times of foods, *J. Food Eng. 9*:275 (1989).

23. Q. T. Pham, A fast unconditionally stable finite difference scheme for conduction heat transfer with phase change, *Int. J. Heat Mass Transfer 28*:2079 (1985).

24. L. Rebellatio, S. Del Giudice, and G. Comini, Finite element analysis of freezing processes in foodstuffs, *J. Food Sci. 43*:239 (1978).

25. H. K. Purwadaria and D. R. Heldman, A finite element model for the prediction of freezing rates in food products with anomalous shapes, *Trans. ASAE 25*:827 (1982).

26. Q. T. Pham, The use of lumped capacitance in the finite element solution of heat conduction problems with phase change, *Int. J. Heat Mass Transfer 29*:285 (1986).

27. C. Bonacina and G. Comini, On the solution of the nonlinear heat conduction equations by numerical methods, *Int. J. Heat Mass Transfer 16*:581 (1973).

28. D. J. Cleland, *Prediction of freezing and thawing times for foods*, Ph.D. thesis, Massey University, New Zealand, 1985.

29. P. Mallikarjunan and G. S. Mittal, Heat and mass transfer during beef carcass chilling—modelling and simulation, *J. Food Eng. 23*:277 (1994).

30. P. Mallikarjunan and G. S. Mittal, Prediction of beef carcass chilling time and mass loss, *J. Food Proc. Eng. 18*:1 (1995).

31. V. O. Salvadori and R. H. Mascheroni, Prediction of freezing time and thawing times of foods by means of a simplified analytical method, *J. Food Eng. 13*:67 (1991).

32. D. Wang and E. Kolbe, Analysis of food block freezing using a PC-based finite element package, *J. Food Eng. 21*:521 (1994).

33. R. C. Hsieh, L. E. Lerew, and D. R. Heldman, Prediction of freezing times for foods as influenced by product properties, *J. Food Proc. Eng. 1*:183 (1977).

34. T. Mihori and H. Watanabe, An on-line method for predicting freezing time using time/temperature data collected in early stages of freezing, *J. Food Eng. 23*:357 (1994).

35. T. Mihori and H. Watanabe, A two-stage model for on-line estimation of freezing time of food material: A one-dimensional mathematical model which has analytical solution, *J. Food Eng. 23*:69 (1994).
36. J. P. Miller, The use of liquid nitrogen in food freezing, *Food Freezing: Today and Tomorrow* (W. B. Bald, ed.), Springer-Verlag, London, 1991.
37. C. Bailey and R. P. Cox, *The Chilling of Beef Carcasses*, The Institute of Refrigeration, London, 1976.
38. J. D. Daudin and M. V. L. Swain, Heat and mass transfer in chilling and storage of meat, *J. Food Eng. 12*:95 (1990).
39. B. M. Drumm, R. L. Joseph, and B. M. McKenna, Line chilling of beef, 1: The prediction of temperature, *J. Food Eng. 16*:251 (1992).
40. B. M. Drumm and B. M. McKenna, Line chilling of beef, 2: The effect on carcass temperature, weight loss and toughness, *J. Food Eng. 15*:285 (1992).

12

Separation Processes for the Food Industry: Process Models for Effective Computer Control

K. Niranjan
The University of Reading, Whiteknights, Reading, England

N. C. Shilton
University College Dublin, Dublin, Ireland

I. INTRODUCTION

Selective separation of one or a mixture of components from a bulk liquid or mass is one of the most common operations encountered in the food industry today. There are a wide variety of processes available, including centrifugation, filtration, leaching, distillation, and use of synthetic membranes. In the context of food technology, separation processes provide an aid to processing, and they generally have no preservative effects. Applications in the food industry include production, purification, or concentration of foods or components, for example, juices, enzymes, proteins, and beer clarification [1]. One other major use of separation processes in the food industry is that of food analysis, whether of flavors or nutritional components.

In the analysis and development of computer control systems for separation processes, a variety of process models need to be used. The accuracy and the level of complexity required in developing a computer control system depend on the processing objectives, the control objectives, and the degree of automation required to enforce these objectives. Both analog and digital computers are used for process control. The choice is governed by task and the cost of use. In general, the costs of analog computer control systems depend more on job size than do those of digital control systems. Significant developments in the chemical industries, which have taken place in the recent past, can be applied to food processing plants. In the food context, however, the control strategy does not depend on the real-time interaction between the process parameters and product quality; it is often based on setting and controlling the process parameters at values (or a range of values) deduced from past experience. As a result, the descriptions of process models are not integrated with the control strategy. In this chapter, we illustrate the principles of mathematical models describing three selected separation processes: filtration, extraction/leaching, and pressure-driven membrane processes. The chapter does not provide a comprehensive survey of all relevant published information on these topics; the coverage is restricted to principles. The depth of coverage is more a reflection of the extent of use of these operations in the food industry rather than the depth of knowledge available in the literature. It is hoped that readers can draw ideas for using these principles in their specific applications.

II. ILLUSTRATIVE SEPARATION TECHNIQUES

A. Filtration

Filtration can be defined as the separation of solids from a suspension in a liquid by means of a porous medium or screen that retains the solids and allows the liquid to pass [2]. A major use of filtration in the food industry is the filtration of beer during brewing. This is done by adding a compound called kieselguhr during mashing. This causes precipitation, and the precipitate forms a filter cake as the beer is drained off. Kieselguhr addition was done manually in the past; however, automation of this process has shown savings in filtration time, has decreased the amount of kieselguhr required, and has improved the process efficiency [3, 4]. This illustrates the common practice of using filter aids, which are added to increase cake permeability. Filtration can be divided into two broad categories: cake and depth filtration. In cake filtration, the particulate phase is largely stopped at the surface of a supporting porous medium while the fluid passes through. In depth filtration, the particles in the slurry are captured in the interstices of a porous solid phase. In practice, depth filtration may well precede cake filtration in a given process. Furthermore, both types of filtration may be employed in a given plant. For instance, water drawn into a plant from surface sources is often treated using deep-bed sand filters, whereas effluent water may have to be treated using pressure, vacuum, or gravity filters that involve cake formation. In general, the solid content of the slurry will determine the choice of method adopted: depth filtration is used for taking out small quantities of contaminants, whereas cake filtration will be employed for more concentrated slurries.

1. Theory of Cake Filtration

The common assumption made in developing theories for cake filtration is that the solid particles deposited on the cake remain without migration. The other important assumption relates to the cake characteristics, i.e., whether it is hard and incompressible, or otherwise. In general, the flow of fluid through the cake can be well described by Darcy's equation:

$$\frac{1}{A}\frac{dV}{dt} = k'\frac{\Delta P}{\mu L} \tag{1}$$

which indicates that the flux through the cake is directly proportional to the pressure gradient ($\Delta P/L$) across it, and inversely proportional to the fluid viscosity (μ). Here, A is the cake cross-sectional area, V is the fluid volume flowing at any time t, and k' is a factor reflecting the cake permeability. For hard and incompressible cakes, k' is constant independent of the pressure gradient. During a given operation, L, the filter cake thickness, is a function of time, which can be easily be related to V since the solids concentration (C) in the feed slurry is known. Thus

$$L = \frac{VC}{A\rho_c} \tag{2}$$

where ρ_c is the density of the cake. A general differential equation can be obtained for fluid flow through the cakes by eliminating L between Eqs. (1) and (2), which results in

$$V\frac{dV}{dt} = \frac{k'A^2\rho_c\,\Delta P}{\mu C} \tag{3}$$

Practical filtrations can either be approximated as taking place (1) at a constant rate or (2) at a constant pressure. For constant rate filtration, dV/dt in Eq. (3) is uniform, given by V/t, and if Q denotes the constant flux ($Q = V/At$), the pressure drop required to maintain this flux can be easily be shown to be

$$\Delta P = \frac{\mu C Q^2 t}{k' \rho_c} \tag{4}$$

In the case of constant pressure filtration, the pressure drop across the cake is maintained constant, either by increasing the pressure of the cake side of the filter or by lowering it on the other side. It follows from Eq. (3) that the flow rate will vary with time. Solving Eq. (3) assuming ΔP to be constant, and with $V = 0$ at $t = 0$ as the initial condition, we get

$$V^2 = \frac{2k' A^2 \rho_c t \, \Delta P}{\mu C} \tag{5}$$

The pressure drop caused by equipment, piping, and the filter medium resistance has been ignored in the above analysis. This is generally justifiable, except when the cake thickness is relatively small. For a more accurate expression for flow, especially during start-up, one must account for equipment and septum resistances. In some cases, there may be an appreciable bleed of solid material through the filter medium at the start of the process until the solid particles bridge across the medium openings. Once again, end corrections may have to be applied under such conditions.

Considerable complications can arise if the cake is compressible. For such situations, one must determine k' under various applied pressures using, say, a compression permeability cell, and this factor has to be taken into account while solving the differential equation (Eq. 3).

Filtration of non-Newtonian fluid–solid mixtures is highly relevant to food processing. The theory of cake filtration for power law–exhibiting fluids has been described in detail by Shirato and coworkers [5], who also describe other quantitative models for cake filtration.

2. Theory of Depth Filtration

The theories developed have almost exclusively been concerned with deep-bed sand filtration. As mentioned, this type of filtration is only applied when small quantities of contaminant, e.g., those causing turbidity in water, are to be removed. The drop in the contaminant concentration with depth is assumed to be exponential. Thus if a liquid containing contaminants at a concentration level of C_0 passes through a bed of length L, the final concentration is given by

$$C = C_0 e^{-\lambda t} \tag{6}$$

where λ is a constant known as the filter coefficient of the bed. Clearly, λ will not remain constant. Its value for a clean bed will indeed be much higher than for a bed in which the contaminants have been deposited. In practice, λ must be correlated with the nature of the contaminants as well as with the nature of the packing of the bed, and then incorporated into Eq. (6).

B. Leaching/Extraction

Leaching and extraction processes are concerned with the separation of soluble constituents from liquid or solid mixtures by use of a selective liquid (aqueous or organic) or supercritical solvent. The process may be used either for the production of solution from materials (e.g., apple juice from apples) or to separate substances such as proteins, lipids (edible oils), or pigments. An excellent introduction to the use of extraction of liquids from liquid mixtures (liquid–liquid extraction) in the food industry is given by Hamm [6].

Extraction as a process, therefore, depends on the transfer of components across interfaces from one phase to another. The two phases must therefore be contacted, and separated when the transfer of the component approaches equilibrium. The simplest extraction system comprises three components: the solute or the material to be extracted, the solvent, and the nonsolute portion of the feed mixture.

It is convenient to model continuous liquid–liquid extraction processes on an equilibrium stage basis [7]. In general, if an extraction system consists of two phases, three variables are available to satisfy performance specifications: the feed rate, the solvent rate, and the number of stages required to achieve a given rate of transfer. Before designing an extraction device, one must ensure that reliable equilibrium data is available. A typical ternary equilibrium diagram is shown in Fig. 1. In the absence of experimental data, predictions can be made using UNIQUAC or NRTL methods. A sound introduction to the principles underpinning the design of liquid–liquid extractors is given by Khoury [8]. The lengths of the segments of the tie line passing through M' (Fig. 1) determine the ratio of the phases according to the lever arm rule, which follows from simple material equations. If L', Q', and M' represent the masses at the points shown in Fig. 1, and x is the mass or mole fraction, the following component balances hold:

$$M'x_{M'S'} = (L' + Q')x_{M'S'} = L'x_{L'S'} + Q'x_{Q'S'}$$

$$M'x_{M'R'} = (L' + Q')x_{M'R'} = L'x_{L'R'} + Q'x_{Q'R'} \tag{7}$$

$$M'x_{M'E'} = (L' + Q')x_{M'E'} = L'x_{L'E'} + Q'x_{Q'E'}$$

where the first subscript refers to the mixture M', phase L', or phase Q'; and the second refers to the solvent (S'), raffinate (R'), or extract (E')

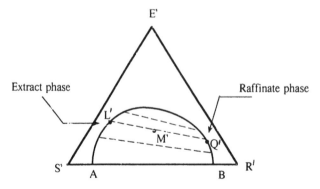

FIGURE 1 Liquid–liquid equilibrium diagram for a ternary system at a given temperature and pressure. E': extract; S': solvent; R': raffinate (feed stream minus the extracted material). The dashed lines are the tie lines connecting the equilibrium phase compositions and thus providing the basis for selectivity. The equilibrium ratio (or partition coefficient) is the ratio of the concentration of the components in the two phases, the extract phase composition forming the numerator. Each of the binaries E'R' and E'S' forms a single liquid phase, while the binary R'S' forms two liquid phases between the compositions represented by points A and B, and a single phase outside this range. The point M' represents the combined composition of a mixture that separates into two phases at equilibrium.

components. It follows that the ratio of the phase can be expressed as

$$\frac{L'}{Q'} = \frac{x_{M'S'} - x_{Q'S'}}{x_{L'S'} - x_{M'S'}} = \frac{x_{M'E'} - x_{Q'E'}}{x_{L'E'} - x_{M'E'}} = \frac{x_{M'R'} - x_{Q'R'}}{x_{L'R'} - x_{M'R'}} \tag{8}$$

Given a feed and a solvent, Q_0' and L_0', respectively, their compositions are first of all plotted on the ternary diagram (Fig. 1). The point M' is determined as the one dividing the line $L_0'Q_0'$ such that the ratio of the lengths of the segments $(L_0' M')/(Q_0' M') = Q_0'/L_0'$. The tie line through M' then determines the composition of the extract and the raffinate.

Let us now consider calculations for a multistage extraction column operating, say, countercurrently. While the ternary triangular diagram can be used to determine the number of theoretical stages required to achieve a given level of separation, the concept of solute free solvents passing countercurrent to each other can be used to simplify calculations. The concentrations are then given as the ratio of the solute to extraction solvent, Y, and the ratio of the solute to feed solvent, X. Equilibrium data in terms of these concentrations are nearly straight lines.

Consider a stagewise extraction scheme, shown in Fig. 2. If the two solvents are assumed to be completely immiscible, F' is the same as R',

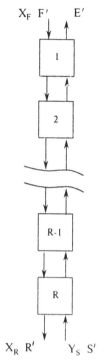

X_F F' E'

X_R R' Y_S S'

FIGURE 2 Stagewise countercurrent extraction: a theoretical scheme.

the flow rate of the raffinate phase; and S' is the same as E', the flow rate of the extract. If X and Y are the mass fractions of the solute in the raffinate and extract phases, respectively, on a solute free basis, a material balance around the feed end of the extractor down to any stage, n, gives

$$Y_{n+1} = \frac{F'}{S'} X_n + \frac{S'Y_E - F'X_F}{S'} \tag{9}$$

An equation similar to this will result if a material balance is taken around the raffinate end of the extractor up to stage n, except that this equation will relate the extract concentration in the nth stage with the raffinate concentration in the $(n - 1)$th stage. Thus,

$$Y_n = \frac{F'}{S'} X_{n-1} + \frac{S'Y_S - R'X_R}{S'} \tag{10}$$

Equations (9) and (10) are known to describe the operating line on a plot of Y against X, which will have a slope of F'/S'. The overall material balance around the extractor is given by

$$Y_E = \frac{F'X_F + S'Y_E - R'X_R}{E'} \tag{11}$$

The operating line and the equilibrium relationship must be solved simultaneously to obtain the number of theoretical stages to achieve a given level of separation. The solution can either be achieved graphically (Fig. 3) or numerically, by stepping between the equation of the operating line and that representing the equilibrium relationship. Analytical expressions are available to determine the number of stages if the solvents are completely immiscible and the equilibrium relationship is a straight line. The situation when the solvents are miscible can be quite complicated. Related information is available [7]. The methods used to design solid–liquid extractors or leaching devices are not conceptually different; Schwartzberg [9] reviews these methods.

In supercritical fluid extraction, the solvent is used at temperatures and pressures above the vapor–liquid critical point. The density is extremely sensitive to temperature and pressure changes near the critical point. For instance, the density of carbon dioxide (the most commonly used supercritical solvent) at the critical point is more similar to organic liquids than gases, and this property leads to similar solubilities, thereby resulting in it being a very good solvent. The solute is also known to diffuse more rapidly through a supercritical solvent than through a liquid solvent; in other words, supercritical solvents exhibit enhanced transport properties. Moreover, the supercritical solvent can be readily separated

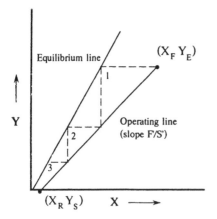

FIGURE 3 Determination of the theoretical number of stages when mass fractions are expressed on a solute free basis.

from the solution merely by reducing the pressure and/or by increasing the temperature. These advantages are sometimes offset by the need to operate the extractor under unfavorable conditions of pressure and temperature. For example, caffeine is removed from coffee using supercritical carbon dioxide at pressures as high as 160–220 atm and temperatures around 90°C.

Conceptually, the design procedures for liquid–liquid extractors, discussed earlier, ought to be applicable to supercritical fluid extractors. However, phase equilibrium models are not readily available, which render process design, scale-up, and control rather difficult. King and Bott [10] have discussed the relevant concepts in detail.

C. Membrane Processes

Certain natural membranes, or membranes made of polymeric materials, are known to be selectively permeable to certain substances. This property of membranes can be exploited to separate permeable components from a multicomponent fluid mixture. The discrimination is based on the molecular shapes, sizes, and charges of the components. Areas of membrane applications can be divided according to the "driving force" causing component transfer across any membrane. For example, hydrostatic forces control *ultrafiltration* (UF) and *reverse osmosis* (RO); electrostatic forces control *electrodialysis*; and osmotic pressure controls *osmodehydration*. Gutman [11] has produced an exhaustive classification of all current areas of membrane applications, primarily based on driving force considerations. The following discussion will be restricted to pressure-driven membrane separation processes, in particular *microfiltration*, UF, and RO, which are more relevant to the food and drink industry.

Concentration of solutions using membrane technology is an attractive alternative to evaporation because components can be separated at fairly low temperatures (this advantage can be exploited to maximize quality retention), and it is economically viable to operate even on a smaller scale. Membrane technology has been tested in the large-scale manufacture of almost all kinds of food materials, but the most significant application is in the dairy industry, where this technology is used in the processing of wheys. Membranes are also being used to concentrate skimmed milk directly. In the production of milk powder, skimmed milk is first concentrated by reverse osmosis. The concentrate is then subjected to evaporation before spray drying. A considerable amount of soft and semihard cheese is now being made by ultrafiltration of skim milk, full cream milk, and formulated milk. Membranes are also used in the beverage industries. Fruit and vegetable juices including sugar beet juice, tomato juice, and lemon juice can be concentrated by reverse osmosis. Applications of reverse osmosis in the wine industry include production of low alcohol wine, must concentration, and more promisingly, tartar stabilization (i.e., removal of tartar, which tends to precipitate in a stored bottle or gives a hazy appearance). By subjecting wine to reverse osmosis, the concentration of tartar can be increased above its saturation value at a given temperature, resulting in its precipitation. After separating the precipitated tartar, the concentrate and permeate are reconstituted [12].

Consider the situation shown in Fig. 4, where a concentrated solution at high pressure (milk) is separated by a membrane from a dilute solution at lower pressure (water). The diffusion of the permeate through the membrane is (1) aided by the transmembrane pressure and (2) opposed by osmotic pressure developed because of the concentration difference. Further, if it is assumed that Darcy's law can be applied to the flow occurring through the membrane, the net permeation velocity, over time intervals so short that neither volume nor concentrations have changed, is given by

$$v = A(\Delta P_m - \sigma \Delta \pi) \tag{12}$$

where A is the membrane permeability, ΔP_m is the pressure differential across the membrane, and $\Delta \pi$ is the difference in osmotic pressure between the solutions on either side of the membrane. The term σ is a

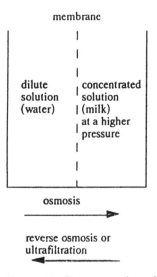

membrane

dilute solution (water) | concentrated solution (milk) at a higher pressure

osmosis

reverse osmosis or ultrafiltration

FIGURE 4   Representation of osmosis and reverse osmosis.

membrane characteristic, introduced to account for the relative permeabil-
ities of the solvent and solute: if the membrane is permeable to the solvent
and completely impermeable to the solute, $\sigma = 1$; likewise, if the mem-
brane is only permeable to the solute and lets no solvent through, $\sigma = 0$. In most practical cases relevant to food processing operations, high
rejection membranes are used; i.e., the membrane can be assumed to be
largely impermeable to the solute. Therefore, with $\sigma = 1$, Eq. (12) reduces
to its more commonly used form:

$$v = A(\Delta P_m - \Delta\pi) \tag{13}$$

For high rejection membranes, the permeate is predominantly solvent,
and therefore $\Delta\pi = \pi_i$, the osmotic pressure of the feed solution. Further,
when the molecular weight of the solute is low (which would be the case
in many reverse osmosis applications), the solution osmotic pressure is
high and significant in comparison with the applied pressure differential
ΔP_m. On the other hand, when solvent is separated from solutions containing high-molecular-weight solutes such as lactose and other proteins (such
separations being more common in ultrafiltration or microfiltration applications), the solution osmotic pressure π_i is low and negligible in relation
to ΔP_m. Thus Eq. (13) further simplifies to

$$v = A \, \Delta P_m \tag{14}$$

It is also evident from Eq. (13) that the minimum value of applied transmembrane pressure for any process to be feasible is $\Delta\pi$; clearly, the upper
limit depends on the membrane tolerance. To a first approximation, this
information can also be used to estimate the maximum factor by which
a feed solution can be concentrated using a given membrane: for example,
if the feed has an osmotic pressure of 100 psi and the membrane can
tolerate a maximum pressure of 600 psi, the solution can at most be concentrated by a factor of 6.

In reverse osmosis applications, one must know the values of osmotic pressure, especially as a function of solution concentration and
temperature. For food substances such data are not readily available;
some experimental data on food components are reported by Cheryan
[13]. Note that in the case of protein solutions, osmotic pressure varies
with pH as well as with concentration and temperature. To get a feel for
osmotic pressure values, note that for substances such as milk and fruit
juices, with total solids in the range of 5–20%, the osmotic pressure takes
values between 100 and 500 psi.

In addition to permeate flux discussed above, one must consider the
degree of concentration achieved in a given process. It is customary to
define the latter in terms of membrane rejection or retention:

$$R = \frac{C_f - C_p}{C_f} \tag{15}$$

where C_f and C_p are the concentrations of the solute in the feed and the permeate, respectively. It is also possible to express R in terms of the fluxes. The behavior of membranes is described by both *permeate flux* and *rejection values*. It is important to appreciate that although reverse osmosis, ultrafiltration, and microfiltration are all pressure-driven membrane separations, the *intrinsic mechanism of transport* through reverse osmosis membranes and, hence, the flux and rejection values, are different from those occurring in the latter two applications.

One must appreciate that the intrinsic performance capabilities of membranes are seldom realized in practice. Theoretical models for permeate flux have only been experimentally validated in the case of fresh membranes, and that only for low values of ΔP, low feed concentrations, and high feed velocities; significant deviations are observed otherwise. Typical variations of flux with ΔP_m are shown in Fig. 5. It is clear that the flux, after increasing initially with ΔP_m, attains a limiting value; in certain cases, it even decreases subsequently as ΔP_m increases. The deviation from linearity is attributed to a phenomenon known as *concentration polarization*: the accumulation of the rejected solute at the membrane surface, which causes higher solute concentration to prevail near the membrane than in the bulk solution. In its simplest form, concentration polari-

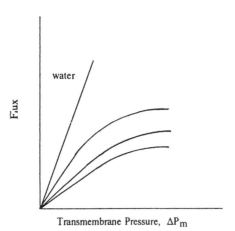

FIGURE 5 Variation of flux with transmembrane pressure. In the case of water, the plot is linear; for most other solutions, the flux attains a limiting value due to concentration polarization.

zation can have two possible effects:

1. The higher solute concentration adjacent to the membrane can cause solute to diffuse back and limit the net volumetric flux reaching the membrane.
2. The higher concentration adjacent to the membrane (often significantly higher) can also set up an even higher osmotic back-pressure.

Although the former effect is widely accepted, the effect of applied pressure on permeate flux is offset in either case, and the flux variation deviates from linearity. In addition to concentration polarization, *fouling of membranes* can also cause significant departure from any theoretical model that considers only intrinsic membrane characteristics. These extrinsic factors—fouling and concentration polarization—can occur in reverse osmosis as well as ultrafiltration, and they are therefore general phenomena associated with membrane transport. Unlike other separation operations, it is advisable to incorporate transient variations in performance at the design stage itself.

1. Membrane Modules and Their Operation

The heart of any membrane separation equipment is its module, which contains the membrane. The membrane is generally made of cellulose acetate polymer; other polymers such as polyamides, polysulfones, and polyvinyl chloride (PVC), cast individually or as a composite, are being increasingly used. In addition, inorganic membranes such as those made of ceramic are also available. Both hydrodynamics (or fluid mechanics) and mass transfer play a crucial role in module design; other important considerations are easy access for cleaning and cost-effective manufacture and membrane replacement. The role of fluid mechanics in membrane processes has been discussed in great detail [14, 15]. The relationships necessary for calculating friction losses and mass transfer parameters will depend on the module type; four types of modules are in wide use: the tubular module, the plate and frame module, the hollow fiber module, and the spiral wound module [12].

Membrane modules are either operated batchwise or continuously. When the volume of liquid to be processed by a module is small, it is generally operated batchwise. In a simple batch-operated plant, the liquid feed is taken in a tank and circulated through the module. The retentate is either fully or partly returned to the feed tank (Figs. 6a and b). As the permeate is withdrawn, the concentration of the rejected components in the tank increases with time. Often, in practice, a fraction of the retentate is recycled around the membrane, as shown in Fig. 6b.

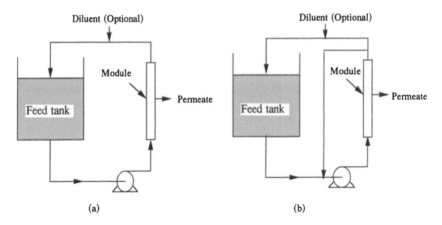

FIGURE 6 Simple batch-operated module (a) with the retentate being fully returned to the tank, and (b) with a fraction of the retentate recirculating. Diluent
addition is optional. When diluent is added in an ultrafiltration process, it is known
as diafiltration.

An analysis of a batch module with the retentate fully returned to the
feed tank will now be discussed. If C is the concentration of the rejected
component at any time t and V is the volume of the tank contents, then
assuming that the membrane totally rejects this component,

$$CV = C_i V_i \tag{16}$$

where the subscripts i refer to initial values. Equation (16) simply states
that the mass of the totally rejected component is constant at all times.
The rate of change of volume is obtained by differentiating Eq. (16) w.r.t.
t. Thus

$$-\frac{dV}{dt} = \frac{V_i C_i}{C^2} \frac{dC}{dt} \tag{17}$$

Moreover, if v is the permeate velocity and A is the surface area for
permeate flow,

$$-\frac{dV}{dt} = vA \tag{18}$$

Substituting for $-dV/dt$ from Eq. (18) into (17), the processing time for
increasing the concentration of the rejected component is given by

$$t_p = \frac{V_i C_i}{A} \int_{C_i}^{C_t} \frac{dC}{C^2 v} \tag{19}$$

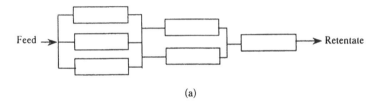

(a)

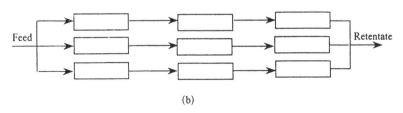

(b)

FIGURE 7   Single pass module arrangements for continuous operation: (a) tapered arrangement and (b) squared-off arrangement; permeate streams are not shown.

The integral can be evaluated if the relationship between permeate velocity and solute concentration is known.

Continuous membrane plants are laid out in the form of parallel trains of modules, which are themselves arranged in series (Fig. 7). The retentate from each set of parallel trains is often subjected to pressure boosting before introducing it to the next set of trains. These arrays of modules can either be "tapered" or "squared off," as shown in Figs. 7a and b, respectively. The tapered arrangement is favored so that the cross-flow in each module can be maintained at a value that will minimize concentration polarization.

The arrangements described in Fig. 7 are "single pass" units. It is also possible to have a "feed and bleed" type of arrangement, in which most of the retentate from any module is recirculated after pressure boosting. The recirculation flow rate around each stage is much larger than both feed and bleed rates. A typical three-stage feed and bleed continuous plant is shown in Fig. 8.

The performance of a cascade of modules is studied in terms of "module separation factor," and for the nth stage, it is defined as

$$\lambda_n = \frac{y_n}{x_n} \tag{20}$$

where y_n is the mole fraction of the component in the permeate and x_n is

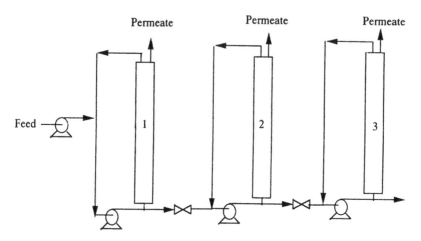

FIGURE 8   Three-stage feed and bleed continuous plant.

that in the retentate. In addition, the term split factor is also defined as

$$\theta = \frac{P}{F} \tag{21}$$

where P is the mass flow rate of the permeate from a given module that is fed at a rate F. Consider a cascade of single pass modules, all arranged in series, as shown in Fig. 9. A mass balance around the nth module can be written as follows:

$$P_{n+1} = P_n + R_n \tag{22}$$

The component mass balance equation is given by

$$P_{n+1}y_{n+1} = P_n y_n + R_n x_n \tag{23}$$

Eliminating R_n between Eqs. (22) and (23), it follows that

$$P_{n+1}y_{n+1} = P_n y_n + (P_{n+1} - P_n)x_n \tag{24}$$

Equation (24) can also be expressed in terms of λ and θ using Eqs. (20) and (21):

$$y_n = y_{n+1} \frac{\lambda_n}{1 + \theta_n(\lambda_n - 1)} \tag{25}$$

Similar equations can be written relating component mole fraction in the permeate of any stage with that in the corresponding feed stream. Combin-

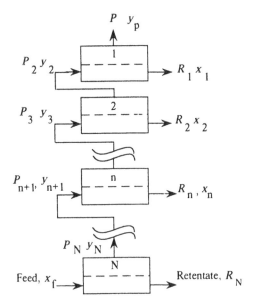

P y_p

P_2 y_2 1 R_1 x_1

P_3 y_3 2 R_2 x_2

P_{n+1} y_{n+1} n R_n x_n

P_N y_N

Feed, x_f N   Retentate, R_N

FIGURE 9 Cascade of modules without reflux.

ing all such relationships, the mole fraction in the final permeate stream is given by

$$y_n = x_f \frac{\prod_{i=n}^{N} \lambda_i}{\prod_{i=n}^{N}[1 + \theta_i(\lambda_i - 1)]} \tag{26}$$

It follows from the definition of the split factor (Eq. 21) that the mass flow rate of the final permeate is given by

$$P_n = F \prod_{i=n}^{N} \theta_i \tag{27}$$

If λ_i and θ_i values are the same for all the modules in the cascade, Eqs. (26) and (27) simplify to

$$y_n = x_f \left[\frac{\lambda}{1 + \theta(\lambda - 1)} \right]^{N+1-n} \tag{28}$$

and

$$P_n = F\theta^n \tag{29}$$

When the expression for x_n is substituted from Eq. (20) into Eq. (25), the following equation is obtained, which is known as the equation of the operating line:

$$y_{n+1} = [1 + \theta_n(\lambda_n - 1)] x_n \tag{30}$$

If, for a situation, x_f, F, and the desired y_n are given, and θ and λ values are known, it is possible to solve the above equations to determine n, the number of stages. The approach described earlier to determine the number of stages in extraction and distillation columns can be used here; i.e., the equation of the operating line, Eq. (30), can be solved simultaneously with the "equilibrium curve." In the case of membrane separations, however, there is no "thermodynamic equilibrium." Instead, it would be represented by the separation characteristics of the module, which depend on (1) membrane selectivity, (2) fluid dynamics in the module, (3) component concentrations, and (4) overall flow pattern, i.e., whether the retentate and permeate flow co-current or countercurrent in relation to each other, and whether these streams follow plug flow or back-mixed flow. For simplicity, only a limiting case will be considered in which λ tends to 1. The operating line given by Eq. (30) reduces to $y = x$. Moreover, the equilib-

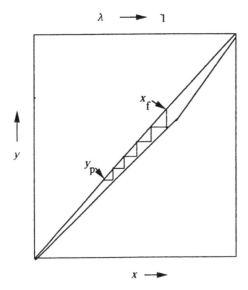

FIGURE 10   Theoretical stagewise diagram for the limiting case $\lambda \to 1$.

rium line will also be very close to this line. This case represents a small separation factor and hence a large number of module stages is needed for a given separation; see Fig. 10.

III. CONCLUDING REMARKS

The principles underpinning process design of three specific separation operations practiced widely in the food processing industries have been described with a view to encourage integration of process modeling with computer-based control strategies. It is evident that all of these processes have a viable background for exploitation; however, current practice largely centers on the control of process variables that have been set at specific values rather than being controlled and manipulated by the transient events taking place during the process. By integrating phenomenological models with control equipment, the efficiency of separation processes can be vastly improved.

REFERENCES

1. P. J. Fellows, *Food Processing Technology*: *Principles and Practice*, Ellis Harwood, New York, 1988.
2. J. M. Coulson, J. F. Richardson, J. R. Backhurst, and J. H. Harker, *Coulson and Richardson's Chemical Engineering*, vol. 2, Pergamon Press, Oxford, 1991.
3. J. Voborsky and T. Struma, Automated metering of keiselguhr during filtration of beer, *Kvansny Prumsyl 40*:107 (1994).
4. E. Rawlinson, Prototype promises a bonus for brewers, *Process Industry J. 7*(10):40 (1992).
5. T. Murase, E. Iritani, J. H. Cho, and M. Shirato, Determination of filtration characteristics of power law non Newtonian fluid–solid mixtures under constant pressure conditions, *J. Chem. Eng. Japan 22*:65 (1989).
6. W. Hamm, Liquid–liquid extraction in the food industry, *Handbook of Solvent Extraction* (T. C. Lo, M. H. I. Baird, and C. Hanson, eds.), Wiley Interscience, New York, 1983, p. 593.
7. R. E. Treybal, *Mass Transfer Operations*, McGraw-Hill, New York, 1980.
8. F. M. Khoury, *Predicting the Performance of Multistage Separation Process*, Gulf Publishing, 1995.
9. H. G. Schwartzberg, Leaching of organic materials, *Handbook of Separation Process Technology* (R. W. Rosseau, ed.), Wiley, New York, 1987.
10. M. B. King and T. R. Bott, *Extraction of Natural Products Using Near Critical Solvents*, Blackie, Glasgow, 1993.
11. R. G. Gutman, *Membrane Filtration: The Technology of Pressure Driven Crossflow Processes*, Adam Highler, Bristol, 1987.

12. R. Rautenbach and R. Albrecht, *Membrane Processes*, John Wiley, New York, 1989.
13. M. Cheryan, *Ultrafiltration Handbook*, Technomic Publishing, Lancaster, Pennsylvania, 1986.
14. G. Belfort and N. Nagata, Fluid mechanics and cross flow filtration: Some thoughts, *Desalination 53*:57 (1985).
15. G. Belfort, Membrane modules: Comparison of different configurations using fluid mechanics, *J. Membrane Sci. 35*:245 (1988).

13

Computerized Food Warehouse Automation

J. Pemberton Cyrus and Lino R. Correia
Technical University of Nova Scotia, Halifax,
Nova Scotia, Canada

I. INTRODUCTION

A food warehouse is a building where food and kindred products are stored before being distributed to retailers. Thus warehouse activities essentially involve flow and storage of goods, distribution methods, and stock control [1].

Food warehouses have unique requirements. Every food has an ideal shelf life, after which product quality may slowly or rapidly deteriorate. Meat, fish, and dairy products must be stored at desired frozen or refrigerated temperatures. Consumption of such foods, subject to temperature abuse during storage and handling, may lead to food poisoning. Certain products need to be stored under controlled atmospheric (CA) conditions with specific humidity levels to slow down the ripening process. Spices need to be stored in a dry environment. Thus food quality is a function of time, temperature, and perhaps humidity during storage and transportation.

The need for cost reduction and the availability of digital computer–based technology has lead to the proliferation of warehouse automation. The growth of multiple retailing and the advent of grocery supermarkets with large turnovers of goods at marginal profits have produced a desperate requirement for efficient distribution. Digital computers can easily integrate all aspects of data processing from receiving to shipping activities in warehouses and number crunching in offices. Automation can also reduce the personnel required to work under the adverse environmental conditions of subfreezing temperatures in some warehouses.

Benefits of warehouse automation are

Decreased product loss or damage
Decreased handling, inventory, and cycle or rotation time
Decreased human resources cost
Increased throughput and efficiency
Improved accuracy
Improved inventory and resources management, reporting, and communication

A major drawback of automation can be prohibitively high costs. Hence it is necessary to consider the pros and cons of automating a warehouse before selecting the best method for a situation [2].

The purpose of this chapter is to examine computerized automation in food warehouses by reviewing information requirements and computer use in warehouses, discussing the planning process, identifying system design issues, describing system architecture, and presenting some applications.

II. FOOD WAREHOUSE INFORMATION REQUIREMENTS

One must define specific information requirements prior to design and computerized automation of food warehouses. Some requirements are monitoring and control of product location, product quality, sanitation, occupational safety, and energy efficiency. A longer information requirement list requires a more complex warehouse management system (WMS); therefore one should limit the requirements to only those needed to meet the most important goals of the organization.

A. Monitoring and Control of Product Location

Product placement is critical for food and other products that can emanate and absorb strong odors. The knowledge of product location is helpful in simulating material handling and in determining lead times and stock levels. If product overcrowding results in lower productivity, then house-keeping is a concern. The traditional approach of preassigning products to a specified location may not be necessary for an adequately automated warehouse, but the juxtaposition of products must be controlled by the WMS.

B. Product Quality

Product quality of some foods can be ascertained by recording time, temperature, and perhaps humidity. Most food products have a "best before" consumption date. Thus from a legal perspective, stock rotation is critical. For perishable commodities it may be necessary to measure product quality. For example, regimented stocks of bagged cocoa were "mapped" according to a precise floor pattern and constantly sampled to check for any possible deterioration. Also the bags were frequently restacked so that all bags were regularly probed [3]. Product or package damage can occur during storage in aisles. Cartons storing liquid products are especially susceptible to damage, and they may leak [4]. Thus a WMS can set up rules to rotate stock, to restack for sampling, and to prevent avoidable product or package damage.

C. Sanitation

Rodents, insects, and birds can contaminate food produce. The U.S. Food and Drug Administration (FDA) and the Canadian Department of Health conduct appropriate sanitation inspections of food warehouses. Hence, any WMS must monitor rodent exclusion and extermination, bird and insect control, inbound inspection, and cleanup methods and equipment [4, 5].

D. Occupational Safety Hazards

The U.S. Occupational Safety and Health Act (OSHA) applies to the warehouse industry. For example, cluttered aisles and slippery floors due to product spills can cause safety hazards [4]. Dust control of particulate food is important to prevent explosions. A computerized warehouse facility can monitor dust levels and storage in aisles; it also typically carries less inventory, hence reducing safety hazards.

E. Energy Consumption

Frozen and refrigerated warehouses are high energy consumers. A computerized warehouse typically increases throughput, thus lowering the refrigeration load. Also, minimizing truck movements can conserve energy. A WMS can track energy consumption.

III. COMPUTER USE IN WAREHOUSE AUTOMATION

In automating a warehouse, computer use appears in the following areas:

Warehouse management systems
Automated material handling
Warehouse design
Distribution optimization

The level of automation can be classified by the degree of capital investment, the overall material flow capacity of the system, and the labor productivity. We may consider three major automation levels: a WMS involving forklift operators interfacing with a remote computer; a WMS with automatic guided vehicles (AGVs); and a WMS with an automated storage and retrieval system (AS/RS) [6].

A. Warehouse Management Systems

A warehouse management system is a software package designed to meet the information requirements of a warehouse. A WMS is usually purchased from a software vendor, and it may be extensively customized by either the vendor or customer before final implementation. A typical WMS may have the following features [7–9]:

Shipping and receiving
Put-away
Storing and retrieving pallet-loads
Picking from split cases
Sequencing goods at the dock for staging onto trucks
Generating pick-lists on paper
Generating pick-lists on radiofrequency data collection (RFDC) terminals
Generating bar code labels
Packaging
Inventory management
Resource management
Controlling automated material handling equipment
Sequencing and scheduling the use of dock doors
Warehouse layout and material handling system design evaluation
Truck routing and scheduling
Order entry and sequencing

The typical modules and relationships of a WMS to the organization's information system (IS) are shown in Fig. 1. This figure was developed from details and diagrams in [7] and [10]. The various subsystems are shown with a solid outline, and external sources of information are shown with dotted outlines.

In [11], the success of Ritter Food Corp. in automating its food distribution business is discussed. The implementation cost of the WMS was in the region of $1 million. Among other improvements, the system allowed minimization of out-of-stocks to less than 1.5% of items ordered, and closer management of worker incentive schemes. The WMS is interfaced with the RoadNet vehicle routing system, which produces daily vehicle routes, to balance truck loading and ensure that products are loaded in reverse stop sequence.

1. Inventory Control, Order Processing, and Receiving/Put-Away

These processes involve a number of activities, such as age tracking, ensuring first-in, first-out (FIFO) shipping of goods, determining locations of goods for put-away, and automatic order entry. One function of the WMS is to automate these functions to the maximum possible degree. If age tracking and FIFO are fully automated, then food spoilage can be reduced or eliminated. Automatic location of goods for put-away can ena-

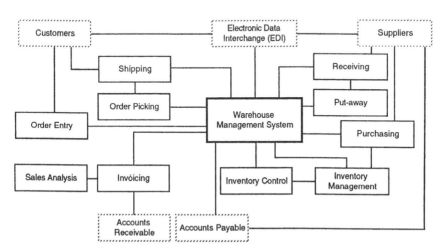

FIGURE 1 The modules and relationships of a warehouse management system (WMS).

ble optimization of picking sequences to minimize the travel time of, and interference between, order pickers. Automation of ordering may be accomplished with the help of links between the warehouse, suppliers, and customers by the use of electronic data interchange (EDI). This may allow shorter order cycles, and enable the warehouse to achieve just-in-time operations strategies.

2. Order Picking

Order picking operations may be optimized by use of optimal pick-lists. The objectives may be to minimize travel distance while picking, or to optimize stacking of the pallets, for example. However, the primary use of IS in order picking is to create the pick-lists and communicate the picking schedule to the order pickers. A simple pick order may be used; for example, an order may be picked in the sequence in which the items will be stacked on the customer's shelves [12]. The scheduling of orderpicking also allows detailed monitoring of orderpickers' efficiencies.

The advantages of paperless order picking are reported in [13] to be

Increased productivity of the order pickers (20% to 50% improvement)
Reduction in picking errors (90% to 95% reduction)
Pick exceptions corrected directly on the host computer
Reduction or elimination of post-pick data entry for pick exceptions
Reduced inventory
Reduced cycle counting
Expedited (hot) orders obtained directly from the host computer
Stock replenishment done on-line
Picker productivity tracking
Stockkeeping unit (SKU) activity tracking
Orderpicking workload forecasting and balancing
On-line order status reports
Packing lists with pick exceptions generated automatically
Interfaces to other warehouse information systems

The typical performance ranges for different picking systems are given in [14] and are summarized in Table 1. In this table, lines per hour are calculated as total lines shipped annually divided by total annual labor hours.

B. Warehouse Design

Computers may be used in determining optimal layout of the warehouse. Many software packages are available for layout [15]. The FactoryFLOW

TABLE 1 Typical Pick Rates for Various Picking Systems

Picking system	Pick rate (lines per hour)
Pick to cart	25–125
Pick to tote	100–350
Horizontal carousel	50–250
Vertical carousel	50–350
Miniload automated storage and retrieval system (AS/RS)	25–100
Automatic picker	500–1000

package [16] operates in conjunction with AutoCAD, which makes it convenient for analysis and production of final layouts. Very rarely will a package be able to produce an optimal layout, but they can usually produce very good layouts in a short time.

Plant layout requires detailed knowledge of material flows (origins, destinations, volumes, handling equipment), and the relationships among flows (e.g., are there two products whose paths must not cross?) and among departments. In addition to these considerations, a food warehouse design must consider the location of freezer and refrigerated spaces, separate storage of liquids, sanitation, and occupational health and safety.

C. Material Handling Systems

Material handling systems often incorporate computer-based control systems, and the WMS can sometimes be interfaced with these to provide an integrated operating environment. For example, if an automated storage and retrieval system (AS/RS) uses computer-controlled positioning of its truck, then it is possible to provide real-time feedback of the truck's location, and therefore take snapshots of the system state: how many orders are finished, which orders are on time, and estimates of completion time for outstanding orders.

The various options for automated material handling are described in [17]; some options include automated storage and retrieval systems, automated guided vehicle systems, and carousels.

1. Automated Storage and Retrieval Systems

An automated storage and retrieval system (AS/RS) is a narrow-aisle, high-bay storage system with an automated driverless truck system for storing and retrieving items/pallets under direct computer control. The items are stored randomly in the AS/RS, with their locations recorded in the computer system. When an item is requested, the truck moves to the appropriate location, retrieves the item, and deposits it on a conveyor or transfer station. An AS/RS has the advantages of high storage density, high throughput, high security, automated picking and placing, very accurate picking, low labor cost, and the ability to handle a large number of SKUs. The major disadvantage is high capital cost.

For a dairy cold storage facility with a 330,000 gal total storage capacity, an AR/RS computer control system performed the following activities: continuous and up-to-date inventory control, balanced product rotation, improved product protection, effective space utilization, reduced operating costs, automated handling, reduced errors, and more efficient distribution. Finished products were logged into the computer and transported via carts and dollies into the AS/RS [6].

The AS/RS is popular in retail supermarket distribution centers and is seldom used in manufacturers' regional distribution centers or in raw material warehouses [18].

In [19], an AS/RS with a case handling capacity of 4000 cases/h using radio-controlled rack entry modules (REMS) is reported.

2. Automated Guided Vehicle Systems

An automated guided vehicle system (AGVS) is a system of autonomous driverless vehicles with programmable routes. The vehicles usually follow either a buried cable emitting a radiofrequency (RF) signal, or a painted line on the floor. Some newer AGVs are guided by bar-coded station tags or use "dead reckoning". In the dead reckoning systems, sometimes called self-guided vehicles (SGVs), the vehicle maintains a record of its location based on calculations of distance traveled and turns executed. An AGVS is expensive, but it has route flexibility and low labor cost.

A soft drink bottling and distribution plant used 24 AGVs to transport 120 pallets of product per hour from the plant to its warehouse and distribution building, and ultimately to its shipping docks [6]. In a soft drink warehouse and distribution center having a capacity of 12,000 pallets, nine automatic stacker cranes patrolled the warehouse aisles and about 20 AGVs moved pallets to the warehouse from the bottling lines and from the warehouse as necessary [20].

A cereal processing company used 11 AGVs to transfer product from the bakery to a six-aisle AS/RS in its workshop. Computers assigned a storage location and automatically transferred the product into the assigned space. When a pick order was entered into the computer, the AS/RS cranes retrieved the appropriate product in the correct order according to preprogrammed instructions (code, date, FIFO, mixed pallet configuration, truck weight distribution, etc.) [6].

3. Conveyors and Carousels

Conveyors are the traditional workhorses of warehouse automation. They provide high-speed transport of items between fixed locations, but they are more limited in their routing than AGVs. Conveyors are typically used in loading trucks and in some types of order picking.

Carousels are vertical or horizontal circulating storage racks that can be moved on demand.

4. Computer Simulation

Detailed computer simulation models may be developed to mimic the operation of the warehouse design, highlighting possible bottlenecks and allowing precise determination of resource requirements such as the numbers of lift trucks, racks, loading bays, and order pickers. Experiments with different layouts of racking may be performed, as well as different picking and put-away policies. Law and Kelton [21] provide comprehensive coverage of the theory and application of simulation modeling and analysis.

D. Distribution Optimization

Deliveries may be scheduled to minimize the lateness of orders, or to ensure that all orders are picked and delivered with the minimum use of resources. Software may also be employed for routing and scheduling of delivery trucks. Optimal or near-optimal routing of trucks may save 10% or more in travel costs. If routing is integrated into the WMS, then orders can be loaded onto the trucks in reverse stop order, further improving performance of the shipping process.

IV. THE PLANNING PROCESS

The success of an information system project depends greatly on the quality of planning and control exerted. The process usually can be broken into a number of phases, as shown in Fig. 2. This model of system development is loosely based on the system development life cycle (SDLC). The SDLC model is an effective tool for visualizing the system development process. The phases of the project should not be considered mutually exclusive, but rather as general guidelines to activities that must be performed as part of a continuous, highly interdependent, and somewhat

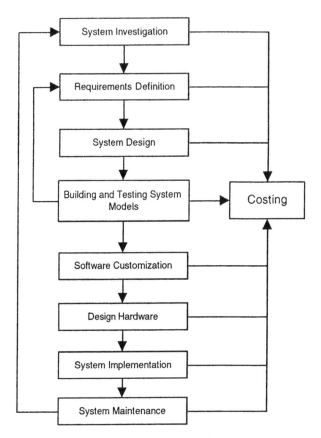

FIGURE 2   An overview of the planning process.

iterative process. As shown in Fig. 2, cost is a major factor in the decision–making process.

A.   System Investigation

The system investigation lays the foundation for your information system. Every aspect of the system must be interrogated, and the system analyst must gain complete familiarity with all operating procedures as they exist.

1.   Goals

Typical goals of a warehouse are high throughput, low labor costs, high accuracy, outstanding customer service, easy access to information, minimal inventory costs, smooth surges of activity, and a short learning curve for workers [22].

Computerization may help in achieving many of the goals of an efficient warehouse, but computerization itself is not an acceptable goal. The goals must be defined in terms of the productivity measures of the warehouse. Examples of goals may be reduced cycle time, higher inventory accuracy, or reduced inventory. Goals may also be based on customer requirements such as bar-coding standards, or on supplier standards such as electronic data interchange (EDI).

Possible goals for warehouse computerization are

Reduce response time to picking or shipping errors
Increase productivity of workers
Reduce cycle time
Reduce operation costs
Increase customer satisfaction
Meet supplier standards
Meet customer standards
Reduce inventory levels
Eliminate shrinkage
Eliminate stale-dated inventory items
Reduce cycle counting
Increase turnover (inventory turns)
Eliminate picking errors
Reduce truck unloading time
Eliminate truck waiting time for unloading
Reduce truck loading time
Reduce put-away time
Reduce picking time (order picker travel time)
Reduce customer unloading time
Increase number of full pallet loads
Improve operator training with less training time
Produce timely and informative management reports
FIFO inventory

2. Project Team

The project team manages the information systems project and implements the resulting system. Requirements for a project team [10] are

1. At least one member from distribution and at least one from the management information systems department
2. Full-time commitment of team members to the WMS project

3.   A project manager, possibly part-time
4.   A sponsor at the executive level

In assembling the team, take an interdisciplinary approach, selecting members who know the details of the system's operation. Necessary skill sets include management of information systems, distribution management, and food warehouse operations. Executive sponsorship is necessary to ensure the cooperation of organization members at all levels.

The project team is also responsible for system integration. The WMS must interface with material requirements planning, accounting, and distribution information systems. The task of system integration is therefore quite demanding. Help may be available from a number of external sources [7]:

1.   Computer and software suppliers help to integrate the flow of information throughout the warehouse. They supply software packages and customization services, and in some cases they can provide turnkey solutions.
2.   Suppliers of material handling systems can integrate material flows throughout the warehouse. These suppliers may also provide the computer systems needed to integrate material handling hardware into the WMS.
3.   Automatic identification equipment suppliers provide bar coding and radiofrequency data collection systems and software for linking data collection into the WMS.
4.   Systems integrators may play the role of project managers, for example, designing software, purchasing all equipment, and managing the installation.
5.   Engineering consultants design the warehouse, doing a requirements analysis and giving complete specifications for material handling equipment, layout, site planning, and building design.

3.   Systems Audit

Review the business processes and prepare a description of how data and materials are generated and used in the warehouse. Data flow and material flow diagrams allow analysis of current operations to determine possible improvements in business processes.
Diagramming also helps to identify sources of inefficiency that may be eliminated by automation. The system audit typically involves the following activities:

1.   Review business processes.
2.   Prepare data flow and material flow diagrams.

3. Identify areas of inefficiency in the business processes.
4. Determine the value and cost of each business process.
5. Rank the processes from most to least important based on value and cost.

The business process review involves interviewing potential system users and documenting their data inputs, information processing, and outputs. From these interviews and documents, the WMS requirements may be extracted. Finally, discuss the requirements with the user to determine if anything was omitted or misunderstood.

Data flow diagrams show the origin, processing, and destination of data. An example of a data flow diagram for material receiving is shown in Fig. 3. This diagram uses standard data flow symbols, where data sources and sinks are shown as rectangles, a file as an open-ended rectangle, and processes as rounded rectangles. In Fig. 4 is a simplified material flow diagram for receiving pallets and storing them in an AS/RS. Flow-charting software, such as MacFlow for the Macintosh or ABC Flow-charter for Windows, is available; these packages simplify the process of chart drawing by maintaining logical links between the nodes of the chart.

B. Requirements Definition

The requirements definition is the skeleton of the IS. Every potential user of the system must be interviewed, and each user's requirements documented. Although some requirements may be ignored to produce a manageable IS project, one must be conscious of all possible user requests. The main activities are [23] identification of all information sources and uses, and searching for areas of potential improvement in information flow.

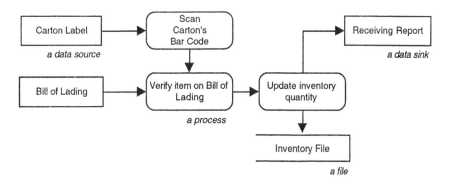

FIGURE 3   A data flow diagram for part of the receiving process.

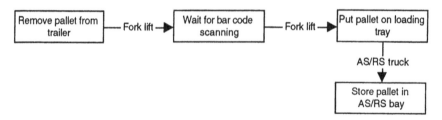

FIGURE 4 A material flow diagram for receiving pallets.

C. System Design

The objective of the system design is to produce a blueprint for an IS
that will meet all user requirements. System design proceeds in stages:
database design, interface design, and algorithm design.

1. Database Design

Database design is the specification of a database structure that allows
production of all outputs as specified in the user requirements. This design
is detailed within the confines of a data model. The relational data model
is the most popular at present, because of its simplicity and power. A
good database design will eliminate certain types of errors, referred to as
anomalies in the relational database jargon, and minimize data redundancy. Pratt and Adamski [23] is a good introductory text on database
design.

2. Interface Design

Interface design specifies the "look-and-feel" of the information system:
every report, screen, and label must be designed with the objectives of
maximum usability. Some features of good interface design are the following:

1. Data entry screens accept data in the order that it is available,
    as the screens are traversed from left to right, and top to bottom,
    with no sudden jumps required.
2. For maximum data entry speed via keyboard, it should not be
    necessary to move the hands from the keyboard to control the
    screen.
3. Reports should provide just the right amount of information
    needed to do the user's job: a good guideline is that the report
    must provide just enough information to make the correct de-
    cision.

4. Every abbreviation on the report should be explained either on the report or in a quick-reference card.
5. Every input/output (I/O) screen or report should be fully documented as to purpose, interpretation, and operation.
6. The language (jargon) of the interface should be that of the users.

Graphical user interfaces (GUIs), such as Windows or the Apple Macintosh operating system, allow significant savings in training time and reduce the intimidation of first-time users of information systems. A typical data entry screen in the Microsoft Windows operating system is shown in Fig. 5; this screen was written for a WMS in Microsoft FoxPro. The use of a GUI also makes available easy report customization and data queries. Use of GUI-based operating systems on cheap fast microcomputers makes high-performance interfaces available throughout the organization.

3. Algorithms

Algorithms define the computations that convert inputs to outputs. For example, algorithms may be used to produce pick-lists from orders and

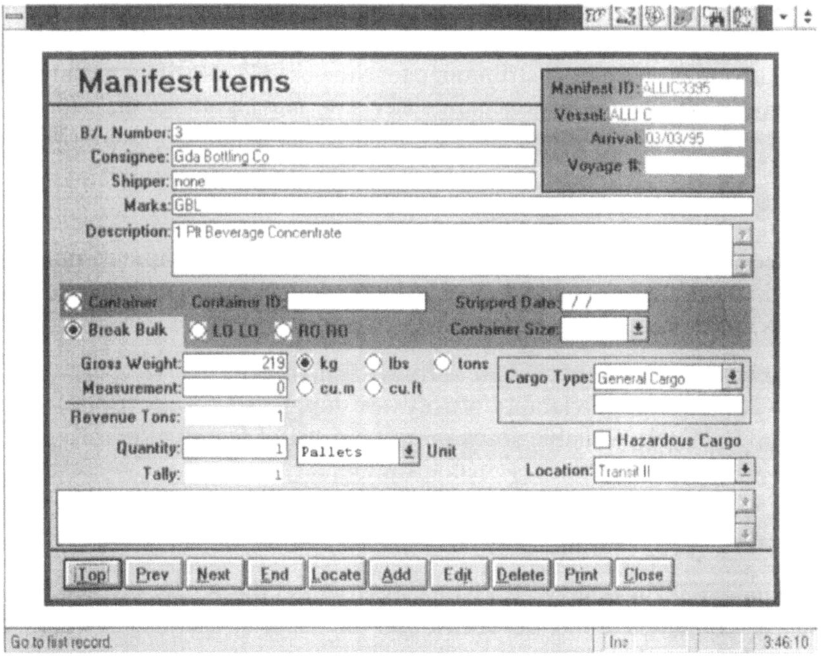

FIGURE 5 A data entry screen in Microsoft Windows.

customer information, or to determine the sequence in which orders will be processed. Proper algorithm design is critical to the correct operation of the IS, and every algorithm should be exhaustively tested before final implementation. Algorithms for sequencing and scheduling may be based on complex mathematical theory and can usually be proved correct before programming.

D. Building and Testing System Models

The complete system design can be modeled with simulation software. This allows for detailed evaluation of each design alternative in terms of the performance goals. Quite often, the process of building the system model will highlight errors in the design or show potential problem areas in operations. Building animations of the models allows the system users to view and verify correct operation of the models. Overall, simulation modeling of the system provides greater confidence in the final solution. There are many simulation packages that provide effective modeling and animation capabilities. Available software includes SLAMSYSTEM, SIMAN/Cinema, Taylor II, AutoMod II, GPSS/H with Proof Animation, Micro Saint, and ProModel; these packages all allow the analyst to build stochastic simulation models.

In stochastic simulation, a probabilistic model of the operations is constructed. This model is used in a large number of random instantiations to generate probability distributions for the possible outcomes. Analyzing these distributions allows accurate decision making about the capacity, structure, and operating policies of the material handling system (MHS).

E. Software Customization

Based on the performance of the system models, it may now be possible to choose the software functions for the final system design. The following steps are involved:

1. Determine necessary software functions.
2. Determine viability of software suppliers.
3. Match available software to your list of features.
4. Develop plans for customization.

F. Hardware Design

Equipment changes may also be necessary, and these may involve computers, communications networks, and the material handling system. At this point, detailed design of hardware changes is done.

G. System Implementation

System implementation includes all stages of design, construction, and installation of the IS. Effective implementation requires

> Top-management commitment to the project; this is usually encouraged by positive economic justification.
> Constant feedback from users.
> Effective project management and cost control.
> Users' commitment to the project, through perceptions of ownership of the design; this can be achieved by frequent meetings with users or user representatives to discuss design changes and to determine if the design meets users' needs.
> Economic justification of design decisions.
> Justification of design decisions by productivity gains.

The plan for final installation of the system should consider the following issues [10]:

> Testing of the WMS and all peripheral systems
> Training all staff in the new processes
> Staffing the information systems department
> Documentation of all procedures
> Installation of material handling system changes
> Installation of terminals, scanners, printers, and other computer peripherals
> Development of storage maps for the WMS, and bar code labeling of storage locations
> Entry of initial inventory into the WMS database
> Documentation of maintenance and disaster recovery procedures
> Planning of postinstallation user support
> Phased installation of the WMS
> Cut-over to the live WMS

H. System Maintenance

System maintenance via remote control is a popular feature of warehousing software; however, stringent security measures must be taken to ensure that unauthorized access to data and programs is denied. The major areas for system maintenance are hardware, software, and data.

Hardware maintenance can usually be covered by a long-term contract from the system vendor or a third party; alternatively, an in-house maintenance department may be established.

The objective of software maintenance is to correct programming errors (bugs) and to add new features to the WMS as the warehouse organization changes. Software maintenance is best performed by the original programmers of the WMS, whether in-house or outsourced.

Data maintenance can be done in-house and involves correcting data entry errors and any data errors caused by system failures. Historical data must also be periodically archived to off-line or near-line storage. Occasionally, the data or file structures may be sufficiently badly damaged to require the use of professional data recovery specialists or the system's programmers to recover the data.

I. Costing

The cost of the system must be determined as accurately as possible in the early planning stages. This allows a complete benefit–cost analysis. The cost of borrowing, or a discount rate, should be used in the analysis because different components of the system may have widely varying life spans. For example, computer systems are usually obsolete after 5–10 years, but some material handling equipment can serve effectively for 20–30 years.

The cost of maintenance and support is the one most likely to be overlooked in the planning stages, but these can be the largest cost components of the system.

Typical cost factors are

Consulting fees for design of software, computer network, and material handling system
Computer hardware (computers, printers, and other peripherals)
Computer software (off-the-shelf purchase price, development costs, or turnkey price)
Software installation (installation, configuration, interfacing to existing systems, testing)
Network software (priced per user)
Network hardware (hubs, concentrators, media converters, interface cards, and other)
Network cabling (cable and terminators)
Network installation (cable installation, termination, and testing; network hardware and software installation, configuration, and testing)
Material handling equipment (AS/RS, AGVs, conveyors, and other)
Material handling equipment installation and testing
Training
Cost of down time for installation, testing, and training
Automatic data collection equipment and software

Installation of automated data collection (ADC) systems, interfacing to other software, and testing

Maintenance contracts (computer software, hardware and network, and material handling system)

V. SYSTEM DESIGN ISSUES

The major system design issues are accuracy, speed, security, reliability, and cost effectiveness.

A. Accuracy

In warehouse automation, accuracy is paramount. Some warehouses expect to operate with error rates well below 0.5%, and the IS must be able to support this in receiving and shipping operations. Accuracy must be judged relative to an industry benchmark, or to existing performance in the firm. In a 1992 survery of warehouse operations [14], picking errors amounted to 1%.

B. Speed

Processing speed includes not only the speed of number crunching, but printing and data entry as well. High technology can give speed improvements, but it is not a substitute for sensible design. Areas to consider in evaluating the speed of the WMS are

Total time to complete a process, e.g., time to process and ship a complete order

Data entry time, e.g., time to enter an order into the system

Report generation time, e.g., time to produce a pick list, from time of initial request until the printed list is available

Analysis time, e.g., time to produce a report of all items sorted by inventory turns

Exception response time: the lag time to produce a warning of an information exception, e.g., the delay before producing an automatic report of a late-shipped order

For example, in the survey [14], the time to put away an order ranged from 0.5 to 96 h, and as shown in Table 1, the performance of the picking system can vary dramatically, depending on the type of facilities and operation policies. The choice of a speed target must therefore be made in consultation with the material handling system designers and operations analysts. Detailed computer simulation models of the existing and proposed systems can also provide accurate information on attainable speeds.

C. Security and Reliability

The security of a WMS is critical. There must be no way to access the system without user names and passwords. All terminals, including RFDC nodes, base stations, and bar code scanners, must have a secure log-on procedure for system access.

Many warehouses are 24-hour operations, and there is little margin for error or down time of the computer systems. However, all computers have down time; therefore, computer systems in those environments must be designed with "mission critical" specifications: redundant CPUs, hard disks, and power supplies, and control systems for automatic switching to backup CPUs or peripherals if there are failures or during preventive maintenance.

D. Cost Effectiveness

The cost effectiveness of the system, i.e., production rate per investment cost, is of great importance. One must measure the cost of the system against the potential savings in operating costs and the possible increase in throughput of the warehouse. Although a WMS may provide great productivity advantages, there is no need to spend any more than necessary to get the job done within specified requirements.

The cost of an information system is mainly in the maintenance and support after installation. The capital costs of acquiring computer equipment, peripherals, and software are usually the lowest part of the overall system cost. Budgeting must allow for the salaries of IS personnel and long-term support contracts with software and hardware vendors.

A benefit–cost analysis can be performed, considering the discounted cost of capital, comparing the total costs for various system designs with their possible benefits.

VI. SYSTEM ARCHITECTURE

A. Computer Networks

Local area networks (LANs) link personal computers typically within a building. If the network connects many buildings, it is referred to as a wide area network (WAN). The most popular networking protocol is Ethernet, and many cabling systems are available for implementation. A network structure is shown in Fig. 6. The network includes LANs interconnected by a fiber optic interrepeater link (FOIRL) into a WAN, and an RFDC system operating off one network node. A network hub is a signal repeater that interconnects a number of 10baseT (unshielded

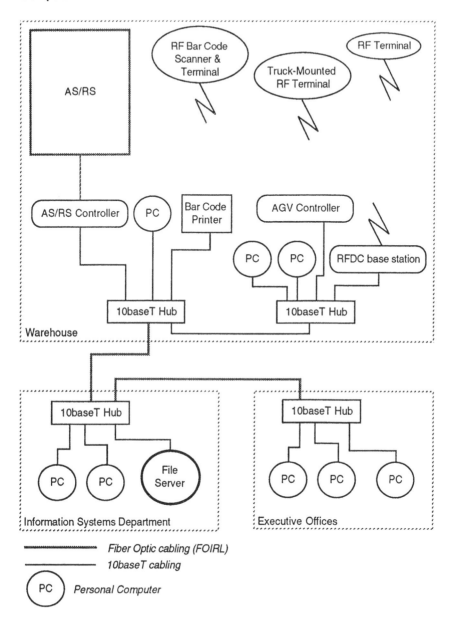

FIGURE 6 An Ethernet network for warehouse management.

twisted pair wiring), FOIRL, and other types of links. Each hub may connect 24 or more computers. Every PC has an internal network interface card.

1. Ethernet Cabling Systems

Ten Base-2 (10base2) networks use coaxial cabling and are limited to segments of 185 m between repeaters. Ten Base-T (10baseT) networks use unshielded twisted pair wiring and are limited to segment lengths (from computer to hub) of 100m.

Fiber optic systems (FOIRL) use multimode fibers in duplex optical cabling to create networks that are immune to electromagnetic interference, but at a high per-node cost. Networks using optical fibers also have high capacity and can operate over long distances: each link between two repeaters is limited to 1 km.

Information on Ethernet protocols and standards is given in [24].

B. Database Architecture

A database is a collection of files containing the data and data structures of the information system. A database management system (DBMS) is a program that performs basic storage, retrieval, and update functions on the database. The DBMS usually has a "fourth generation" programming language (4GL) associated with it, which is used to access the DBMS functions. An application is written in this 4GL, and the application accesses the DBMS. For example, Microsoft FoxPro is a DBMS that uses a superset of the dBase language and a subset of the SQL language to perform its functions. An application can be written in FoxPro. The relationships between application, DBMS, and database are shown in Fig. 7.

1. Client–Server Technology

In a traditional mainframe- or minicomputer-based information system, all processing and storage is done on the central computer, and the user

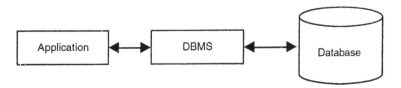

FIGURE 7 The relationships between applications, DBMSs, and databases.

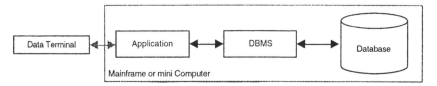

FIGURE 8 Traditional mainframe IS architecture.

accesses this system through a "dumb" data terminal (Fig. 8). This type of system architecture provides high security, but makes sophisticated GUIs difficult to develop. There is very low traffic between the terminal and the mainframe, allowing very fast response times.

In a system using a network of microcomputer workstations, all data is stored on a file server, but all processing is done at the user's workstation, as shown in Fig. 9. This allows sophisticated GUI development, but somewhat less security and speed than the mainframe architecture shown in Fig. 8. The main reason is the bottleneck between the workstations and the server. All data required by the application must travel over the network. As a result of this bottleneck, high-speed networks must be installed, and system response is often slow. The computational requirements for the server are low, and this type of server is usually orders of magnitude cheaper than a mainframe computer.

In a client–server system (Fig. 10) the DBMS resides on a server and sends the result of a data query to the client, thereby reducing network traffic, hence giving faster response times compared with the system in Fig. 9; but it is slower than the mainframe systems. The major advantage of the client–server architecture is that GUI-based systems can be developed, but need not be slowed down by network bottlenecks as in the case of the file server in Fig. 9.

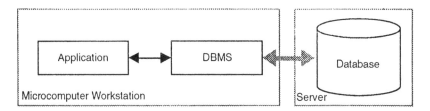

FIGURE 9 Network of workstations with file server.

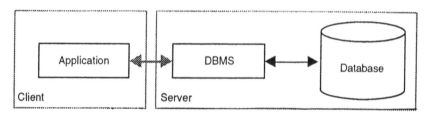

FIGURE 10   Client–server architecture.

2.   Distributed Databases

A distributed database system splits the database over many servers (Fig. 11), but the database is always perceived to be a single structure by any user. The advantage of a distributed database is that multiple organizations may share their data without making a copy for each server. Disadvantages include problems with data security, data integrity, and maintaining consistent data descriptions (referred to as data dictionaries).

C.   Data Entry Technologies

Developing automated data collection (ADC) strategies requires careful planning. Some recommended steps in the planning process are [8]

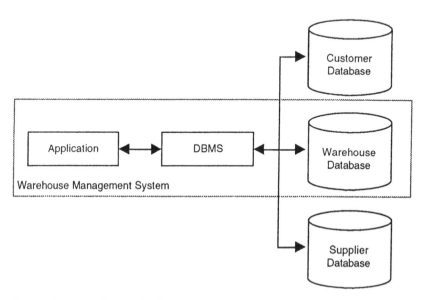

FIGURE 11   Distributed databases.

1. Identify all steps in the warehousing process.
2. Identify areas for use of ADC.
3. Evaluate internal and external constraints and requirements.
4. Determine where to use ADC.
5. Develop system specifications.

A list of suppliers of ADC equipment is provided in [25]. Some of the more advanced data entry hardware technologies are radiofrequency terminals, bar coding, and radiofrequency identification tags.

1. Radiofrequency Terminals

Radiofrequency data collection (RFDC) systems are a network of terminals or microcomputers, communicating to base stations by high-frequency radio. RFDC terminals are usually attached to the lift trucks or carried by the operators throughout the facility. RFDC allows instant inventory updating of the central WMS database, and instant changes in order picking lists. Benefits of RFDC systems are [8]

1. Reduced inventory levels
2. Labor savings
3. Elimination of physical inventory level counting
4. Increased accuracy
5. Enforced operation discipline
6. Reduced paperwork
7. Enforced FIFO
8. Easy to identify and retrieve expedited orders

2. Bar Coding

A bar coding system prints a one-dimensional pattern of narrow and wide bars to represent alphanumeric codes. Every item or carton is labeled with a bar code. These labels can then be read by laser scanners with very high accuracy. Recent advances in bar coding include the development of two-dimensional bar codes, very cheap scanners, and scanners which can read codes from up to 1.5 m away.

3. Radiofrequency Identification Tags

Radiofrequency identification (RFID) tags are read/write devices that allow a limited amount of data (typically 256 bits) to be stored in the tag. The tag can be read and updated via RF signals. This technology allows some database information, e.g., inventory count or product age, to be stored with the item.

VII. APPLICATIONS

In this section, we describe some typical WMS software, give an example of an implementation, and give some references to computer applications.

A. Examples of WMS Software Packages

1. PC/AIM

PC/AIM is a WMS that is based on a personal computer platform. The system comes in paper-based or paperless versions and includes among its features the following functions: receiving, put-away, forward pick/replenish, and shipping [26].

2. Warehouse Control System/400

The software package Warehouse Control System/400 is a WMS that runs on the IBM AS/400. It is designed for medium to large distribution systems, and among its many features are on-line receiving, directed put-away and replenishment, paperless real-time order processing, quality control, cycle counting, physical inventory, and shipping verification [27].

B. A WMS Implementation Example: Sobeys

Sobeys is a supermarket chain with stores throughout Canada and New England [12]. Sobeys owns a number of food warehouses and provides daily shipments to its stores. They use a highly automated ordering and replenishment system, with EDI for orders, advance ship notices (ASNs), and electronic funds transfer.

Their WMS is a highly customized version of Worldwide Chain Store Systems' software product. The software runs on a mainframe computer, but there are many networked workstations dedicated to specific tasks. The system uses EDI to communicate purchase orders (POs) to suppliers, and among the benefits they report are no discrepancies between invoice and purchase order prices, and no order errors. EDI is also used by suppliers to send ASNs at the time of shipment; this allows the warehouse to tightly schedule the unloading of trucks. The warehouses use paper-based pick-lists sorted in reverse order of storage on the supermarket shelves, and forklifts for materials handling. In the warehouses, achievement of 99.8% inventory accuracy is typical, and 99% of shipments arrive at the customer on time.

For upper-management users of the system, a data warehouse maintains two years of historical data for instant access, and SAS is used as a front end to build executive information systems. Also included in the system is a sophisticated labor management system that allows accurate forecasting of daily labor requirements and close monitoring of labor efficiency.

Some of the benefits of the WMS in use at Sobeys are very fast and accurate rotation of food products with almost zero spoilage; very high picking and inventory accuracy; close management of the workforce; improved supermarket put-away efficiency; and high customer satisfaction (the supermarkets are the customers of the warehouses).

REFERENCES

1. D. W. Baker, *Not So Much a Warehouse: The CWS Automated Grocery Distributed Centre, Birtley*, Her Majesty's Stationery Office, London, 1972, p. iii.
2. Anon., Pros and cons of automated warehousing: Finding the best method for a situation, *Quick-Frozen-Foods 41* (6):60 (1979).
3. R. Clark, Up-to-date storage, *Coffee-&-Cocoa-Int. 19* (5):19 (1992).
4. K. B. Akerman, *Practical Handbook of Warehousing*, 1st ed. The Traffic Service Corporation, Washington, D.C., 1983, p. 223.
5. J. F. Cruz and A. Diop, *Agricultural Engineering in Development: Warehouse Technique*, Food and Agriculture Organization of the United Nations, Rome, 1989, p. 49.
6. S. Berne, The ins and outs of automated warehousing, *Prepared Foods 163*(4):77 (1994).
7. C. Trunk, Software spurs warehousing action, *Material Handling Eng. 47* (3):67 (1992).
8. Anon., Smooth data flow ties it all together, *Mod. Materials Handling 47*(4): 24 (1992).
9. Anon., Evaluating warehouse management software, *Mod. Materials Handling 48*(3):43 (1993).
10. R. Firth, Steps to successfully installing a warehouse management system, *Industrial Eng. 27*(2):34 (1995).
11. R. Civin, D. Tanyeri, C. Casper, and D. Harrison, Automation puts the (ware)house in order, *Institutional Distribution 26*(9):184 (1990).
12. B. Jardine, Vice President of Information Systems, Sobeys, Stellarton, Nova Scotia, Canada, verbal communications, November 1994.
13. M. Diehl, Warehouse efficiency with paperless orderpicking, *Material Handling Eng. 46*(11):92 (1991).
14. K. A. Auguston, Compare yourself with the best and worst! *Mod. Materials Handling 47*(6):48 (1992).
15. *Handbook of Industrial Engineering* (Gabriel Salvendy, ed.), John Wiley and Sons, New York, 1992, p. 581.
16. *FactoryFLOW*, CIMTECHNOLOGIES Corp., ISU Research Park #700, 2501 North Loop Drive, Ames, Iowa 50010.
17. Anon., Innovation boosts material handling, *Chilton's Distribution 90*(7):10 (1991).

18. M. J. Russell, Putting it all under one roof, *Food Engg.62*(2):67 (1990).
19. Anon., Automatic storage/retrieval system doubles warehouse Output, *Food Engg. 53*(7):113 (1981).
20. J. Gledhill, Carters kegworth, *Soft Drinks Mgmt. Int. 192*:22 (1989).
21. A. M. Law and W. D. Kelton, *Stimulation Modeling and Analysis*, McGraw-Hill, New York, 1991.
22. J. Wagner, Fire up the terminals: Let's move some product, *Food Processing 55*(3):77 (1994).
23. P. J. Pratt and J. J. Adamski, *Database Systems: Management and Design*, Boyd and Fraser, Boston, 1991.
24. H. -G. Hegering and A. Läpple, *Ethernet*, Addison-Wesley, New York, 1993.
25. Anon., Eleventh annual automatic identification buyers' guide, *Industrial Engg. 25*(4):BG1 (1993).
26. *PC/AIM*, Ann Arbor Computer, 1201 E. Ellsworth, Ann Arbor, Michigan 48108.
27. *Warehouse Control System/400*, Gateway Data Science Corp., 3410 E. University Drive, Suite 100, Phoenix, Arizona 85034.

14

Computer-Based Control Systems in Food Packaging

Harold A. Hughes
Michigan State University, East Lansing, Michigan

I. INTRODUCTION

Packaging is an essential part of the food distribution system in a modern society. Very few food products can be harvested, handled, manufactured, distributed, or sold without being packaged at some time. Even the few items of produce offered without a package are placed in a bag (package) at the checkout station.

Food packages, like food products, come in many forms, styles, and sizes. Products are offered in bottles, boxes, cartons, pouches, tubs, jugs, and many other containers. Packages are made of glass, paper, metal, plastic, and combinations of materials and may be rigid, semirigid, or flexible in form. Food packages range in size from the tiny glass or plastic vials of bakery coloring to gallon jugs of milk and 40 lb bags of potatoes, onions, or citrus fruit.

In spite of the variety of form, style, and size, all food packages have certain functional characteristics in common. Every food package, like every other package, is intended to satisfy one or more of the following four basic functions:

1. *Containment.*   Products are contained in packages so that they can be moved and handled conveniently. Containment is particularly important for liquids and free-flowing powders, such as granulated sugar.
2. *Protection.*   Food products must be protected against a wide array of hazards, including microbial contamination, physical damage, oxidation, and undesirable loss or gain of moisture. Another important issue is the prevention of unauthorized tampering with the contents.
3. *Utility.*   Certain package forms and attachments add value to the product in various ways, such as by making it easier to dispense the products or to reclose the package. A package with higher utility improves the competitive position of a product.
4. *Communication.*   Packages are the primary selling tools in a modern food market. For a product to be successful, the package must communicate an image of desirability, quality, and value to shoppers by prominently displaying the brand name, illustrations of contents, and other information. In addition, the package should communicate a variety of other information, such as the manufacturer's name, "use by" or "sell by" dates, instructions for preparing and serving the product, and the nutritional content of the product.

Food companies commit large investments to design and specify food packages that satisfy these essential functions and are also economical to manufacture and distribute. Researchers and designers carefully select the packaging material and package form to satisfy the requirements of a product and marketing scheme. However, a good package design is of little value unless the package manufacturing processes are equally effective. The packaging operation must consistently and reliably form, fill, close, and seal the packages that perform the basic functions during all phases of distribution and marketing, including a shelf life period that often extends to six months or more.

In the past, line speeds were relatively low and most of the packaging operations were under the direct control of human operators. Operations were monitored and machines were adjusted by the operators. Product inspection was also a function of human workers, using scales, gauges, and other simple tools. Packages were randomly selected and removed from the packaging line to be inspected. Packages were checked by eye to ensure that the package had been filled, labels had been applied correctly, closures were screwed on straight and tight, there were no obvious leaks, and other appearance aspects of the package were acceptable. Packages were weighed to check that the product weight fell within predetermined limits. Depending on the particular product and the package type, other checks were sometimes included, such as measurement of cap removal torque or seal quality.

The traditional approach to inspection just described had two major disadvantages:

1. Compared with modern food packaging systems, processing speeds were slow, but the line still moved too fast to allow inspection of every item. The human eye simply could not effectively evaluate details of the appearance of a package when several hundred packages were processed per minute. Even when line speed was low enough to allow each item to be individually checked by eye, the repetitive nature of the activity caused fatigue. As a result, it was necessary to either institute regular rest periods or accept reduced accuracy toward the end of the day. Further, technology was not available to check the weight and other physical characteristics of every package. As a consequence, the product stream had to be sampled so that the product-package evaluation could be conducted off-line.

2. Many of the inspection methods used in the past destroyed the items being inspected. For example, a cap removal torque test required that the cap be removed, thereby opening the package and exposing the contents to the outside environment. The package and the contents were both scrapped after the test.

Today, electronic instrumentation and computerized controls are available to make a wide variety of operational, quality, and safety checks at different points on the packaging line [1]. Modern instrumentation is fast enough to inspect every package that passes down the line, and the testing processes, even at high line speeds, are nondestructive. Items that fail a test are automatically ejected from the line. A much higher level of inspection is achieved with fewer people actively involved in the process.

Information from sensors can be automatically plotted on statistical process control (SPC) charts to determine when a system is in control. The same information can be channeled to a management information system (MIS). Sensor information can also be fed into programmable logic controllers (PLCs), which can make decisions and issue instructions to adjust the operation of the packaging machinery.

In addition, sensors are now available to make checks that could not be conducted by human operators. Metal detectors are a prime example [2]. The human eye obviously cannot see the inside of most packages. Modern packaging operations, in contrast, routinely include one or more metal detectors. These devices examine the interior of packages, even the interior of some packages made of metal, to detect unwanted metal objects. The information from the sensor device is sent to a control unit, which rejects packages containing unauthorized metal, ensuring that it is not allowed to reach the end of the line and be sent to market.

The next several sections describe and discuss several of the types of sensing equipment used on packaging lines to monitor and evaluate various characteristics of packages and packaging machines. The discussion addresses factors that make automatic control and evaluation a necessity and factors that affect the effectiveness of the sensors. The chapter closes with a discussion of the use of computerized control units to interpret the information provided by the sensor units and manage operation of the line.

II. METAL DETECTORS

Metal can accidentally enter a food product or package in several ways [2]. The raw materials for the food constituents may have metallic contaminants, such as metal tags and lead shot in meat, hooks in fish, or field machinery parts in vegetables. Workers can lose personal effects, such as buttons, jewelry, coins, or keys. Tools, wire scraps, metal shavings, and similar miscellaneous items may be left behind by maintenance workers, and metal fragments can be shed from broken screens in milling and mixing equipment or worn parts of other machines. Metal items may even be deliberately introduced by an individual intending to contaminate the product.

No matter what the source of the metallic contaminant, it is critical that every package that improperly contains metal be identified and rejected. Effective metal detection requires that every item be inspected. For example, a bakery will often have equipment to inspect every loaf of bread before it is placed into a flexible wrapper. A second inspection may take place at a later point in the packaging process.

A. Metal Detection

Depending on the source of the metal contamination, a food package may contain copper, aluminum, lead, iron, steel, or stainless steel. Of these, iron and steel are the easiest to detect and stainless steel is the most difficult. Copper and aluminum fall somewhere between the other metals [2].

Metal detectors using a three-coil arrangement can detect even small pieces of stainless steel and nonferrous metals. The three coils can be arranged in various ways. The encircling coil arrangement (Fig. 1) is commonly used for packaging inspection. The three parallel coils are wound on an electrically insulated frame. The middle coil, the transmitter, is energized by a high-frequency oscillator and generates an inductive field. The field, in turn, induces a voltage in the outer coils, the receivers. If the receiver coils are the same size and parallel to each other, the voltages will be equal. The receivers can be connected in opposition (Fig. 1) to cancel out the voltages and give a null output. However, if a piece of metal passes through the coils, the shape of the electromagnetic field will be altered, changing the voltages generated in the receiver coils. This phenomenon can be used to detect the presence of unwanted metal.

The extremely sensitive coil arrangement is mounted inside a metal case for protection against stray voltages and extraneous inductive fields, which can be generated by nearby machinery. Generally, aluminum is used for the case, but other metals may be used to satisfy special requirements. In addition to providing a screen to protect against false signals, the metal case adds strength and rigidity to the assembly and protects against vibration.

B. Detecting Metal in Aluminum Packages

The balanced coil arrangement just described cannot be used when the product to be inspected is packaged in aluminum because the detector

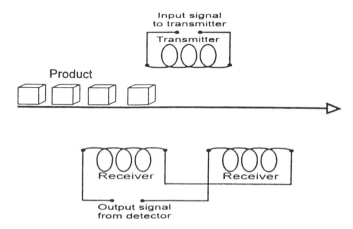

FIGURE 1   Metal detector using a balanced three-coil arrangement.

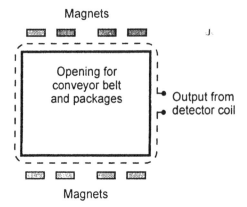

FIGURE 2 Detector of ferrous metals for foil applications.

would respond to every package. An alternative arrangement that is insensitive to aluminum is available for this application [2]. The arrangement (Fig. 2) uses powerful magnets to create a stationary magnetic field. When a piece of ferrous metal moves through the field, cutting the lines of magnetic force, a current is induced in the particle, creating a second magnetic field that moves along with the particle. As the lines of magnetic force from this secondary field cut across the detector coil, a voltage is induced in the coil, indicating the presence of the metal. The primary magnets do not induce an equivalent field in the detector coil because the lines of magnetic force from that source do not move. The single-coil detector works only on ferrous metals, iron and steel, and has no ability to detect even large pieces of nonmagnetic material, such as aluminum, copper, and stainless steel.

Metal detector sensitivity can be influenced by several factors. Certain food products, such as fresh meat, cheese, and pickles, can conduct electricity and appear similar to metal to the detector. It is essential that tests be conducted with the actual product to determine the effect on detector performance before conductive products are packaged. Other factors that influence the sensitivity of the detector include belt speed, environmental conditions, size of metal fragments, size of the detector aperture, and the frequency of the input signal to the transmitter. All of these factors should be evaluated and considered if performance is unsatisfactory.

III. CHECKWEIGHERS

Automatic in-line checkweighers are used to perform basic weight inspection of packages. Checkweighing is used to assure that a packaging operation is meeting weight targets. In addition, checkweighers provide information used for process control and production reporting functions.

The equipment used for checkweighing consists of a weigh platform, conveyor, electronics, and a reject device. The weigh platform is the key component. Packages pass in single file over the platform on a conveyor. Each package is individually weighed as it passes over the weigh platform. Strain gauge load cells are commonly used, but systems that measure displacement and techniques based on resonant frequency of the scale system are sometimes used. The strain value measured by the load cell is electronically converted to determine the weight of the package. Packages of food that weigh less than the setpoint are rejected from the line.

In addition to the basic operation, sorting out underweight packages, some checkweighers can perform more complex functions. For example, a checkweigher with computer control can scan the bar code on incoming packages to identify the contents and intended weight. Then, after the package has been weighed, the computer can compare the weight with the setpoint for that particular package, which has been stored in memory. In this fashion, a single checkweigher can simultaneously verify the weight settings for multiple product streams.

IV. MACHINE VISION

Machine vision systems are available to perform several inspection and verification functions. Machine vision equipment consists of a digitizing camera and a computer with software to analyze and evaluate the data from the camera. Packages that fail one of the preset requirements are rejected under the control of the computer.

Machine vision can be used to verify that package components such as caps, labels, and outserts are in place [3]. The system can scan labels to determine that the correct label has been applied and can also verify that the proper fonts, colors, and logos or other symbols have been applied. The check can be based on a single point of identification or on several points.

Vision systems can also check to determine that a label has been applied properly. The system identifies two or more points on the label and compares the positions relative to each other and with a standard preset in the computer. Then, if the location of one or more points falls outside the allowable range of locations, the package is rejected. Similarly, the system can examine multiple locations on the package to evaluate multiple labels, such as neck labels and body labels on wine.

Machine vision systems are also used to provide information needed to control machine operations. For example, robotic palletizing equipment is flexible and able to stack boxes of differing sizes on pallets, using several different stacking patterns. The machine vision system enables a robot to identify each particular type of box and guide the operation of the arm and the end-of-arm tooling as each box is positioned on the pallet.

The computational load can be heavy when a computer analyzes the appearance and location of a package label, or performs other similar functions on modern packaging lines. Computers are needed to develop the necessary data from the image, compare the data with standardized information, reach a decision, and issue an instruction to keep or reject the item. Today, computers and imaging equipment are available that can perform these complex operations quickly and reliably, examining packages passing at speeds as high as 2000 packages per min, equivalent to one package every 0.03 s. Obviously, the human eye is not capable of inspecting at the same rate.

V. LEAK TESTING

Leaks can occur in any type of package. Two-piece or three-piece cans often leak if the seal at the point where the top (or bottom) is attached to the body of the can is inadequate. Bottles leak because of improper manufacturing of the finish, which prevents proper mating with the cap, because of flaws in the body of the container, or because of cracks, chips, or other damage.

Flexible packages are more susceptible to leaks than rigid sealed packages. Flexible packaging is thinner than most other packaging materials and may be made of aluminum foil or other fragile materials. Packages can leak because of pinholes, cracks, and other manufacturing flaws, incomplete or inadequate seals, or punctures.

A package with any of these flaws can allow external moisture, oxygen, or other contaminants to gain entry. Alternatively, moisture that should be kept inside the package can be lost, allowing the contents to dry out. In either case, the product quality declines and shelf life is shortened. Manufacturers use leak testers to protect product quality by preventing packages with leaks from being sent to market. Modern equipment is fast enough to test each package as it passes down the line. The testing systems are similar, but they vary in details to suit the container type and whether the containers have been filled when the test is performed.

A. Leak Testing Empty Containers

Containers manufacturers commonly test bottles, cans, and similar containers for leaks before delivering them to the filling operation (Fig. 3). A simple leak testing system pressurizes the container and then immediately monitors the resulting pressure to detect changes that would indicate a

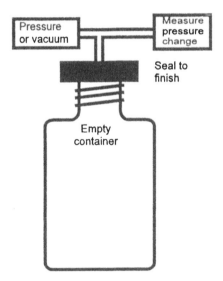

FIGURE 3   Testing empty containers for leaks. An alternative system uses a chamber.

leak [4]. Unfortunately, the system is sensitive to pressure changes from any cause and cannot differentiate a pressure drop due to a leak from a pressure drop due to an increase in the volume of the container. The system can be made more accurate and reliable by providing a finite time interval for stabilization after the container has been pressurized. The container is pressurized by a predetermined amount of air and then held for a short period of time for stabilization to take place [4]. The rate of pressure change is then measured by a sensitive electronic device. If the rate of pressure change exceeds a preset standard, the equipment makes the interpretation that the container has a leak, and it is rejected. The time required for each of the three steps in this process can be controlled by the user, but the total time requirement is usually about 1.5 s. Testing rates adequate to match line speeds of several hundred bottles per minute are achieved by using multiple head testers, either in an in-line or rotary arrangement.

B.   Leak Testing Filled Containers

The equipment for leak testing containers that have been filled and sealed must be more sophisticated because of the wide variances in product weight, volume, viscosity, moisture content, and other physical character-istics [5]. A variation of the method for empty containers, just described,

Testing cups, etc.

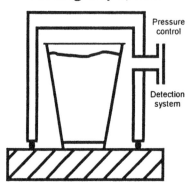

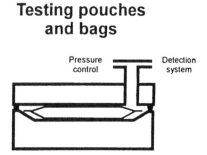

Testing pouches and bags

FIGURE 4 Leak testing filled containers.

is often used (Fig. 4). The filled container is sealed inside a chamber, and the surrounding pressure is either increased or lowered, creating a differential pressure between the inside and the outside of the sealed container. Then, after a time interval to provide for stabilization, the pressure change is monitored. If the rate of change of pressure exceeds the allowable (preset) limits, a signal is generated that the container has a leak and causes it to be rejected. The time required to complete an individual test usually ranges from 1 to 3 s.

VI. OTHER TESTS AND SENSORS

Modern packaging lines make use of numerous other sensors and detectors, including various types of location sensors, such as microswitches that work on mechanical action, or switches that are activated by the completion or interruption of a light beam. Proximity sensors are used to initiate or terminate actions on individual machines, such as controlling the flow of empty containers into an in-line filling machine. Other switches determine whether there is space for a machine to send a processed package down the line to another machine and to determine if a machine should be activated to process a waiting package. Switches can be configured to count the number of items placed into a package or the number of completed primary food packages placed into another package for transportation to market.

In recent years, the machines involved in packaging have become more sophisticated, using sensors and controls to manage each step of the operation. Traditional drive trains consisting of heavy shafts, bearings, clutches, brakes, belts, chains, and gears, which transmitted power and timed the operation of various machine functions, have largely been replaced by individual electric, pneumatic, or hydraulic drives and actuators that power sections of the machine and programmable logic controllers (PLCs) that receive information from the sensors and issue instructions for the various drive motors and actuators to be energized [6]. In effect, the typical packaging machine has been converted into a series of smaller machines that each work as directed by a computer, the central electronic controller, on the machine. The result has been machines that are lighter, more flexible, and easier to set up and operate.

At the same time, other PLCs are used to integrate and coordinate the working of the various machines in a packaging line [7]. In this way, the flow of materials, packages, and products is controlled, and individual machines are monitored to ensure that the assembly of equipment operates as effectively and efficiently as possible. In effect, the entire set of packaging machines is made to operate like a single large machine.

The PLCs, computers, and microprocessors have largely replaced mechanical and electric controls on new machines and packaging lines [6]. The capabilities of these devices, and the machines they control, are being continuously upgraded and expanded as new software and techniques are developed and as the computers are equipped with more memory and are made to operate faster. Through the use of these controls, the packaging machines not only perform their individual functions and work together effectively, but they detect problems, advise the operator about repairs or adjustments, keep track of the production history, respond to external conditions and events, send information to remote computers, monitor automatic inspection processes, and store changeover information.

Computers and electronic controls empower the operator by providing information and by making available the means to initiate actions based on that information. Modern management systems give machine operators greater authority and responsibility. The machine and system controls provide for the practical requirements to implement that authority. The computerized controls enable capable operators to enhance productivity because they keep the lines operating and require less support from mechanics and other service personnel. Putting an intelligent machine in the hands of an intelligent operator is a strong move toward higher productivity [6].

A. The Future

The use of automatic, on-line, inspection will continue to increase as food processors find new techniques and as machinery manufacturers continue to develop new equipment. In some cases, as in the pharmaceutical field, regulations requiring the inspection of every packaged item may be instituted in the food industry, particularly for biologically sensitive processes, such as asceptically packaged products, and for market sensitive products, such as baby food. Even if not required by the government, many companies will voluntarily institute or expand manufacturing practices that involve inspection of every package as a means to ensure the highest possible quality and as a defensive approach to cut down the threat of liability lawsuits. Companies will apply any device or approach that improves the ability to ensure that every package is properly sealed, that the processing and packaging processes were conducted properly, that the package is free of foreign materials, and that the labels and other printing and decorations are complete, correct, and placed properly. These new approaches will be intelligent devices, controlled and managed by computers.

Future packaging lines will be fully integrated [1]. The concept of a set of single machines will be replaced by the concept of a system of machines that operate as a coordinated entity. The central system computer will communicate with the computers operating each component machine, giving instructions to each machine, accumulator, conveyor, and inspection device and receiving back information to be used for statistical process control and similar management information system applications.

VII. SUMMARY

Modern packaging machines use sensors to feed information to computerized controllers, which issue instructions to various parts of the machine. The individual machines are organized into systems to exchange information and receive instructions from a central computerized controller. Metal detectors, vision systems, checkweighers, and leak detection systems are used to inspect every package. Because of the high line speeds, inspection equipment must rely on computers to interpret data and make decisions to reject faulty packages; human vision and other senses cannot react fast enough.

In the future, packaging machine manufacturers will continue to apply electronic sensors and computerized controls to individual machines, and to organize those machines into integrated assemblies that function in a fashion similar to a single large machine.

REFERENCES

1. J. Cudahy *Using Computer Technology to Achieve Packaging Line Integration and Automation*, Institute of Packaging Professionals, Herndon, Virginia, 1993.

2. A. Lock *The Guide to Reducing Metal Contamination in the Food Processing Industry*, Safeline, Inc., Tampa, Florida, 1990.
3. J. Jackson, Manufacturer's Literature, Label Vision Systems, Inc., Peachtree City, Georgia, 1994.
4. T. Stauffer, Non-destructive in-line detection of leaks in food and beverage packages—an analysis of methods, *J. Packaging Technol* 2(4):(1988).
5. T. Stauffer, *Wilco Precision Testers, Testing flexible containers for leaks, Food engineering, 60(6)*:70 1988.
6. Technical Association of the Pulp and Paper Industry, *Intelligent Packaging Lines: Trends in Automation*, TAPPI, Atlanta, 1995.
7. G.G. Davis, *Programmable Logic Controllers (PLCs) on Packaging Machines*, Packaging Machinery Manufacturers Institute, Arlington, Virginia, 1995.

15

Computerized Automation Controls in Dairy Processing

Sundaram Gunasekaran
University of Wisconsin–Madison, Madison, Wisconsin

I. INTRODUCTION

The purpose of automation is to increase process efficiency, safety, productivity, and product quality. This is generally achieved by means of a control system that has been "programmed" with a set of instructions. The control system communicates with every component critical for successful operation of the process or function being manipulated. The dairy process industry, just like the rest of the food industry, traditionally lagged behind other manufacturing industries in introducing automation and computer controls. Some difficulties encountered in automation and computer utilization are lack of suitable sensors, low profit margins, use of batch/continuous operations, and installation of equipment that is not integrated into the whole process. However, the trend is now changing rapidly as more and more dairy operations are being automated [1–3]. Among all food industry categories, the dairy industry is uniquely positioned for easy adoption of computer automation. This is because [4]:

> The dairy plant has significant external record-keeping requirements.

Finished dairy products are homogeneous, with relatively few major ingredients.

Fluid operations in a dairy plant are often of long duration, sequential, and adaptable to software systems developed for continuous processes.

Moreover, for dairy processors, computer automation and control strategy facilitate gathering of critical process information as systematically as the process itself. Thus almost every dairy plant operating today has at least some automatic equipment or process control system in place in scattered areas. The next step will be to integrate most of these scattered "islands" into an overall system with state-of-the-art computer automation and process control systems [5].

Perhaps a critical step in dairy process automation is the introduction of clean-in-place (CIP) systems. The CIP systems do not require disassembly for cleaning. They are designed so that cleaning is performed by circulating detergent solutions through product lines according to a fixed cleaning program. This entails opening and closing valves and regulating temperatures, pressures, and flow rate from a remote location. This was followed by automation of other processes, such as mixing, filling, emptying, heating, and cooling. Because of increased processing capacities, mistakes became costlier, and it became crucial to control these operations more closely. Thus more and more operations are entrusted to programmable controllers that execute a preprogrammed set of instructions at exact times with little or no operator assistance. Computer programs are merely translations of logic into a "language" understandable by the computer [6]. In a plant control setting, logic refers to the sequence and timing of each of the interrelated operations. Thus a set of signals is sent in a certain order to actuate or shut off various components involved in the controlled process in accordance with the logical conditions applying to the process. The components acknowledge the signals by sending feedback signals, which are used to time the execution of a subsequent set of instructions. Some of the key requirements of a control system are that it must be flexible, reliable, safe, and economical.

II. TYPES OF CONTROL SYSTEMS

A process consists of an assembly of equipment and material that relates to some manufacturing sequence. Any control system should be built around the process components so as to link their operations sequentially and enable communication among them. In general, a process control loop consists of four elements: process, measurement, controller, and control element. The sequence of information flow between these elements is illustrated in Fig. 1. To control a dynamic variable such as the temperature of a fluid level in a tank, information must be available on the variable itself, which is obtained via an appropriate measurement. Measurement, for control purposes, is the transduction of the variable into an analog signal. This signal can be a pneumatic pressure, electrical voltage, or current. In the evaluation step, the measured signal is examined and the corrective action, if any, is determined. The evaluation may be performed manually by an operator (manual controller) or automatically by electronic signal processing, pneumatic signal processing, a computer, or a combination of all these. The final control element exerts a direct influence on the process by accepting an input from the controller and transforming it into some action regulating the process.

An example of the basic process control is presented in Fig. 2. The purpose of this control system is to maintain the fluid level at some prescribed level (H) from the bottom of the tank, a fairly common operation in any dairy plant. The level detector and transducer is a device that senses and measures the fluid level in the tank. The level controller evaluates the measurement, compares it with a desired setpoint (H), and produces a series of corrective outputs. The valve controls the flow of fluid in the outlet pipe. In this example, there is no apparent connection between the fluid outlet and the fluid inlet. However, the control valve often will be regulating the fluid inlet, thus completing what is known as a "control loop."

There are three major types of process control loops. They are briefly described below [8].

A. Feedback Control Loop

This is the most widely used control loop. In this, sensors measure actual values of controlled variables. The measured values are compared with

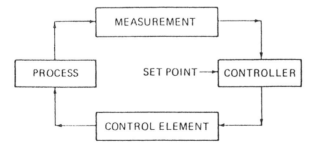

FIGURE 1 Block diagram of a process control loop. (Reprinted by permission, © 1988 Instrument Society of America, from *Measurement and Control Basics* [7].)

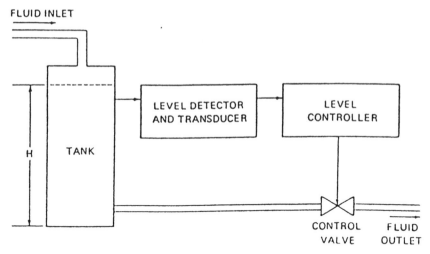

FLUID INLET

LEVEL DETECTOR AND TRANSDUCER

LEVEL CONTROLLER

TANK

H

CONTROL VALVE

FLUID OUTLET

FIGURE 2 Basic process control example. (Reprinted by permission, © 1988 Instrument Society of America, from *Measurement and Control Basics* [7].)

the desired values (the setpoints) of the corresponding controlled variables. The error (difference between desired and measured) is used as input to the feedback controller to regulate the control variable. A schematic of the feedback control concept implemented by an automatic feedback controller is represented in Fig. 3. The feedback control loop is popular because it allows for controlling the process without any advance knowledge. Control is achieved, in this and other control loops, through one of the following predetermined algorithms [7]: on-off, proportional, integral (reset), derivative (rate), and proportional–integral–derivative (PID). The emphasis in the feedback control concept is to eliminate or minimize errors.

B. Feed-Forward Control Loop

This type of control is significantly different from the feedback control. The operator has advance knowledge of the disturbance and adjusts the control variables in accordance with the disturbance. The schematic of the automatic performance control concept is represented in Fig. 4. Unlike the feedback control, which aims at eliminating errors, feed-forward control aims at preventing errors. The trade-off is that the information on the disturbance and the scheme for how the variable is to be controlled should be known beforehand. This entails knowledge and understanding of the exact effects the disturbances will have on the controlled variable. Thus,

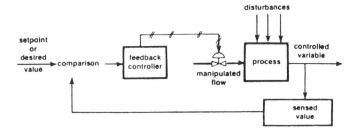

FIGURE 3 A single feedback loop. (Reprinted by permission, © 1988 Instrument Society of America, from *Application Concepts of Process Control* [8].)

it is used only for systems that are well understood (e.g., heat exchangers). Often, a process may be controlled by combined feed-forward and feedback control loops.

C. Sequential Control Loop

This is defined as the step-by-step execution of time-ordered events. Each step is a simple action such as opening a valve, starting a pump, or stopping a motor drive. These discontinuous control functions are quite different from the continuous nature of the feedback and feed-forward loops. The simplest sequential control involves use of electromechanical relays. This was popular during the past few decades. However, the present trend is

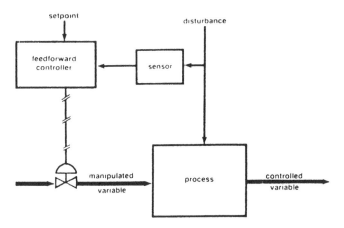

FIGURE 4 A feed-forward control loop. (Reprinted by permission, © 1988 Instrument Society of America, from *Application Concepts of Process Control* [8].)

the digital implementation of sequential control using a programmable logic controller (PLC).

A PLC is designed to control a process unit that is part of a total sequential or batch process. A solid state PLC permits arbitrary interconnection among a number of elementary devices such as switches, relays, timers, counters, registers, and register arrays. The PLC output switches may be connected to I/O (input/output) devices, which can switch equipment on and off, transmit values to and from internal registers, and so forth. The major convenience is that PLCs may be reprogrammed; that is, unlike electromechanical relays, they need not be rewired. Most of the currently available control mechanisms in the dairy process industry are based on PLCs.

III. PROCESS CONTROLLERS IN A DAIRY PLANT

Process parameters such as temperature and pressure were originally controlled by human operators. In the 1940s pneumatic instrumentation was developed to sense process parameters and provide feedback control [5]. Signal transmission to a remote control room was also possible. The control equipment usually remained with the piece of equipment or was routed to a control panel. Pneumatic instrumentation is highly reliable and hence is still in use in many dairy plants. However, these systems are based on single-loop control, namely, one measurement, one control algorithm, one actuator, and one process variable. In the 1960s, digital computer usage began in process control. Analog instrumentation became available, and electronic instrumentation became more stable and repeatable than comparable pneumatic controllers. Signals could be transmitted over longer distances. With traditional control, operating conditions are predetermined to maintain certain levels or temperatures. The PLCs replaced many of the discrete relays, offering much flexibility compared with the traditional, hardwired relay-logic control systems.

Because of concerns over possible system failure, widespread adoption of computer technology was slow. By taking advantage of PLCs in distributing computing over a wider area, concerns over large-scale failures were alleviated [5]. Digitally controlled systems are widely used control mechanisms today. Once computer-based control systems are interfaced with sensors, valves, and switches, the measurement devices can provide input for the computer, while signals from the computer provide input for the control device. A key feature in this system is the ease with which process changes can be made by reprogramming the computer.

Computer automation/control begins with various parameters being controlled. Process control equipment includes transmitters, controllers, and an actuator or control device. The transmitter is the sensor that is continuously sensing the parameter being controlled and, as the name implies, transmits a signal corresponding to the magnitude of the measured parameter. The controller receives this information and compares it with a signal corresponding to the present value of the parameter. If the magnitude of the difference between the desired and measured signals is greater than an allowable limit, the controller sends a signal to the actuator physically responds to the controller signal and carries out the instructed function. The cycle of comparison and correction of the measured value is repeated until the measured value is within the preset limits of the desired value.

The most widely controlled parameters in a dairy plant are pressure, temperature, level, and flow. A schematic of a pressure transmitter is shown in Fig. 5. In this, a diaphragm (a thin metallic film) senses the pressure and moves accordingly. This movement is transmitted via a set of levers and linkages in a direct control transmitter. In an indirect device, strain gauges or other secondary transducers are used to send an electrical signal corresponding to the pressure sensed. A pressure transmitter can also function as a level transmitter if positioned properly. When placed at the bottom of a tank, the pressure sensed is directly proportional to

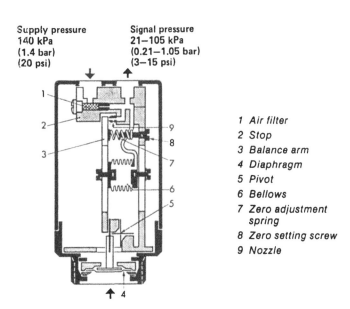

Supply pressure
140 kPa
(1.4 bar)
(20 psi)

Signal pressure
21—105 kPa
(0.21—1.05 bar)
(3—15 psi)

1 Air filter
2 Stop
3 Balance arm
4 Diaphragm
5 Pivot
6 Bellows
7 Zero adjustment spring
8 Zero setting screw
9 Nozzle

FIGURE 5 A pressure transmitter. (Reprinted by permission, Alfa Laval [6].)

the liquid level, and hence the pressure transmitter senses the level and level changes. Other types of level transmitters use ultrasonics and are positioned at the top of the tank (see Chapter 2 for more details). Ultrasonic sensors may be preferred because they will not interfere in any stirring and/or emptying operations taking place in the tank. Most common temperature transmitters use a resistance temperature device (RTD) or a thermistor as the sensing element. In these, the sensor resistance changes due to an applied temperature change. An electrical signal is generated corresponding to the temperature change. In Fig. 6 the principle of automatic temperature control for a plate heat exchanger (PHE) is illustrated. The RTD temperature transmitter sends an electrical signal to a pneumatic controller. The regulating valve responds to the controller signal and adjusts the steam supply to the PHE.

Once the necessary parameters are sensed, control systems can work their way up to the next level—control of individual unit operations. Some of the most common unit operations in any dairy plant include blending, pumping, heating, and storage operations, as well as clean-in-place (CIP) operations [5]. Blending operations using least-cost formulation and computer-aided optimization are well established in the food industry. Metering of milk and other ingredients is often computer controlled. The casein/fat ration for cheese milk is determined on-line using an infrared multicomponent analyzer. The speed of the cream meter is adjusted as needed. Computers are used in standardizing cheese milk to give the correct casein/fat ratio. Yield prediction formulas are used in the calculations. Batch movement and formulation are fully automated. Quantities of raw materials and finished product are tracked and reported. Pumping systems have been designed to avoid problems associated with centrifugal pump cavitation. The systems include a PLC, a sound sensor, and a variable-speed drive at the pump motor. The sensor can detect preliminary cavitation impulses, signaling the variable-speed drive to adjust the pump's revolutions [9]. During pasteurization, a computer can monitor all the process parameters. These include temperatures, valve positions, liquid levels, and feed rates. Some control systems are limited to data acquisition without real-time controllability. In these cases, process control adjustments still depend on an operator. A retort management system that provides temperature and pressure control has been developed. The system monitors these signals along with all facets of retorting including cooking, venting, and cooling. If a temperature deviation occurs, the system automatically recalculates a new process time and makes all necessary adjustments. Liquid levels in tanks are monitored simultaneously and used in production tracking and inventory control. Enclosed cheese vats are designed for automated control. During cheesemaking,

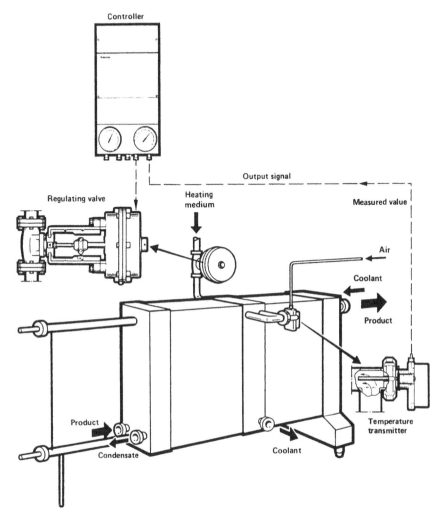

FIGURE 6 Temperature and control circuit for a pasteurizer. (Reprinted by permission, Alfa Laval [6].)

operations such as adding ingredients, controlling temperature, cutting, stirring, draining, and washing are commonly computer controlled. Clean-in-place systems have long been established as major computer process control systems in the dairy industry. Recent developments include computer-controlled detergent dosing systems, environmentally friendly CIP systems, automatic analyzers, and CIP data logging systems.

Beyond unit operations, computer control systems currently available can supply data to a higher-level host computer for further data manipulation [5]. Devices can be mixed and matched and integrated into plantwide control schemes. The use of single-loop controllers involves only configuration without dedicated software. Configurable software is used to sequence control systems so that the process operates as a sequential series of linked operations that can operate independently of each other (fill tank, empty tank, sterilize, etc.) but are still related in terms of order and integrity to process operation. For example [5], consider diverse processes such as manufacturing regular, flavored, and evaporated milk, storing products, and cleaning. In this case, the status of the process will be shown by means of matrices at the operator station. Some of the linked computers will control manufacturing, while another acts as a management computer. With a computerized control system, the desired functions can be monitored and the system programmed for the next product. Computers can chart equipment conditions and locate developing problems at any point in a process. Future systems will include the ability to network with other systems; to communicate with a wide range of protocols including intelligent sensors, to bar code readers, PLCs, and PC software packages; and to provide a fault-tolerant operation.

IV. ESSENTIAL ELEMENTS OF A DAIRY PROCESSING PLANT

Regardless of the type of processing operation, there are certain elements common to all dairy processing plants. An overview of various operations in a typical dairy processing plant [10] is represented in Fig. 7. The common elements in various dairies include [6, 11] (1) tanks for holding, processing, or storing milk, (2) heat exchangers for heating and cooling, (3) pumps, pipes, fittings, and valves for fluid handling, (4) process control equipment, and (5) utility components for water supply, heat generation, refrigeration, air supply, and electric installations.

A. Tanks

There are numerous types of tanks serving different purposes [6, 11]. There are large vertical tanks known as silo tanks in which milk is received and held until further processing. These are often located outside the plant for lack of space. Hence they are specially insulated to maintain milk temperature at about 5°C. Buffer tanks are those used as intermediate storage tanks between subsequent process steps in a plant. For a given process, milk may be held in what is known as a process tank or a mixing

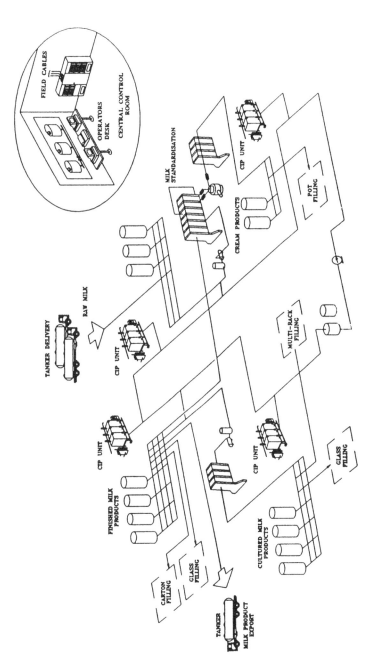

FIGURE 7   A typical dairy process plant. (Reprinted by permission, © 1994 Chapman & Hall, from *Modern Dairy Technology,* Vol. 1, R. K. Robinson (ed.), Fig. 5, p. 70 [10].)

tank. Examples include ripening tanks used for yogurt making and coagulation tanks used for cheesemaking. Due to hygiene and CIP requirements, the tanks are usually made of stainless steel and are equipped with special provisions for filling, emptying, stirring, and monitoring (temperature, level, etc.), as shown in Fig. 8.

B. Heat Exchangers

Heating and cooling are essential steps in all dairy processing operations. Selective control of microbial growth or lethality is achieved by appropriately heating and/or cooling the product. Examples are pasteurization, culture preparation, cooking or scalding while cheesemaking, and so forth. The most popular equipment used for this purpose is the plate heat exchanger, in which both hot and cold fluids flow adjacent to each other, separated by thin stainless steel plates (Fig. 9). Other heat exchanger types include shell-and-tube heat exchangers and scraped-surface heat exchangers. Heat exchangers are very energy intensive and, therefore, must be

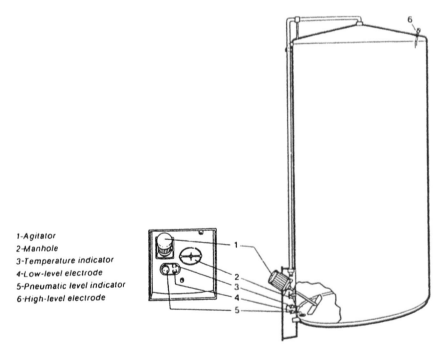

1-Agitator
2-Manhole
3-Temperature indicator
4-Low-level electrode
5-Pneumatic level indicator
6-High-level electrode

FIGURE 8   A silo tank with control panel. (Reprinted by permission, Alfa Laval [6].)

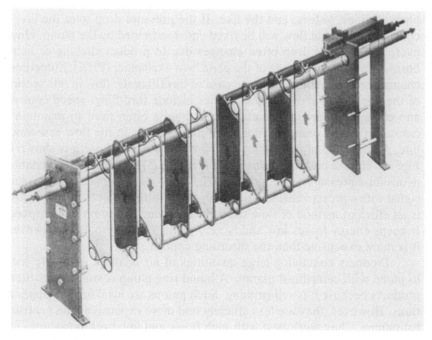

designed for efficiency, as well as for flexibility and cleanliness. Before selecting a heat exchanger for a particular operation, the following should be considered [11]:

1. Materials of construction
2. Performance requirements—temperature, pressure drop, flows
3. Scaling/fouling tendencies
4. Ease of cleaning, inspecting, and servicing
5. Physical changes in the product during the heating and cooling processes
6. Overall efficiency

C. Fluid Handling Components

These include pumps, pipes, fittings, valves, and other accessories that are a part of the system in moving the product from reception to fillers or to other areas where further processing takes place. Centrifugal pumps are most commonly used for moving large volumes of fluid milk across pipes, fittings, valves, and the like. If the pressure drop over the line is constant, the liquid flow will be fixed and determined by the pump. However, the pressure drop often changes due to product sticking or being burnt onto the heating side of the plate heat exchanger (PHE). Sometimes changes are intentional, as in the case of throttling the flow in one section of the line. Control methods in dairies include throttling, speed control, and changing impeller diameter. Throttling is often used to maintain a constant flow by compensating for minor changes in the flow resistance line. An example of a constant-pressure flow control system is shown in Fig. 10. Here a pressure transmitter (see Fig. 5) communicates instantaneous fluid pressure to a controller. The controller compares the pressure signal with a preset value and regulates the throttling valve. Speed control is an efficient method of flow control by varying the pump motor speed. It keeps energy losses low and is very gentle on the product. However, it is more expensive than the throttling control.

Products containing large quantities of air or other gases are hard to pump with centrifugal pumps. A liquid ring pump is suitable for these products because it is self-priming. Such pumps are ideal for CIP applications. However, they are less efficient and more expensive than centrifugal pumps. They work best with high flows and low back-pressures.

A positive displacement pump can handle high-viscosity fluids and can be used as a metering device. Applications include pumping cream,

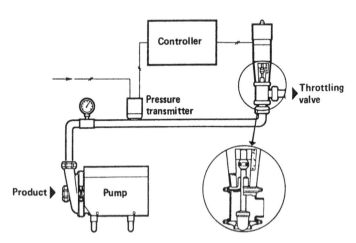

FIGURE 10 Constant-pressure flow control system. (Reprinted by permission, Alfa Laval [6].)

cultured milk products, curd and whey, and so forth. For this pump, use of a throttling valve for flow control on the delivery side of the pump should be avoided, as the pressure increases sharply when the pump outlet is throttled. There usually is a safety relief valve in a positive displacement pump to avoid any damage.

A balance tank is introduced on the suction side of a pump to keep the product at a constant level above the pump inlet; that is, it maintains a constant pressure on the suction side. This ensures a uniform flow.

D. Pipes, Fittings, and Valves

Pipes are the conduits for product flow. All the pipes coming in contact with dairy products are made of stainless steel. Pipes used for handling cleaning solutions and waste may be of other materials. Some typical fittings and valves included as part of the pipe system are (1) bends, tees, and reducers; (2) inspection sight glasses, instrument ports, or bends; (3) valves for stopping, directing, and controlling pressure and flow; and (4) pipe supports.

Normally there are hundreds of valves in a dairy plant. They constitute an essential part of automatic controls. Some commonly used valves are (1) seat valves—widely used for routine open and shut operations, (2) check valves—to prevent the product from flowing in the wrong direction, (3) air-flow valves—to clear pipelines with a blast of air stream, and (4) regulating valves—to accurately control the flow and pressure at various points.

V. DAIRY PROCESSING OPERATIONS

Dairy processing begins with raw fluid milk received at the plant. Due to the highly perishable nature of milk, special care is required in handling and processing fluid milk. Extraordinary efforts must be taken to keep the milk cold (below 5°C) at all times to prevent growth of spoilage microorganisms. As the milk is received, it is tested for its physical, chemical, and microbial quality; measured for volume or mass; chilled to ensure that the temperature remains below 5°C; and stored in large vertical tanks called silo tanks until further processing. While held in silo tanks, milk is gently agitated to avoid cream separation.

Regardless of the final product for which the milk is intended, it goes through a series of processing steps. For example, raw milk is pasteurized, separated, standardized, and homogenized before it is ready for other processes.

A. Pasteurization

The objective of pasteurization is to inactivate spoilage microorganisms so that the milk will be safe and will have a longer shelf life than its unpasteurized counterpart. The particular requirements for pasteurization are specifically designed to kill *Coxiella burnetii*, the most resistant pathogenic organism commonly associated with cow's milk [12]. The extent of microbial inactivation is a function of heating temperature and duration over which milk is held at that temperature.

1. High-Temperature, Short-Time Pasteurization

Some minimum temperature–time treatments recognized by the US Public Health Service for pasteurization in indirect heat exchangers follow [13]:

62.8°C	30 min
71.7°C	15 s
88.3°C	1 s
90.0°C	0.5 s
93.9°C	0.1 s
95.6°C	0.05 s
100°C	0.01

These temperature–time combinations result in high-temperature, short-time (HTST) pasteurization. Higher processing temperatures and/or times result in increased shelf life.

A secondary purpose of pasteurization is to destroy enzymatic systems to safeguard product quality, especially rancidity (enzymatic breakdown of fat) or bitterness (breakdown of protein) [12]. The adequacy of pasteurization is evaluated by testing for inactivation of the enzyme phosphatase. Inactivation of phosphatase (and other enzymes) requires more intensive heat treatment than is necessary to kill the pathogenic bacteria. In Fig. 11, temperature–time curves for lethal effect to bacteria and destruction of enzymes are presented.

Heat treatments during HTST pasteurization and most other dairy processing is accomplished in a PHE, which consists of a pack of stainless steel plates clamped in a frame. The plates are corrugated in a pattern designed for optimum heat transfer. Figure 12 shows a simplified process for milk pasteurization. The PHE with cooling, holding, and heating sections is also shown. Milk is heated to pasteurization temperature (~75°C) in the first section. The heated milk then flows through a holding section, where temperature is held constant. Then it goes to a cooling section, where it is cooled down to about 4°C. The pasteurization requires both heating incoming milk and cooling pasteurized milk. Using the pasteurized

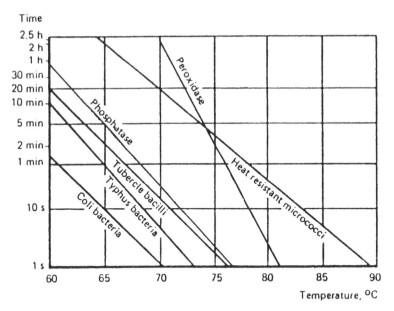

Time

FIGURE 11   Temperature–time combination for lethal effect for bacteria and for destruction of enzymes. (Reprinted by permission, Alfa Laval [6].)

milk to heat the incoming milk, the pasteurized milk is cooled and an improved energy efficiency results. This technique is known as regeneration and is widely used during heat treatments with PHEs.

A constant pasteurization temperature is maintained by a temperature controller acting on a steam regulating valve. Any tendency of the milk temperature to drop is immediately sensed by a temperature transducer, and a signal is sent to the controller, which adjusts the steam regulating valve to allow additional steam. It is a typical feedback control loop. A complete PHE pasteurizer with accessories for process supervision and control is presented in Fig. 13.

2. Ultrahigh Temperature Pasteurization

The ultrahigh temperature (UHT) process involves application temperatures much higher than the HTST process. The temperature–time combinations used for the UHT process are in the range of 135–149°C and 2–8 s. The milk is heated directly by steam injection or infusion, or indirectly in a heat exchanger. In the direct process, steam infusion dilutes the milk. Therefore, the diluted milk is evaporatively cooled to restore its original composition. A typical flow chart of a direct steam injection UHT plant

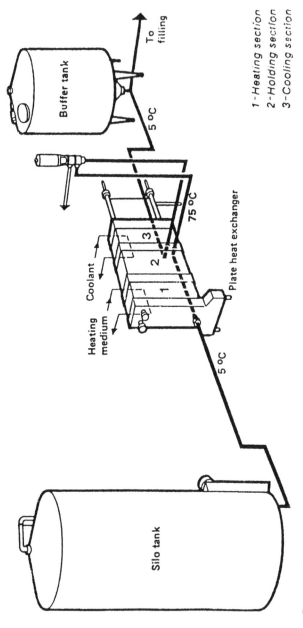

1 - Heating section
2 - Holding section
3 - Cooling section

FIGURE 12 Simplified process for milk pasteurization. (Reprinted by permission, Alfa Laval [6].)

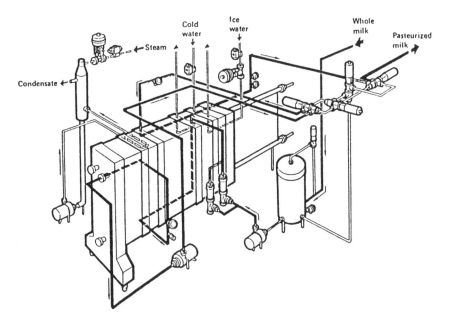

FIGURE 13 Plate pasteurizer with accessories for process supervision and control. (Reprinted by permission, Alfa Laval [6].)

is presented in Fig. 14. Milk is supplied from the balance tank and pumped by the feed pump to the preheating section of the PHE. After it has been preheated to about 80°C, the product pressure is increased by the positive displacement pump to about 4 bar. The pressure should be high enough to prevent boiling of the product and the separation of dissolved air [14]. Too low a pressure will make it difficult to maintain processing temperature and holding time—pressure regulation here is critical. Then the product continues to the steam injector, where the product temperature increases to about 140°C. It is held at that temperature for a few seconds and then is flash cooled in an expansion chamber kept at a partial vacuum by a vacuum pump. The vacuum is regulated in such a way that it corresponds to a boiling temperature some 1–2°C higher than that of the milk coming out of the PHE preheater. This temperature difference is needed to ensure that the correct amount of water vapor is removed to compensate for that added as condensed steam during heating [14]. Therefore, precise control of this differential temperature is a way of controlling the milk composition. A centrifugal pump pumps the milk to the homogenizer. After homogenization, it is cooled to about 20°C in the PHE and then continues directly to an aseptic tank for intermediate storage before filling.

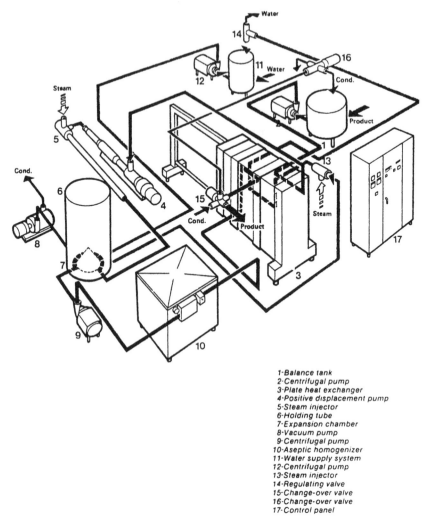

1-Balance tank
2-Centrifugal pump
3-Plate heat exchanger
4-Positive displacement pump
5-Steam injector
6-Holding tube
7-Expansion chamber
8-Vacuum pump
9-Centrifugal pump
10-Aseptic homogenizer
11-Water supply system
12-Centrifugal pump
13-Steam injector
14-Regulating valve
15-Change-over valve
16-Change-over valve
17-Control panel

FIGURE 14 UHT plant with direct steam heating. (Reprinted by permission, Alfa Laval [6].)

B. Cream Separation

Cream separation is a process whereby an essentially fat-free portion (skim milk) is separated from a fat-rich portion (cream) [15]. It is a physical process based on the fact that cream is lighter than the rest of the milk constituents. Until the introduction of the centrifugal separator, cream was obtained by spontaneous or natural separation caused by gravity in deep containers. When milk is imparted a centrifugal force, the separation of cream takes place at a much faster rate than that of natural separation.

A centrifugal separator or a centrifuge consists of a series of inverted conical disks aligned vertically around a cylindrical shaft in a bowl. The disks have small perforations, known as the distribution holes, for the cream to move through. The disks are spaced typically on the order of a few millimeters. Figure 15a shows how fat globules have only a limited

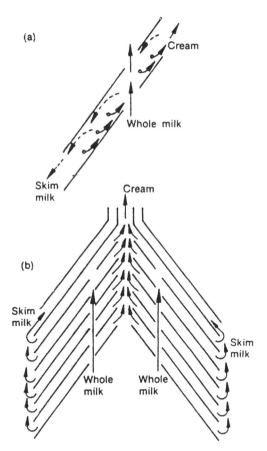

FIGURE 15   (a) Flow of cream and skim milk in the space between disks in a centrifugal separator; (b) a disk stack. (Reprinted by permission, © 1994 Chapman & Hall, from *Modern Dairy Technology*, Vol. 1, R. K. Robinson (ed.), Fig. 5, p. 443 [10].)

distance to move before being directed inward on the upper surface of a disk, while skim milk flows outward on the lower surface of the adjacent disk. The milk normally enters the centrifuge from the bottom and moves up. As soon as the milk enters the centrifuge, it is accelerated by the bowl and disk rotating at 4000–6000 rpm. The heavier milk is thrown toward the wall of the rotating bowl, and the lighter cream accumulates near the center (around the shaft). Figure 15b shows the arrangement of a complete disk stack with relevant flows. The skim milk and the cream are then collected via separate outlets.

The efficiency of separation depends on a number of factors, including [15] the conditions of the incoming milk, milk flow rate, milk temperature, bowl speed, and disk configuration. The volume of cream discharged from the separator is controlled by means of a throttling valve in the cream outlet [6]. If the valve is completely closed, all the milk will discharge through the skim milk outlet. Progressively larger amounts of cream, with a progressively diminishing fat content, will discharge from the cream outlet if the valve is open. Any change in the cream discharge will be compensated for by an equal but opposite change in the skim milk discharge. Thus the pressure in the downstream line will be changed. Therefore a control circuit is installed at the skim milk outlet to keep the counterpressure at the outlet constant, regardless of changes in the cream flow rate [6] (Fig. 16). The control unit consists of a diaphragm valve, and the required product pressure is controlled by compressed air above the diaphragm. During separation, for example, if the pressure of the skim milk drops, the preset air pressure will force the diaphragm down. This action enables the skim milk pressure to increase to the preset value.

C. Standardization

The purpose of standardization is to control the fat content in the milk. During standardization by a batch process, predetermined amounts of cream (separated previously using a cream separator) are added back into the skim milk. However, in the continuous method, cream can be added to skim or whole milk because of ease of control. In fully automated systems, clarifier–separator–standardizer combinations are used. When programmed properly, these systems remix the cream and skim milk that were initially separated when the original milk was partially or totally skimmed.

Figure 17 shows the complete process for automatic, direct standardization of milk. The pressure control system at the skim milk outlet maintains a constant pressure, regardless of fluctuations in the pressure drop over downstream equipment. The cream regulating system maintains a

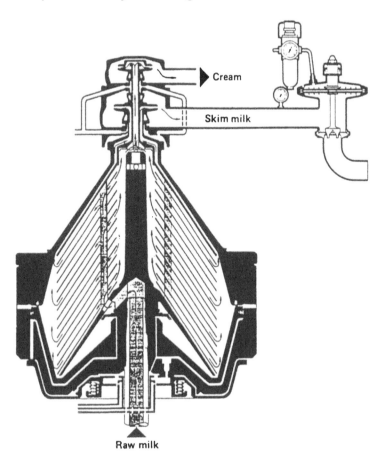

FIGURE 16 Control of constant counterpressure in the skim milk outlet. (Reprinted by permission, Alfa Laval [6].)

constant fat in the cream. The fat content in the cream in the outlet from the separator is determined by the cream flow rate. Some standardization systems, therefore, use flow meters to control the fat content. This is the quickest method. It is also accurate provided that the temperature and fat content in the milk before separation are constant [6]. Different types of instruments for continuous measuring of fat content in the cream can also be used. For example, fat content of the cream can be measured by measuring its density. However, the control is slow, and it takes a long time for the system to adjust when a disturbance occurs. Thus, a combination of accurately measuring the fat content and rapid flow metering is

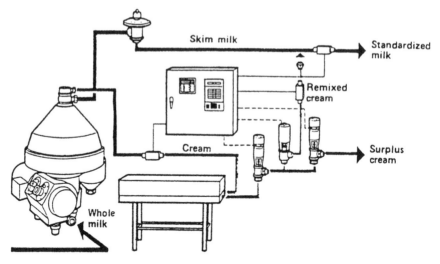

FIGURE 17 The complete plant for direct standardization. (Reprinted by permission, Alfa Laval [6].)

used in controlling the fat content in the cream. Finally, the ratio controller mixes cream of constant fat content with skim milk in the necessary proportion to give standardized milk a desired fat content.

D. Homogenization

Homogenization is the process by which the stability of milk fat emulsion is improved by decreasing the average diameter of milk fat globules. This prevents creaming or fat separation (due to the natural separation described previously). In homogenized milk, fat globules are more uniform in diameter and form a polydispersive system of milk fat with a narrow distribution [16]. The diameter of fat globules before homogenization can vary from 0.1 to 15 μm. In homogenized milk, about 85% of the fat globules are 0.1–2 μm and the rest are under 3 μm. Homogenization is performed at a very high pressure. Thus the main component of a homogenizer is a high-pressure pump that increases the pressure of milk from about 80–220 kPa at the inlet to a homogenizing pressure of 10–20 MPa. The homogenization temperature is normally 60–70°C. When the milk is forced through a narrow opening in the homogenizer head, its velocity becomes very high (about 200–300 m/s). As the milk leaves the homogenizer head, it makes impact at a high velocity on the inside of the homogenizer ring, and the fat globules are shattered.

Homogenization efficiency can be improved by a two-stage process with two homogenizer heads in series. However, the two-stage process is generally used in conjunction with UHT pasteurization because the two-stage process helps to disintegrate certain salts formed during the UHT process.

VI. MANUFACTURE OF CERTAIN POPULAR DAIRY PRODUCTS

In this section, manufacturing procedures for certain popular dairy products are briefly described. The emphasis is on providing an overview of the process and yet highlighting some unique features of the processes and equipment and/or control requirements.

A. Fermented/Cultured Milk Products—Yogurt

Fermented/cultured milk represents an array of products including yogurt, kefir, and cultured buttermilk. Among these, yogurt is by far the most popular. Cultured products are so named due to the addition of a starter culture to milk during their manufacture. Table 1 presents different bacteria used as starter cultures for a variety of dairy products.

Bacterial cultures of *Lactobacillus bulgaricus* and *Streptococcus thermophilus* are used as starters for yogurt manufacture. The ratio of *bacilli* to *cocci* in the culture and in the yogurt is about 1:1 or 1:2. This balance must be actively maintained to assure the desired qualities. Dairy plants manufacture bulk starter cultures from commercially available concentrated cultures. A block diagram of starter manufacture is shown in Fig. 18 and the equipment used for this process is shown in Fig. 19.

Yogurt is usually classified as [6]

Set type yogurt, which is filled immediately after inoculation with bulk starter and incubated in packages.
Stirred type yogurt, which is inoculated and incubated in a tank. After incubation, the product is cooled before filling.
Drink type yogurt, which is based on the stirred type. The coagulum is broken down to a liquid before filling.

The stages in the production of set, stirred, and drink types of yogurt are depicted in Fig. 20. The production line for stirred type yogurt is shown in Fig. 21. Stabilizers and sweeteners are added to milk after standardization. Then the milk is homogenized and heat treated (90–95°C for 5 min). After appropriate heat treatment, the milk is cooled to incubation temperature and starter culture is added. Then the milk is kept in incuba-

TABLE 1 Bacteria Used in Starters

Bacterium	Effect	Product
Propionic bacterum shermanii	Flavor/aroma	Emmental cheese
Lactobacillus bulgaricus	Acidity/flavor/aroma	Yogurt, kefir
Lactobacillus lactis	Flavor/aroma	Cheese
Lactobacillus helveticus	Flavor/aroma	Cheese
Lactobacillus acidolphilus	Acidity	Acidophilus milk, cultured milk
Streptococcus thermophilus	Acidity	Yogurt, Cheddar and Emmental cheese
Streptococcus diacetilactis	Acidity, flavor/aroma	Butter, cultured cream, cultured milk
Streptococcus lactis and *Streptococcus cremoris*	Acidity	Cheese, butter, cultured cream, cultured milk
Leuconostoc citrovorum and *Leuconostoc dextranicum*	Flavor/aroma	Cheese, butter, cultured cream, cultured milk
Streptococcus durans and *Streptococcus faecalis*	Acidity, flavor/aroma	Cheddar cheese, Italian soft cheese

Source: From Ref. 6, with permission.

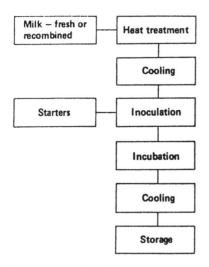

FIGURE 18 Block diagram of starter manufacture. (Reprinted by permission, Alfa Laval [6].)

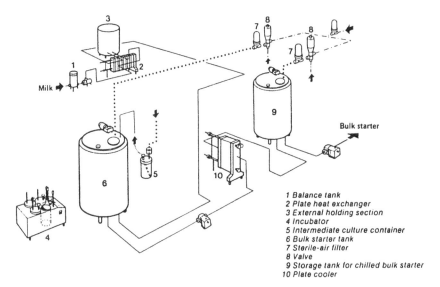

FIGURE 19 Starter culture line with substrate heat treatment in a plate heat exchanger. (Reprinted by permission, Alfa Laval [6].)

tion tanks held at constant temperature. At the end of the process (verified by measuring the pH of the product to be within 4.2 to 4.5), the yogurt is cooled rapidly in a PHE. The cooled product is passed on to filling machines via buffer tanks. Flavorings such as fruits are added when the yogurt is transferred from buffer tanks to the filling machines.

B. Cheese

Cheese is the fresh or ripened product obtained after coagulation and whey separation of milk, cream, partly skimmed milk, buttermilk, or a mixture of these products. The cheese varieties available on the market can be broadly classified according to (1) moisture content—hard, semi-hard, and soft cheese; (2) coagulation method—using rennet, acid, or both rennet and acid (cottage cheese); (3) microorganism used for ripening—lactic acid bacteria or others such as molds (for Roquefort, Gorgonzola, and Camembert); and (4) texture—round eyed, granular, and close-textured. Figure 22 shows a schematic of the simplified cheesemaking process.

Milk is subjected to normal pretreatments such as cream separation, fat standardization, and pasteurization. In addition, bactofugation is performed in cases where the presence of spore-forming organisms such as

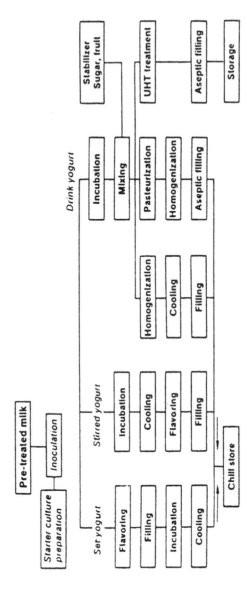

FIGURE 20 Stages in the production of set, stirred, and drink types of yogurt. (Reprinted by permission, Alfa Laval [6].)

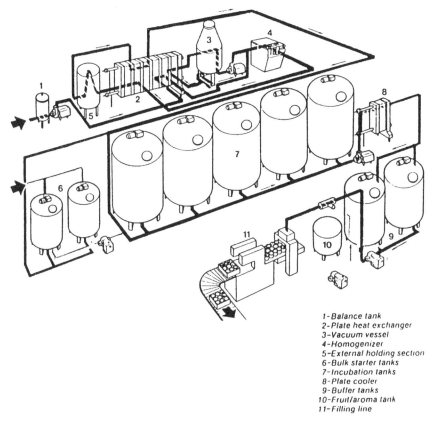

1-Balance tank
2-Plate heat exchanger
3-Vacuum vessel
4-Homogenizer
5-External holding section
6-Bulk starter tanks
7-Incubation tanks
8-Plate cooler
9-Buffer tanks
10-Fruit/aroma tank
11-Filling line

FIGURE 21 Production line for stirred type yogurt. (Reprinted by permission, Alfa Laval [6].)

Clostridinum sp. in milk can lead to product loss. During bactofugation, 2–3% of the total volume of milk is separated in a centrifuge. This bactofugate contains both the undesirable microorganisms and the fat content of milk [17]. The bactofugate is sterilized by live steam injection at 130–140°C for a few seconds. After cooling, it is added back to the pasteurized cheese milk.

Often the cheese milk is preripened by adding the starter culture and keeping it at low temperature for several hours. This is known to improve the properties of the cheese. Certain additives, such as calcium chloride ($CaCl_2$), coloring agents, and cheese ripening agents, can be added at this stage. The $CaCl_2$ addition (5–20 g/100 kg of cheese milk) improves the firmness of the coagulum.

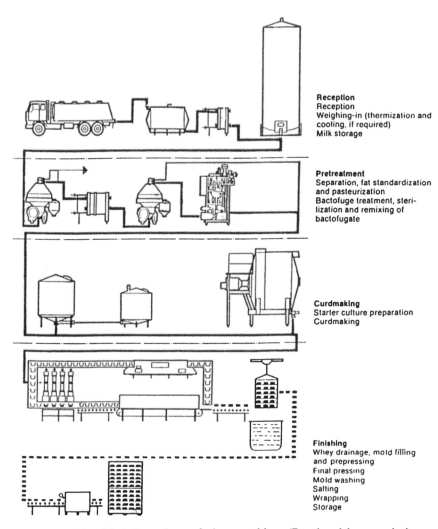

Reception
Reception
Weighing-in (thermization and cooling, if required)
Milk storage

Pretreatment
Separation, fat standardization and pasteurization
Bactofuge treatment, sterilization and remixing of bactofugate

Curdmaking
Starter culture preparation
Curdmaking

Finishing
Whey drainage, mold filling and prepressing
Final pressing
Mold washing
Salting
Wrapping
Storage

FIGURE 22   Simplified flow chart of cheesemaking. (Reprinted by permission, Alfa Laval [6].)

The cheese milk then enters the curdmaking state. At this stage, upon addition of a coagulant, milk changes from being a liquid to a semi-solid gel, known as the coagulum or curd. The coagulant is added at a specified rate (e.g., 28.4 mL calf rennet/L of milk), and coagulation takes place at 30°C.

After the coagulum reaches a desired firmness, the curd is cut. Stainless steel knives are used for cutting. The knives are also used as an agitation mechanism; the curd is cut with the sharp side of the blades, while the curd–whey mixture is stirred with the blunt side. A typical cheese vat with curd cutting/stirring blades is shown in Fig. 23. The coagulum is cut at a very low speed, and the speed is progressively increased to avoid fat and casein losses in the whey, which reduce the cheese yield.

Whey can be drained off without stopping the stirring tools. Stationary or rotating strainers are used for this purpose. If the stirring is stopped, which is a common practice, the whey must be drained off very quickly before the curd becomes lumpy.

Full automation of cheesemaking has been hampered by the lack of proper sensors to time when to cut the coagulum. This is critical. If a coagulum is cut too soon, cheese yield suffers due to loss of curd fines. Cutting too late results in a high-moisture cheese of less than desired

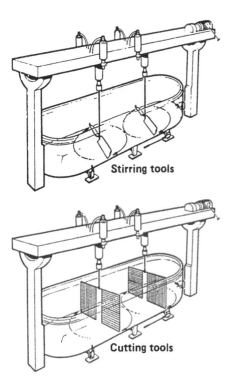

Figure 23   Cheese vat with cutting/stirring tools. (Reprinted by permission, Alfa Laval [6].)

texture [18]. Thus cheesemakers strive to identify the optimal cut time. Currently, it is based mostly on the subjective judgment of the cheesemaker or according to a set time after adding the enzyme. There have been a number of efforts in developing a cut-time sensor. The Stoelting company markets a sensor, OptiSet, which works based on changes in the heat dissipating ability of the coagulum from an electrically heated wire when the coagulum begins to firm [19]. Other efforts still under way with encouraging results use optical or ultrasonic techniques. The optical sensor measures the diffuse reflectance of light (820 nm) by the coagulum during coagulation [20]. The ultrasonic sensor measures the attenuation of the sound energy (at 1 MHz) during coagulation [21]. The signals obtained using these sensors have been used to correlate with the subjective cut-time determinations very satisfactorily. However, a robust and reliable automatic cut-time sensor is still not available. This will continue to pose some problem for the total automation of cheesemaking.

After the initial whey drainage, the curd is heated to accelerate whey removal. The treatment of the curd–whey mixture at this stage varies in relation to the type of cheese produced. The curd handling process is unique to different cheese types. The method used in manufacturing Cheddar cheese (close-textured cheese) is explained here. The close texture is obtained by subjecting the curd to treatment that causes all the lactose to ferment before molding. The Cheddaring process is started when the maximum cooking temperature has been reached and the curd condition is suitable for pitching. After Cheddaring, the blocks are milled, salted, stirred, and pressed into desired shapes. Salting affects the flavor, starter development, and consistency and improves the keeping quality of the cheese.

The cheesemaking operation has been extensively mechanized and automated in keeping with increasing demands of both quantity and quality. Figure 24 shows a schematic of a modern plant for manufacturing Cheddar cheese. Pressing in molds is still the most common method of giving Cheddar cheese its final shape. Handling the molds requires an extensive system. The molds are expensive and need frequent replacements. Therefore several continuous and automatic systems have been devised in recent years to form a cheese without using a mold. In this, the prepared curd is drawn by vacuum to the top of a tower, where it begins to fuse and form a continuous column. A constant vacuum is applied over the entire curd column, which causes the product at the base to be consistent, uniform, and free of air [6]. The columns are automatically lowered, cut (into blocks of about 20 kg each), and sent for packing and ripening.

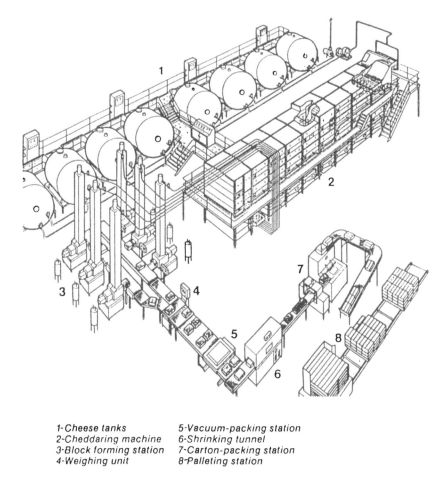

1-Cheese tanks   5-Vacuum-packing station
2-Cheddaring machine   6-Shrinking tunnel
3-Block forming station   7-Carton-packing station
4-Weighing unit   8-Palleting station

FIGURE 24   A modern plant for the manufacture of Cheddar cheese. (Reprinted by permission, Alfa Laval [6].)

C. Concentrated and Dried Dairy Products— Condensed Milk

The main objectives for concentrating or drying dairy products are to increase shelf life, to improve functionally, and to reduce cost of handling, transporting, and storing. The production of concentrated and dried products has the advantage of processing and storing all market surpluses, especially fluid milk and cheese.

Condensed milk may be unsweetened or sweetened. Unsweetened condensed milk is obtained by evaporating water from milk. It is a sterile product used as a substitute for fresh milk. Sweetened condensed milk is basically concentrated milk to which sugar has been added. The high sugar concentration ($\sim$ 45%) increases its osmotic pressure and destroys most of the microorganisms [6].

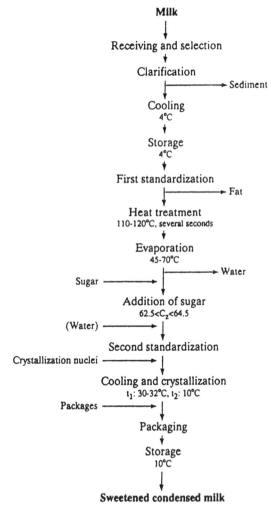

FIGURE 25 Flow chart of unsweetened condensed milk production. (Reprinted with permission by VCH Publishers © 1995 [16].)

Flow charts for unsweetened and sweetened condensed milk production are presented in Figs. 25 and 26, respectively. For unsweetened condensed milk, the steps up to standardization are typical milk processing operations. The preheating step helps to decrease the number of microorganisms present, thus intensifying the sterilization effect and enabling milder sterilization. It also increases the stability of milk during subsequent sterilization.

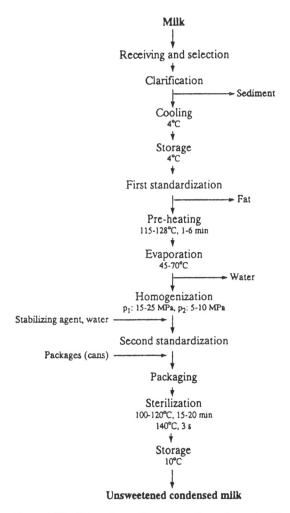

FIGURE 26   Flow chart of sweetened condensed milk production. (Reprinted with permission by VCH Publishers © 1995 [16].)

Plate or tubular heat exchangers are used for preheating. The extent of preheating largely determines the viscosity of the final product and is extremely important for product quality. The time–temperature regime of preheating is usually 93–100°C for 10–25 min or 115–128°C for 1–6 min.

After preheating, the milk enters an evaporator. Because of the undesirable effects of high temperatures necessary to evaporate water at atmospheric conditions, evaporation of water from milk is performed under vacuum. However, to avoid any microbial growth, the evaporator temperature is never below 45°C. The evaporator typically used is of the multiple-stage, falling-film type. This type of evaporator operates at a much lower temperature range (~ 50–60°C), thus offering a several advantages such as decrease of heat-induced changes in milk components, low energy consumption, and easy maintenance. They are simple to operate and maintain a stable regime of operation. Control and adjustment of the regime are easy and accurate; namely, once adjusted, further regulation is automatic.

The dry solids content increases as water is evaporated. Product density is continuously monitored. When the density is about 1.07, the solids concentration is about right. At this point, 2.1 kg of fresh milk has been condensed into 1 kg of unsweetened milk of 8% fat and 18% nonfat solids.

The condensed milk is then homogenized in a two-stage process, cooled to about 14°C, and standardized to the final product composition. To improve the heat stability during sterilization, a stabilizer (disodium or trisodium phosphate) is added. Any addition of vitamins is also made at this time. The milk is then filled in cans and sent to an autoclave for sterilization (100–120°C for 15–20 min).

The manufacture of sweetened condensed milk is similar to that of the unsweetened condensed milk just described. However, the product is neither homogenized nor sterilized. Instead, addition of sugar and cooling with crystallization are included. About 0.18 kg of sugar is added for 1.0 kg of fresh milk. The composition of the sweetened condensed milk is 8% fat, 45% sugar, 20% milk solids, and 27% water.

The most critical stage in manufacturing sweetened condensed milk is the cooling with crystallization. This is done right after evaporation. Because the water in the product can only hold about 50% of the total lactose, the other 50% of the lactose is crystallized. To produce high-quality sweetened condensed milk, it is important that the amount of lactose does not appear as a small number of large crystals but rather as a large number of small crystals (of diameter <10 μm). To ensure this, milk is inoculated with powdered lactose crystals. The crystallization is brought about by rapidly cooling the product while agitating it vigorously. Cooling with crystallization is accomplished by using a double-wall tank-crystallizer, fluid-flow continual coolers, or a vacuum crystallizer.

VII. PLANT SANITATION

Because of the perishable nature of dairy products and the potential proliferation of microorganisms, it is absolutely essential to thoroughly clean
and sanitize the plant. Negligence in this regard can lead to costly consequences. A variety of detergents and disinfectants, in conjunction with
different systems of cleaning, are based on the equipment and operations.

Clean-in-place (CIP) technology is now widely used in the dairy industry. In this, the entire system is cleaned by circulating appropriate
cleaning agents without physically disassembling the system components.
This has greatly reduced plant downtime and associated costs. More im-
portantly, it paved the way for automation and computer controls for the
entire plant operation.

For effective CIP, the equipment must be designed to fit into a cleaning circuit and must also be easy to clean. All surfaces must be accessible
to the detergent solution. Machines and pipes must be installed in such
a way that they can be efficiently drained. The CIP system can be programmed for a specific ampliation. For example, a CIP program for a
pasteurizer circuit can consist of the following stages [6]:

1. Rinsing with warm water for about 8 min
2. Circulating an alkaline detergent solution for about 20 min at
   about 75°C
3. Rinsing out the alkaline detergent with water
4. Circulating (nitric) acid solution for about 15 min at 70°C
5. Gradual cooling with cold water for about 8 min

The CIP systems installed in dairies can either be centralized or
decentralized, depending on the nature of plant layout, extent of plant
operations, and so forth. A schematic of a centralized CIP system is shown
in Fig. 27. The water and detergent solutions are pumped from the storage
tanks in the central station to the various CIP circuits.

A station of this type is usually highly automatic. The tanks have
electrodes for high and low-level monitoring. Routing of the cleaning solutions is controlled by a sensor which measures the conductivity of the
liquid. CIP programs are controlled from a computerized sequence controller. Large CIP stations can be fitted with multiple tanks in order to
provide the necessary capacity.

The effectiveness of the CIP process depends on the flow rate, which
must generate turbulence to remove "soil," the concentration and temperature of the cleaning fluids, and the time allowed for each stage in the
process [22]. All these factors must be closely controlled for a successful
CIP process. Special valves, with double seats and limit switches, are
used in the CIP lines to obtain a positive indication of whether the valve
is fully open or fully shut.

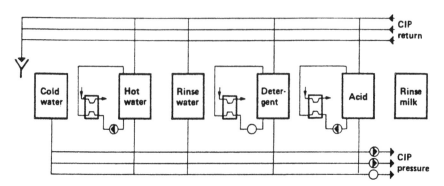

FIGURE 27 Cleaning of several circuits from a central CIP station. (Reprinted by permission, Alfa Laval [6].)

Some CIP systems operate on a once-through basis; others recirculate all the fluids to their respective tanks. The third is a combination of the first two. Recirculation helps save energy and chemicals, but the effectiveness of cleaning may be questionable [22].

As mentioned earlier, the introduction of CIP systems was critical in establishing computer process control systems in the dairy industry. The CIP programs are controlled from a computerized sequence controller. The detergent and water tanks are equipped with sensors for high- and low-level monitoring. Routing of the cleaning solution is controlled by a sensor that measures the conductivity of the liquid. Recent developments include computer-controlled detergent dosing systems, environmentally friendly CIP systems, automatic analyzers, and CIP data logging systems [5].

VIII. SUCCESSFUL DAIRY PLANT AUTOMATION

The technology of dairy processing has changed drastically over the last decade or so in scale and sophistication. Thus effective monitoring of dairy manufacturing processes is now absolutely fundamental to the safe and efficient operation of plant and equipment, and to the production of consistent, high-quality and ultimately trustworthy product for the consumer [23]. The information that a computer-controlled system can provide a plant manager is crucial for making critical decisions. Inefficient resource and materials utilization and other problem areas are more easily discovered when thorough processing records are maintained and compared with ingredients used. Modern automation systems have numerous benefits at affordable costs. They are

1. Improved quality efficiency and productivity because an automated system can facilitate more precise control than an operator.
2. In situations where an operator has to respond, the automated computer-controlled system can provide better information faster.
3. Improved flexibility of operation, that is, changing product lines or runs as necessary.
4. Improved inventory control of finished products as well as ingredients and packaging material supplies.

Perhaps another major reason to automate dairy processing is the need to document. Increased documentation, perhaps tracing all the steps in conversion of raw materials into finished products, is a likely outcome of increased Food and Drug Administration (FDA) scrutiny [4]. Computer-controlled operations can assist with Hazard Analysis and Critical Control Points (HACCP) programs, which, if implemented and executed properly, can assure product safety. Computerized operations increase the monitoring and accountability of an HACCP program. *Listeria* and *Salmonella* are major safety concerns for the dairy industry. Product recalls due to safety problems not only cost dairy processors immediate monetary loss but have a long-term cost associated with loss of consumer confidence. Considering these, dairy plants cannot afford to wait too long to automate with computer controls.

One must approach plant automation in a systematic manner. Before specifying a process control system, it is essential to prepare process flow diagrams showing all the unit processing operations that need to be controlled and the sequence in which the product moves from one operation to another [24]. This provides a "big picture" to plant personnel, which is necessary for gaining a proper perspective on and orientation of the control system. Next, each unit operation should be looked at more critically and in detail to identify and characterize all the specific control points that will require detailed specifications. Some steps to consider for successful automation are [25]

1. *Understand the present plant operations.* Perhaps this appears elementary, but this is a key first step toward successful automation. Begin by examining floor diagrams and operating procedures for each operation in the plant. Evaluate the purpose and value of each step in the procedure, including the peripheral operations such as utilities, warehousing, and distribution.

2. *Involve plant personnel.* Personnel involved in the day-to-day running of the plant know the most about the plant. Their impact as to the condition and requirements for certain systems must be taken into consideration. The workers should review the flow diagram and operating procedures. They should also be involved in all of the other strategies and logistics discussed for automation.

3. *Simplify the operation.* It is recommended that all operations and products should be redesigned to eliminate unnecessary or improper elements. This will eliminate the problem of automating many of the wrong things and thus will further improve plant efficiency.

4. *Create a detailed long-range plan.* This is simply to keep a futuristic view of plant operations, expansions, and so forth in mind. The long-range plan is no place to skimp. Costs of automation systems are dropping steadily, and they get easier to use. So systems that might seem out of the question now could well be commonplace very soon.

5. *Create a short-term plan.* The short-term plan will cover the initial transition to an automated system. It should be compatible with the long-term plan but should include even more detail.

6. *Evaluate the system.* This is to examine (1) adaptability of the particular automation system selected and the plant operations and (2) its suitability for satisfying the requirements of both short- and long-range plans.

7. *Choose a vendor.* This is a purchasing decision to be based not only on cost but on the compatibility of system components, scale-up, technical training and support, software and hardware upgrades, and so forth.

One of the recent trends in plant automation is called computer-integrated manufacturing (CIM). (See Chapter 20 for more details.) CIM brings together manufacturing systems, information systems, and human systems. Manufacturing systems include product design, production process, material flow, machine performance, and plant layout. Information systems include system architecture, databases, communication networks, fault tolerance, and man–machine interfaces. Any successful implementation of a CIM requires close work and communication with the human element [5].

Implementation of CIM techniques is usually a gradual process involving detailed planning, equipment acquisition, training, and implementation. A common practice is to begin with islands of automation that are well suited for eventual computer integration. Stages for a large-scale implementation of CIM might include defining and developing the system concept; functional requirements; functional design; detailed design, coding, and testing at the unit, module, and system level; then installation, start-up, and audit.

Automated factory systems are being integrated with office information system [26]. Specific goals here include improved productivity, reduced inventory costs, improved quality, and more flexibility. As computer-based tools are becoming more available, the competitive advantages of using CIM techniques are increasing. Product analysis from suppliers can be received and processed quicker, reducing inventory holding time prior to shipping. With the ability to deliver timely information, a company can respond to customer demands more readily.

Another major advantage of CIM lies in automatic data acquisition. If handled properly, it results in less duplication of data and effort. There are fewer mistakes, and the data are more accessible. This leads to more consistent decision making. Centralized process control increases the value of individual PC-based stations throughout a plant. By integrating these stations, the data acquisition rate can increase. Improved monitoring of product quality is beneficial for regulatory compliance. Before beginning a CIM project, the application needs to be adequately defined and all of the risks and benefits considered. Mathematical models of the system can be used to simulate performance.

Computer automation and controls in the dairy industry can also help processors to take advantage of yet another recent trend—real-time manufacturing (RTM) or just-in-time (JIT) manufacturing (see Chapter 20 for more details). RTM is the use of computer automation and data acquisition systems to provide information on manufacturing as it happens (real time) [27]. This allows the exact status of cost, efficiency, utilization, and so forth of the manufacturing unit to be known at any time. Thus it allows processors to respond immediately to consumers and retailers in the market. RTM can also assist in control and analysis of unit operations, as well as more holistic measurements. It also can change the entire nature of business from one of manufacturing-to-stock to one that manufactures-to-order.

REFERENCES

1. J. David and P. Filka, Automatic control systems in milk processing, *Automatic Control and Optimisation of Food Processes* (M. Renard and J. J. Bimbenet, eds.), Elsevier Applied Science, New York, 1988, pp. 381–390.

2. P. Demetrakes, Food plants, *Food Processing 56*(3):52 (1995).
3. J. M. Hellman, Cains Foods integrates its information system, *Food Manufacture 69*(6):27 (1994).
4. K. Schexnayder, Why wait to automate, *Dairy Foods 92*(7):30 (1991).
5. R. L. Olsen, Computer applications: Expert systems, *Dairy Science and Technology Handbook, Vol. 3: Applications Science, Technology and Engineering* (Y. H. Hui, ed.), VCH Publishers, New York, 1994, pp. 105–154.
6. Alfa-Laval, *Dairy Handbook*, Alfa-Laval Food Engineering AB, Sweden, 1987.
7. T. A. Hughes, *Measurement and Control Basics*, Instrument Society of America, Research Triangle Park, North Carolina, 1988, pp. 1–20.
8. P. W. Murrill, *Application Concepts of Process Control*, Instrument Society of America, Research Triangle Park, North Carolina, 1988, pp. 11–67.
9. C. Honer, Dairy plants get smart, *Dairy Foods 90*(4):87 (1989).
10. W. M. Kirkland, Automation in the dairy industry, *Modern Dairy Technology, Vol. 1: Advances in Milk Processing* (R. K. Robinson ed.), Chapman and Hall, London, 1994, pp. 433–472.
11. T. Gilmore and J. Shell, Dairy equipment and supplies, *Dairy Science and Technology Handbook, Vol. 3: Applications Science, Technology and Engineering* (Y. H. Hui, ed.), VCH Publishers, New York, 1994, pp. 155–294.
12. P. Jelen, *Introduction to Food Processing*, Reston Publishing, Reston, Virginia, 1985, pp. 125–151.
13. V. Caudill, Engineering: Plant design, processing and packaging, *Dairy Science and Technology Handbook, Vol. 3: Applications Science, Technology and Engineering* (Y. H. Hui, ed.), VCH Publishers, New York, 1994, pp. 295–329.
14. H. Burton, *Ultra-High Temperature Processing of Milk and Milk Products*, Elsevier Applied Science, London, 1988, pp. 77–128.
15. C. Towler, Developments in cream separation and processing, *Modern Dairy Technology, Vol. 1: Advances in Milk Processing* (R. K. Robinson, ed.), Chapman and Hall, London, 1993, pp. 61–105.
16. M. Caric, *Concentrated and Dried Dairy Products*, VCH Publishers, New York, 1994, pp. 1–49.
17. A. Y. Tamine and J. Kirkegaard, Manufacture of Feta cheese—industrial, *Feta and Related Cheeses* (R. K. Robinson and A. Y. Tamine, eds.), Ellis Horwood, London, 1991, p. 70.
18. R. Scott, *Cheesemaking Practice*, Elsevier Applied Science Publishers, London, 1986, p. 193.
19. T. Hori, Objective measurements of the process of curd formation during rennet treatment of milks by the hot wire methods, *J. Food Science 50*:911 (1985).
20. F. A. Payne, C. L. Hicks, S. Madanagopal, and S. A. Shearer, Fiber optic sensor for predicting the cutting time of coagulating milk for cheese production, *Trans. ASAE 36*:841 (1993).
21. S. Gunasekaran and C. Ay, Milk coagulation cut-time determination using ultrasonics, *J. Food Process Eng, 63–73* (1996).

22. T. Greeves, Introduction to food processing, *Automation in the Food Industry* (C. A. Moore, ed.), Blackie, London, 1991, pp. 1–28.

23. G. Godfrey, Introduction, *Food Process Monitoring Systems* (A. C. Pinder and G. Godfrey, eds.), Blackie Academic & Professional, London, 1993, pp. 1–11.

24. A. Teixeria and C. F. Shoemaker, *Computerized Food Processing Operations*, Van Nostrand Reinhold, New York, 1993, pp. 51–100.

25. J. Man, Ten steps to successful automation, *Dairy Foods 93*:45 (1992).

26. E. Keating, S. E. Davis, and D. T. Houck, Snap-together technology for easy systems integration, *Food Processing 55*(4):23 (1994).

27. D. Toops, The real story on real-time manufacturing, *Food Processing 56*(6): 45 (1995).

16

Computerized Automation and Controls in Meat Processing

John W. Buhot

Commonwealth Scientific and Industrial Research
Organisation (CSIRO), Tingalpa D.C., Queensland, Australia

I. INTRODUCTION

The consumption of meat by humans can be traced back into antiquity. However, the meat processing sector has been slow to adopt new technologies. Perhaps this is because of the nature of the work itself and the social connotations in various societies of being involved in the process of killing animals. The variation in shape and size of animals also poses major technical difficulties. It is only in the last hundred years that slaughter, dressing, fabrication, storage, and further processing of meat have become operations for which collective labor can be assigned to specific production tasks in a mass production environment. Figure 1 shows a schematic representation of the industry process flow.

At the turn of the century the majority of livestock were processed manually using the technique of solo-butchering. In some cases a small team of process workers operated with one carcass. A major innovation occurred in the 1950s when the "on-rail" system of slaughter and dressing was introduced to meat processing plants. This system is the basis of the

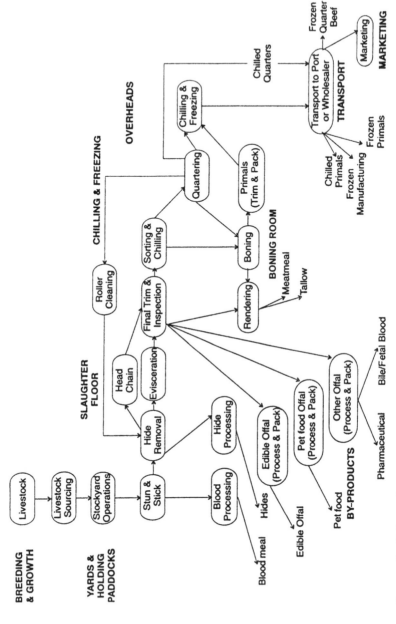

FIGURE 1 Schematic representation of meat industry process flow.

majority of current slaughter and dressing operations. In this system the carcass is suspended by the rear legs and transported to individual process operators by an overhead conveyor. The process operator carries out a unique task or a small number of predefined tasks on each carcass. The number of tasks that an individual operator can perform is primarily determined by the rate of production. The carcass is then conveyed to the next operator, where a further set of tasks is carried out. This sequence is repeated along the slaughter and dressing line until all necessary tasks are completed.

At the time of its introduction, the on-rail system and its components were electromechanical devices and system control was by hard wire relays and timers. The introduction of programmable logic controllers (PLCs) and subsequent computer control techniques have increased the ability to control the processing operations. This has resulted in improvements in product quality and processing efficiency. Various mechanical aids to processing have also been introduced to the industry. These include hydraulically and electrically powered cutting tools and pneumatic flaying knives. Their adoption has enabled further gains in productivity to be made as well as addressing the occupational health and safety issues that are of major importance in operations of this nature.

Spiraling costs and increased competition from other sources of food has provided the incentive for the industry to consider far greater use of computerized automation and controls in meat processing. Considerable research and development have been directed toward improving processing efficiency through the introduction of new technology. A new beef slaughter and dressing method has been developed in Australia [1, 2]. New Zealand has produced a number of systems and equipment for use with sheep processing [3]. In Europe much work has been carried out in the automation of pig processing [4]. Equipment manufacturers have incorporated advanced technologies used in other industries into their product range, and increased employment opportunities now exist for professional engineers, scientists, and skilled technical personnel.

II. SLAUGHTER AND DRESSING TECHNOLOGIES

A. Bovines

1. Overview

The traditional practices of slaughter and dressing of bovines are labor intensive and therefore and substantial cost to the conversion of the live animal to meat. The working environment is unpleasant; process workers carry out tasks that are repetitive, monotonous, unpleasant, and in some instances, dangerous. The Australian meat processing industry has recognized that automation offers the potential to reduce unit costs of production, provide a better working environment for the employees, and increase the efficiency of operation. The industry instigated a research and development program to provide abattoirs with an automated slaughter and dressing system. The Meat Research Corporation and the Commonwealth Scientific and Industrial Research Organisation's Division of Food Science and Technology jointly undertook this work [1, 2].

Initially, all the tasks needed for the slaughter and dressing operations were identified and described. Boundary conditions for the tasks were also defined based on historical production data including weight range, sex, breed, and throughput rate. A sequence of automated actions that could produce the same output as the manual task was then specified for every task. Preprototype mechanisms were built to test the validity of these actions and sequences. Consequently, a full-scale system, known as FUTUTECH, was constructed and evaluated at an abattoir. This system consisted of a number of modules that were interconnected by an overhead conveying system. Each module performed a task or group of tasks automatically using algorithms determined as part of the task description. The large variation in bovine size and weight precluded the use of conventional robotics. Bovine anatomy and physiology proved difficult to describe mathematically. Mechanical replication of the actions of a process worker performing a manual task was also precluded, as the control system could not be replicated in a cost-effective manner.

FUTUTECH demonstrated that automated machines could perform many of the tasks presently being carried out manually. In particular, the more arduous tasks and those involving the greatest risk of personal injury could be automated. Individual modules could be interconnected to form a system that could process all breeds of bovine in the live weight range of 280–800 kg at processing rates of up to 90 animals per hour. The system also offered the potential for labor reduction of up to 50%. These findings were used to specify the commercial system. Figure 2 shows the bed dressing conveyor and scissor lift. Figure 3 shows the head removal module (foreground) and the pubic symphysis module.

2. System Description

a. Lead-Up, Restraint, and Stunning Module. The module provides the means to separate one bovine from a group being led up the race before slaughter. Preslaughter stress is reduced by the controlled manner in which the animal is conveyed to stunning chamber. This is achieved by a moving-floor conveyor combined with a set of pressure sensitive roller gates that follow the contour of the animal. The floor of the chamber

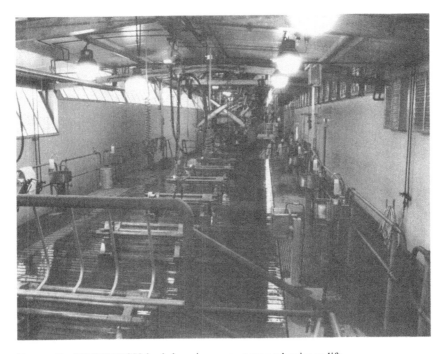

FIGURE 2   FUTUTECH bed dressing conveyor and scissor lift.

is automatically lowered, and the animal is supported by two longitudinal bars under the leg pits. Two vertical bails capture the neck. The animal is electrically stunned by applying a voltage through the neck bails, to the cranium of the fully restrained animal. A cardiac arrest is produced by the application of an additional voltage to the animal via the neck bails to the longitudinal bars under the leg pits. The animal is rendered insensitive to pain before exsanguination in a completely humane manner, and the nervous system is depolarized preventing poststun convulsions and violent kicking. The need for manual intervention in the lead-up race has been removed, and manual stunning has been eliminated.

   b.   Exsanguination Module.   This module permits exsanguination to be carried out automatically. The neck of the stunned animal is fully stretched by the automatic rotation of the neck bails of the lead-up, restraint, and stunning module. An extendible arm housing a pneumatically powered oscillating-blade knife automatically locates the neck above the sternum. The knife penetrates the hide and enters the thoracic cavity severing the brachiocephalic trunk. Bleeding occurs through the neck

FIGURE 3 FUTUTECH head removal module (right foreground), hide pulling module (partly obscured), and pubic symphysis module.

wound, and the blood is collected in a drain below the animal and removed for further processing. The manual tasks of cutting open the neck and severing the brachiocephalic trunk have been eliminated.

    c. Horn and Hoof Removal Module. The carcass, still in the prone position, is conveyed from the exsanguination module to the horn and hoof removal module, where the horns and hooves are removed automatically. Sensors locate the horns and a mechanical arm automatically positions a pair of hydraulically powered cutters at the attachment point of the horns to the skull. The cutters then sever the horns from the skull. Similarly, sensors locate the hooves, and hydraulic cutters attached to a rotating arm sever the hooves from the legs. These automatically controlled operations replace awkward manual tasks using heavy cutting apparatus.

    d. Bed Dressing Module. The carcass is conveyed from the horn and hoof removal module to the bed dressing module, where it is automatically rolled from the prone to supine position and located in a cradle. The cradle is attached to a moving-slat conveyor that also carries process workers. This configuration permits a redesign of manual tasks and allows a process worker to perform a variety of operations on an individual carcass in an ergonomic manner. These include sealing the esophagus, severing the diaphragm, and preparing the hide for subsequent automatic removal. At the end of the module the forelegs and rear legs of the carcass are manually shackled in preparation for transfer to an overhead conveyor. Traditionally, the operator remained stationary while the carcass passed by, allowing only a small number of repetitive tasks to be carried out on successive carcasses.

e.   Overhead Conveying.   The carcass is lifted by pneumatic scissor action to an overhead conveyor. A carcass processed using the FUTU-TECH system is suspended horizontally by all four legs on a dual-rail conveying system, unlike traditional abattoir practice. The mid-sagittal plane of the carcass is orthogonal to the direction of travel of the conveyor. This method of transport was selected to provide greater stability to the carcass. The conveyor design allows the fore and rear legs to be raised or lowered as required. This gives a measure of control over the orientation and position of the internal organs within the body cavity. The first section on the conveyor provides a park station. This allows a process worker to manually separate and seal the rectum.

f.   Pubic Symphysis Module.   The overhead conveyor transports the carcass from the park station to the pubic symphysis module. Optical sensors locate the pubic symphysis, and it is automatically divided using a small circular saw attached to the end of an extendible arm. This eliminates a repetitive manual task that required the process worker to use a heavy and bulky saw.

g.   Hide Removal Module.   The overhead conveyor moves the carcass from the pubic symphysis module to the hide removal module. Sensors locate the portion of hide that has been manually separated from the carcass at the bed dressing module. Clamps mounted at the ends of hydraulic cylinders automatically grip the hide. The cylinders retract downward, pulling the hide away from the body, then over the head, and finally placing it in a disposal shute. This module replaces at least two process workers depending on the type of hide pulling equipment and quality of carcass finish required.

h.   Head Removal Module.   The overhead conveyor moves the carcass from the hide removal module to the head removal module, where the head is automatically separated from the body. A pressure sensor determines the position of the first cervical vertebra (atlas joint) of the carcass by locating the base of the skull. Clamps secure the head to a base plate on the module, and a hydraulically powered cleaver severs the head from the neck. The head is then automatically transferred to a hook on the adjacent head chain. This module has eliminated a number of manual tasks that are extremely arduous. The manual operations of head severance and transfer cannot be carried out in an ergonomically preferred orientation. The weight of a bovine head is around 20 kg, and consequently repetitive strain injuries to the shoulder and arms of the process workers are common in conventional slaughterfloors.

i. Sternum Module. The overhead conveyor then moves the carcass to the sternum module. A set of optical sensors locates the xiphoid cartilage. The sternum and xiphoid cartilage are then automatically divided using hydraulically powered shears mounted at the extremity of an articulated arm, which moves above the sternum form the xiphoid cartilage to the neck. This module eliminates a manual operation that requires a process operator to use a large, heavy reciprocating saw or shears in a manner that transmits large forces to the upper body and arms of the operator.

j. Evisceration Module. The evisceration module follows the sternum module. Here the viscera are automatically removed from the carcass. The overhead conveyor transports the carcass to this module. A hydraulically powered extendible arm with a flat paddle-shaped end enters the thoracic cavity. A pressure sensor on the paddle locates the spine. The arm follows down the spine from the thoracic cavity to the pelvis, pushing the viscera in front of it. The complete viscera rolls over the pelvis between the rear legs and falls onto the viscera table located below the carcass. The tail is also automatically severed from the carcass at this module. An optical sensor located at the end of an extendible arm locates the third coccygeal vertebra, and a hydraulically powered cutter attached to the arm severs the tail at this point. The module eliminates the function of two or more process workers depending on the method of manual evisceration. The use of a paddle to separate the viscera from the body can significantly reduce the damage to the visceral organs and possible carcass contamination through leakage of stomach and bowel contents.

k. Splitting Saw Module. The carcass is conveyed from the evisceration module to the splitting saw module, where it is automatically split into two sides using a bandsaw mounted on an articulated frame. Pressure sensors locate the spine at the pelvis, and four guide rollers attached to the frame engage the carcass beneath the spine. The saw blade then engages the vertebra and begins the splitting procedure. The frame travels beneath the carcass using the guide rollers and pressure sensors to accurately constrain the spine. The saw separates the carcass through the mid-sagittal plane. The sides are then conveyed to manual trim and weighing stations. This procedure has two major advantages. First, it eliminates a manual operation that requires the operator to use a large saw, and second, it is more accurate in cutting along the mid-sagittal plane than the manual operation, particularly when operator fatigue sets in.

3. Hygiene

The provision of automatic sterilization [5, 6] is a feature of the FUTU-TECH system. An automatic cleaning procedure is initiated at the end of every operating cycle. All components that make direct contact with the carcass or come into proximity to the carcass are automatically retracted to fully enclosed sterilization. The equipment is washed with cold water. This is followed by a hot-water (82°C) wash. The procedure is designed to prevent cross-contamination from one carcass to another through physical contact of unwashed equipment, and also to contain water and other airborne particles within the cabinet.

B. Ovines

1. Overview

The New Zealand meat industry has invested heavily in the development of slaughter and dressing equipment of ovines. The program was initiated in response to increased labor costs, more stringent hygiene regulations that had decreased slaughter and dressing productivity, the need for consistent size and shape, and high product quality. Meat processing companies, farmer producer boards, government research agencies, commercial engineering companies, and the Meat Industry Research Institute of New Zealand (MIRINZ) provided input for the equipment development [3]. Figure 4 shows a schematic layout of a mechanized inverted dressing system that incorporates these developments.

2. Equipment Description

a. Stunning Machine.   Electrical stunning is the preferred method for stunning sheep. Two methods of stunning have been developed. These are "head-only" and "head-to-body." Head-only stunning meets Halakic slaughter requirements. It results in an initially still animal that starts to produce a paddling or running movement even after the throat has been cut. Such movement can be reduced by passing an electric current through the carcass, preferably by using running electrodes after shackling. A head-to-body stun results in cardiac arrest, and the current through the body reduces subsequent movement. Both types of stunning have been easily adapted for automation. The first automatic stunner for sheep and

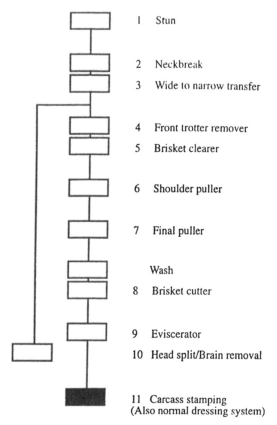

1   Stun

2   Neckbreak

3   Wide to narrow transfer

4   Front trotter remover

5   Brisket clearer

6   Shoulder puller

7   Final puller

    Wash

8   Brisket cutter

9   Eviscerator

10 Head split/Brain removal

11 Carcass stamping
    (Also normal dressing system)

FIGURE 4   Schematic layout of mechanized inverted ovine dressing system.

lambs evolved from a unit developed in Europe for pigs. This unit was then modified for sheep by New Zealand researchers. MIRINZ developed an automatic stunner that used a single "V" restrainer. Each sheep is brought to a position with its head adjacent to two grids of nozzle electrodes. Then the two grids move inward until they contact each side of the animal's head. A voltage is applied to the head from the nozzle electrodes, and the nozzle assembly sprays water onto the contact region to assist electrical conduction. The design of this stunning system was improved by New Zealand's Alliance Freezing Company to handle horned stock. This improved stunning system uses a dual "V" restrainer system in which one conveyer feeds the other. The use of two conveyors allows controlled spacing of the animals. The machine has also been adapted for head-to-body stunning.

b. Hock (Front Trotter) Removal.   A mechanical device grips both the radius and ulna bones of the forelegs. This configuration and method of restraint permit the front hock (or trotters) to be removed automatically. Removing the front trotters before the pelting operation eliminates the possibility of carcass contamination from the trotters.

c. Hide Removal (Pelting).   A considerable amount of labor is required to remove the hide (pelt) from sheep and lambs. Consequently, most of the effort devoted to automate ovine slaughter and dressing has been in the area of pelt removal. MIRINZ has developed a two-stage automated depelting machine. Stage one is a shoulder puller, which removes the pelt from the shoulder and back regions. Stage two is a final puller that pulls the pelt off the rear legs. This requires the carcass to be suspended from the front legs, whereas traditional pelt removal practice requires the carcass to be suspended from the rear legs. Systems using this new carcass orientation are known as "inverted" dressing systems. The inverted dressing system has several advantages over the traditional system. These include less manual labor, reduced operator skills, lower worker injury risk, and lower levels of microbial contamination, particularly in the most valuable hindquarter area. The system produces pelts of a quality at least as good as those from traditional manual systems. New manual butchering techniques have also been developed to enhance the advantages of the inverted dressing system over the traditional system.

d. Head Processing.   The European Community requires the head to be skinned when edible brains and tongues are recovered. The completely skinned head must also be presented with the body for examination as part of the carcass inspection procedure. Several head-skinning machines have been developed. The most successful is a machine that incorporates a small shaft that grips a flap of skin near the nose and removes the skin by a rolling action. This machine is used in most sheep processing plants throughout New Zealand and Australia.

e. Evisceration.   The main objective of the new sheep evisceration system is to maximize the efficiency of the entire offal removal and handling system and to improve the hygiene and product quality aspects. Maximum efficiency is achieved by a combination of mechanization and a revised method of product handling. The two machines that have been developed are the brisket cutter and the eviscerator. The brisket cutter performs the first two operations of evisceration cutting the brisket and opening the belly area. The eviscerator then removes the complete gut set. The method of operation is similar to that of an automatic eviscerator for beef.

C. Porcines

1. Automatic Lairage System

Handling procedures in traditional large lairages can cause stress, as will fighting amongst the more aggressive animals. Skin damage can also occur as a result of such behavior. The Danish Meat Research Institute has developed a fully automatic lairage system [4] designed to keep pigs in groups of 15 to 18. This corresponds to the compartment sizes on most Danish transport vehicles. The system uses normal pig behavior, particularly resting behavior, to minimize stress.

The small pen system consists of a series of flap gates with vertical bars, allowing the pigs to look through the gates. The pig groups are moved from the off-loading bay to the pens. During this stage the live inspection is carried out by a veterinarian. A push-hoist gate automatically moves behind the pigs, allowing a flap gate to close. The push-hoist gate is then raised and moved back to collect the next group of pigs. As the flap gates close, a through is automatically lowered and water made available to the pigs. Built-in ventilation ducts in the pen walls optimize ventilation at pig level. Emptying the pens follows a procedure similar to that of filling the pens.

A comprehensive test of the system has shown a significant improvement in animal welfare, reduced skin damage and eccymosis, and a reduction of labor in the lairage area. The system is in operation at one Danish slaughterhouse operating at 800 pigs per hour.

2. Stunning

The Danish and Swedish Meat Research Institutes [4] have jointly developed a new automatic system for stunning pigs in small groups. Fifteen to eighteen pigs are divided into three groups of five to six pigs using slide gates controlled by a vision system that measures the floor area covered by pigs. The gates are pressure sensitive and cannot harm the pigs during partitioning. The pigs are then transferred to the stunning apparatus, which consists of three chambers that allow continuous loading, stunning, and unloading. Carbon dioxide gas is used as the stunning agent. After stunning is complete, the animals are automatically orientated and transferred to the shackling operator. Full-scale system trials have proven successful at production rates of 400 pigs per hour.

3. Scalding

Traditionally, pigs are scalded by immersing the bled carcasses into scalding vats filled with hot water or by spraying the carcasses with recirculated hot water in cabinets. The water becomes contaminated by dirt from skin, hair, urine, and feces. Scalding with humidified air provides a method that is aesthetically more appealing. The Danish Meat Research Institute, in collaboration with SFK Meat Systems [4], has developed and tested a condensation system for hanging scalding using a special humidifier module, where heat and moisture are transferred to the scalding air by atomization of hot water. This process can maintain a constant temperature and 100% humidity under varying loads.

4. Rectum Sealing and Separation (Fat End Loosening)

A commercial company, Auto Meat [4], has developed a machine for automatically sealing the rectum and separating it from the surrounding tissue. This is referred to as "fat end loosening." The machine initially positions the carcass. An extendible arm with a severing device at the end locates the anus and separates the rectum from the body. A second device also located at the end of the arm cleans the rectum using a combined rinsing and sucking action. Another cutting tool automatically divides the pubic symphysis. Finally, a hot water and a lactic acid solution is used to automatically clean the equipment. Once the rectum has been loosened, it drops into the abdominal cavity of the carcass. Microbiological analyses have shown that the use of this procedure does not result in contamination problems.

5. Carcass Opening

Carcass opening has been successfully automated using a machine developed by Auto Meat [4]. It is capable of a throughput of 360 to 380 pigs per hour. The machine initially locates and positions the carcass in the correct orientation. The front legs are spread. A shearing device then locates the pharynx, enters the thoracic cavity, separates the sternum, and opens the abdominal cavity. The shears are retracted and automatically washed with hot water. The machine has now been certified for commercial operation after a two-year test and commissioning period at a Danish slaughterhouse.

6. Evisceration

Automatic evisceration for pigs has reached the demonstration phase [4]. The processes included in the demonstration equipment are the identification and location of salient points within the carcass; the severing of the connective tissue that attaches the internal organs to the back; the detachment of the diaphragm from the chest wall; and the final removal of the complete organ set, leaving the esophagus and trachea connected to the head at the throat.

The intersection of the diaphragm with the chest, the point at which the kidney tab penetrates the diaphragm, and the point at which the connective tissue supporting the rectal tract commences are each identified and located using a neural network system. This system takes image information from a 3-D camera system and generates the X, Y, and Z coordinates of the required points with respect to a datum point. The carcass is then presented to a position-controlled, hydraulically powered two-axis robot that uses the information derived from the visual scan to carry out the cutting processes. The cutting tool severs the connective tissue from the rectum to the diaphragm. The diaphragm is then severed from the chest walls. A puller grasps the viscera by the lungs and pulls the intact viscera from the carcass, leaving the esophagus and trachea attached to the head.

7. Back Scoring and Finning

Most Danish pig carcasses are back scored and finned before carcass splitting to prevent damage to the loins by the splitting saw. The Danish Meat Research Institute and the Danish firm Danfotech [4] have developed a prototype machine that is capable of positioning the carcass, prescoring, separating muscle from the spinous process, back finning, and finally cleaning the cutting apparatus with hot water.

8. Loosening of the Leaf Fat

A French manufacturer, Durand International [4], has developed a machine for automatic loosening and pulling the leaf fat. The machine locates, positions, and constrains the carcass. A mechanism then grips the leaf fat and loosens it by pulling it away from the carcass.

III. COMPUTER SIMULATION OF SLAUGHTERFLOOR OPERATIONS

A. Overview

Simulation is regarded as a powerful tool in modeling and analyzing complex system in which varying degrees of randomness affect the performance of the tasks, the arrival and flow between processes, the availability of resources, and the competition for shared resources. The advent of modern computer technologies such as the microcomputer, animation, user interfaces, and graphical display facilities have permitted computer simulations to be used as part of a decision support process by those directly involved in the industrial environment. A number of generic systems simulation models have recently been developed. These incorporate the major variables and randomness associated with slaughterfloor operations. It is now possible for managers to audit current operations and quickly develop models of alternative layouts. They can examine the consequences of proposed changes in an operating environment where production cannot be disrupted and safety issues may preclude physical experimentation. These consequences include the impacts of changes to processing rates at given work areas; the global effects of blockages or stoppages at particular points; the performance of various levels of production and manning; the impact of revised equipment availability; and the impact of new technology. The outcomes can be communicated in a simple but highly effective visual manner to operational personnel.

B. Slaughterfloors

Simultech Pty Ltd., in conjunction with the University of Southern Queensland [7], has developed a computer simulation system for bovine slaughterfloors in Australia. The description of the operating components of commercial slaughterfloors provided the baseline for the simulation system specification. Interpretive structural modeling (ISM) methodologies, together with current methodologies used in ecological interface design (EID), were then used to formulate a logical model. The description of operating characteristics of available meat processing technology formed the basis of the decision support systems. Purpose-built data sets and animations were used to verify and validate the prototype models. These models were then generalized across the industry using a logical "linearized" model.

The user can specify the number of carcasses to be processed together with carcasses to be processed together with carcass characteristics such as breed, disease status, and the presence or absence of horns. The nature and behavior of any conveyor can be specified, and the operating parameters of speed, spacing, and distances can be altered. The user can also specify operational input data for any given work area. This includes process time, setup and reset times, sterilization time, number and nature of personnel, equipment failures, physical layouts, accumulated work, damage, and task partitioning. The model will also accept system breakdowns.

The simulation can provide information on the number of stops initiated and the total time for stops attributed to a particular work area; the total stoppage time for each conveyor; total blockage caused by stoppages elsewhere; and the average busy and waiting time for each work area. Graphical displays are available for the distribution of waiting times and work area distances required for particular tasks. The simulation can also provide information on yields.

IV. SLAUGHTERLINE DATA NETWORKS

The Danish Meat Research Institute [4] has developed a new information system to collect, process, and exchange the increasing quantity of data measured and recorded on the slaughterline. The new data system has an open architecture that allows future integration of new equipment without changes to the existing systems. It permits automatic sorting of carcasses in the chillers based on data collected on the slaughterline and gives more effective production planning and forecasting.

The system uses a number of computers that can exchange information from a database. The computers in turn can interrogate the various items of equipment located throughout the plant. These include the PLC systems controlling the line flow; the automatic classification center and manual backup system; the automatic weigh station; the inspection system; automatic identification readers; the manual input of supplier identification numbers tattoed on the carcasses and equipment for the registration of fat samples from entire male pigs. The networks used for connection of the individual subsystems are based on the BitBus standard from Intel Corp., which permits the use of standard hardware and software. Data from the slaughterline can be transferred for production planning or accounting. In addition to production data and analysis, the system can log alarms. This alarm log can be used by the technical staff for fault finding and maintenance.

The system facilitates the automatic sorting of carcasses to specific overhead rails in the chillers. The sorting process is controlled by a dedicated computer linked to an automatic carcass identification system. The identification number is read automatically when the carcass arrives at the chiller. The carcass is physically located on the correct rail in the appropriate chiller using a predefined sorting group function and a location plan. Summary displays and reports allow personnel in the chillers to continuously observe the day's production results.

V. CARCASS CLASSIFICATION, INSPECTION, AND GRADING

A. Overview

Automated carcass inspection offers the ability to objectively grade meat. This provides a more reliable basis for payment based on carcass quality. The aim of automated carcass inspection systems is to measure certain features of the carcass's image and relate that to its yield (fat to lean) ratio. A number of techniques are being applied to the meat processing sector to obtain objective classification. The most advanced of these is video image analysis. Ultrasonic analysis is also being evaluated.

B. Bovines

The Meat Research Corporation has sponsored work in Australia relating to the classification of beef carcasses. Systems Intellect Pty Ltd. has developed a carcass classification system that uses vision image analysis. This system, known as VIASCAN, obviates the need for backlighting to isolate the carcass by subtracting a stored background image from that with the carcass present. The classification also uses color images to segment the carcass into various classes. These include removed bruises, exposed lean, thin fat cover, yellow fat, creamy fat, and white fat.

Beef is currently graded using a subjective technique in Australian abattoirs. A skilled human operator visually inspects a cross-sectional area of the *longissimus dorsi* (ribeye) muscle. The grading score, a major determinant of market value, is based on the ribeye area, surface fat depth, meat color, and fat and marbling within the ribeye area. The method is prone to error, as the boundary between the ribeye muscle and surrounding fat is not always distinct. The image may also contain regions other than the ribeye area that have been incompletely separated by intramuscular fat. The industry is currently evaluating an objective method of assessment using video image analysis of ribeye characteristics. The analysis uses simple Red–Green–Blue color thresholding to separate the ribeye area from the surrounding fat. Meat color and fat color are determined over the ribeye area. Marbling is obtained by thresholding. The ribeye area and surface fat depth are also calculated using this technique, and an objective score is given.

C. Porcines

The Danish Meat Research Institute [4] has developed an automated classification center for carcass classification, grading, and branding. The system is installed throughout Denmark, and it has classified more than 100 million pigs.

The carcass classification system consists of a number of stations. Two position detectors determine the position of the front leg and the pubic bone at the anatomic measuring station. These measurements are then used in the probe measuring station to position the optical probe carriers. Seven optical probes are inserted into one of the sides. The probes record the thicknesses of fat and meat. Finally, health and grade stamps are branded on both sides at the branding station. Originally a conventional pattern recognition system was used to interpret the reflection profile from the optical probes. Neural networks are now used for profile interpretation and the evaluation of recorded meat and fat thickness measurements.

VI. FABRICATION (BONING)

A. Bovines

Computer-controlled mechanical pulling arms provide the simplest form of automated fabrication operation. A process worker (or boner) manually separates the meat from the bone using a knife. However, the manual pulling actions associated with the knifing work are replaced by a machine. A computer-controlled pulling arm grips the meat and pulls it away from the bone. The pulling force is controlled by a load sensor. The process worker can adjust both the speed and force exerted by the pulling arm. Commercial systems using this principle include the PROMAN system and the KARLVENT AV "Meat-O-Matic."

An automatic rib deboning machine has been developed in Australia by CSIRO [8]. The commercial version is marketed by Australian Meat Technology Pty Ltd. Initial manual prework is carried out on the split carcass as it leaves the slaughterfloor prior to chilling. The side is conveyed to the boning room after chilling, where the foreleg is manually removed. The side is quartered, and the forequarter loaded into the machine. The sternum is automatically located and gripped. The gripper mechanism then peels the meat away from the ribs in conjunction with a roller bar. Figure 5 shows the machine. The machine is currently being adapted to "on-rail" side deboning.

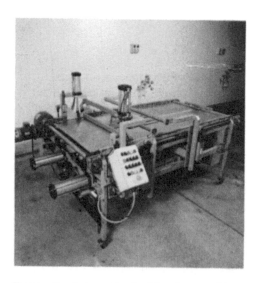

FIGURE 5 Automatic rib deboning machine.

A research project currently being carried out by National Institute of Agricultural Research (INRA) in France uses robotics, vision sensing, and neural networks to fabricate a beef carcass. The computerized cutting trajectory is determined by reference to a large database containing geometric and anatomical information on beef carcasses. The cutting tool is navigated through the meat using the calculated trajectory.

B. Ovines

The New Zealand processing industry has invested considerable effort in the area of sheep and lamb fabrication. MIRINZ has developed two machines that are now finding acceptance in industry.

1. Loin Boning Machine

The loin boner [3] is designed to remove the two loin eye muscles (*longissimus dorsi*) from lamb and mutton short loins (short saddles) that contain the lumbar vertebrae. The machine can process up to six loins per minute with one operator and eight loins per minute with two operators. There is greater than 10% increase in muscle removed compared with existing manual boning methods. Product quality is excellent, as the boned loin has a consistent size and shape and a smooth surface finish. There is a complete absence of knife cuts. This boning process has been extended to long saddles (6–8 thoracic vertebrae plus the lumbar vertebrae).

2. Chine and Feather Bones

The chine and feather boner [3] is designed to remove the chine and feather bones (thoracic vertebrae) from rack saddles, which contain thoracic vertebrae, ribs, and eye muscles (*longissimus dorsi*). This process is performed in such a way that the eye muscles remain attached to the ribs during the removal of chine and feather bones. The machine can process up to nine rack saddles per minute with one operator. There is an increase of approximately 30 g of muscle removed per rack saddle compared with existing manual boning methods. At a production rate of five rack saddles per minute, one less person is required. Product quality is excellent with consistent cut positions and clean smooth cuts.

VII. REFRIGERATION

A. Overview

Meat is an organic product and is subject to biological contamination and biochemical degradation. Cooling and freezing have traditionally been used to prolong the life of the product. Several factors must be considered when refrigerating meat. These include compliance with health and hygiene regulations, weight loss during chilling, the effect of chilling rates on tenderness, the control of microbiological growth, condensation in chillers, energy usage and management, and the effect of temperature on packaging material. Mechanical refrigeration is used by the vast majority of meat processing establishments throughout the world, although liquid nitrogen and carbon dioxide equipment are increasingly finding applications in the industry.

B. Supervisory Control

It is now increasingly common for meat processing establishments to install a supervisory control center with their refrigeration equipment. The hardware and software for these centers have been commercially available for a number of years, and there are many suppliers throughout the world capable of providing, installing, and programming a system for individual requirements. The major requirements of such systems are the provision of alarm and fault detection status and energy monitoring. Alarm and fault detection provide a knowledge of the ability of plant and equipment to function as specified. Major product degradation can occur when refrigeration equipment does not perform in accordance with specification. This degradation includes unacceptable microbial growth due to temperature rises or low air velocity, tenderness reduction in the meat through excessive cooling rates, and the hardening of the external fat cover. In extreme situations it is possible for frozen product to thaw out during storage. Thawing and refreezing are not acceptable practices because of risk of microbial growth. Energy monitoring allows quantifiable data to be obtained for the establishment of energy conservation and management programs, which are critical to the cost-effective operation of a meat processing plant.

Regulatory authorities in many countries specify that the deep muscle temperature of a carcass must fall below a certain value within a certain time. Beef, for example, must achieve a deep muscle temperature of less than 8°C within 24 h. Conventional chilling practice would often result in the hardening of the subcutaneous fat. A process worker would then have difficulty cutting the fat with a knife in the subsequent fabrication or boning operation. Computerized control now provides a means of overcoming this problem. The air temperature and air velocity in the chiller can be altered through the chilling cycle to provide a severe chill at the start of the cycle and a tempering environment at the end of the cycle. The computation is based on the weight and fat covering of the carcass. These measurements have been previously recorded during the weigh and classification stations on the slaughterfloor. This is of particular use with carcasses that have been electrically stimulated during slaughter and dressing to prevent "cold shortening."

C. Dynamic Modeling

The major benefits of dynamic energy modeling are the ability to predict the effects of changing situations and circumstances that can occur in abattoir production and, consequently, the opportunity to analyze different strategies for coping with changing circumstances within the abattoir. Dynamic modeling of refrigeration plants has been carried out for more than 10 years by Massey University and MIRINZ in New Zealand [9] and has resulted in a number of software packages including RADS, LOADS, and the Refrigeration Loads Analyser, which operates in the Microsoft Windows environment. These packages are based on room-by-room analysis and require significant data collection and input to stimulate a refrigeration system's operation. Research is currently being directed at the prediction of refrigeration from production information and basic technical data without considering specific rooms.

The basis of this approach is to break the daily production into a set of product batches, to define the type of conditions (air temperature and velocity) to which these will be exposed, and to define the product form at each stage (carcass, side, carton). From these data, the modeling system estimates what fan power will be necessary, what the resultant door heat loads related to product movement will be, how cooling is needed in the boning rooms, and what the insulation heat loads will be. It also predicts the rate of product heat release with time and, after summing the various heat loads, gives an overall heat load versus time prediction. This is then processed through an engine room calculation module, which, given information on the type of engine room, determines the necessary compressor power inputs.

VIII. CONCLUSIONS

The meat processing sector has traditionally lagged behind most manufacturing industries in the use of automation and computer-based technologies. The rising cost of processing and increased competition from other foods has provided the incentive for the development of computerized automation and controls in the industry. The arrival of modern computers, their computational ability, user-friendly software, and the cost of the systems now make it feasible for these technologies to be incorporated into meat processing operations. Industry survival will mandate their adoption throughout the industry.

REFERENCES

1. J. W. Buhot and R. J. Rankin, Automated slaughter and dressing of bovines, IARP First Workshop on Robotics in Agriculture and the Food Industry, Avignon, France, June 1990.
2. J. W. Buhot, The development of automated slaughter technology for bovines, Conference on Advanced Technologies for the Canadian Meat Industry, Banff, Alberta, 1992.
3. G. R. Longdell, Advanced technologies in the meat industry, *Meat Science J. 36*:277 (1994).
4. J. Zink, Application of automation and robotics to pig slaughtering, Danish Meat Research Institute Manuscript No. 1271E, 1995, Ref No. 31.616.
5. J. W. Buhot, A. F. Egan, D. J. Walker, K. Bowtell, and R. A. Wharton, Automation techniques and contamination during the slaughter and dressing of cattle, World Health Organisation Consultation on Research on New Slaughter Technologies to Reduce Cross Contamination, Riskilde, Denmark, February 1990.
6. *The Production of Chilled Meat for Export*, Workshop Proceedings, CSIRO Division of Food Processing, Meat Research Laboratory, 1991.
7. M. P. McFarlane and G. W. Hood, Systems simulation of slaughter floor operations, Proceedings of Meat '95, The Australian Meat Industry Research Conference, Gold Coast, Queensland, Australia, 1995, pp. 4B19–4B21.
8. D. J. Heidke, Boning technology, Proceedings of Meat '95, The Australian Meat Industry Research Conference, Gold Coast, Queensland, Australia, 1995, pp. 4B13–4B16.
9. R. D. S. Kallu, A. C. Cleland, and S. J. Lovatt, Direct and indirect benefits of dynamic modeling of meat plant energy systems, 28th Meat Industry Research Conference, Auckland, New Zealand, 1994, pp. 307–316.

17

Computerized Process Control in Industrial Cooking Operations

Christina Skjöldebrand

SIK—The Swedish Institute for Food and Biotechnology,
Göteborg, Sweden

I. INTRODUCTION

We are visiting a small town somewhere in Europe [1, 2]. Both households and small companies are located in the town. One of the small companies is called "Local food." It produces fresh bread and bread-based convenience foods. About 400 different items are produced. Most of the products are distributed to nearby households. The products are filled French rolls, crepes, pies, quiche Lorraine, and pizza. New products occasionally are developed. Customers can buy food via TV in their homes. The products are delivered immediately after preparation, within 1 to 2 hours. If a customer has special wishes concerning the filling, quality, and the size of the product, the customer can specify this via the TV screen. The order goes directly into the production computer that controls the processes. The raw materials are based on frozen dough and different fillings such as prawns, minced meat, shellfish, rice, cod, clams, champignons, other mushrooms, vegetables, chicken, meat, and boiled egg.

The storage rooms for raw materials are located at one end of the production unit. There is both cold and frozen storage for fillings and dough. The plant and the processes are fully automated, and the line is flexible and can be tailored to the product. Equipment for sterilization, blanching, pasteurization, boiling, deep fat frying, and different ovens are available. New heating techniques are tested and used when suited. Equipment with microwave, infrared (IR), and minimal processing, that is, nonthermal processing or minimal heating, is also used. At the production line, a computer is installed for operator decision support and process control. The computer programs are easy to use, and the graphic interface and hardware are user-friendly. On-line sensors measure quality-related properties, such as the color of the crust, texture, and aroma. The raw materials are characterized using sensors. When the product is finished, the process stops, the product is packed, and it is sent directly to the customer.

The productivity on the production unit is 90%, and the process equipment is used during 75% of the production time.

Another example: We visit a large-volume soup production unit. They produce vegetable soup that is distributed throughout Europe, called "The quality soup." The soup consists of a viscous fluid with particles. The particles may be peas, carrots, or beans.

The soup is sterilized in a continuous flow using a scraped-surface heat exchanger (SSHE) and then is aseptically packed. The raw material is checked using different sensors to characterize various chemical and physical parameters. Orders come from distributors via computer and go directly into the production computer that controls the process. Sensors for measurement of temperature and quality-related properties, such as the viscosity, tenderness, and softness of the particles, are used in the process. The particle load is also measured and recorded on the computer. The required volume is input as data. From the relationship between the quality parameters and the raw material properties, the process is controlled and the process temperature is set. The process changes when there is a deviation from the required product parameters. The process is manipulated from a control room, where the operator works. An operator has full responsibility for the process, and the process is fully automated. The operator also has a simulator of the process as a control tool.

As the 1990s progress, the food industry is gaining a much clearer picture of the food production processes of the future [3]. This will be a time of change, especially in Europe, which will eventually become one single market. A patchwork of incompatible economies will be transformed into a tapestry of open and unrestricted competition. The compelling domestic industries in the various states are considered to operate on the pan-European platform [4]. New competitors will appear on the market with deregulation and the removal of trade barriers [3]. In the future, the different continents probably will be connected by different information technology tools.

The keys to competitiveness are high product quality and product profile, good service, reliable deliveries, and lower cost [5]. This chapter presents different tools and aids involving computers that are used for industrial cooking operations in food production. Using the tools can increase output and efficiency in the production system. The chapter discusses trends, process descriptions, and developments in computerized process control in industrial cooking operations. Then it gives a more detailed look at both on-line and off-line computerized control systems. Finally, a few thoughts on the future in this field are given.

II. TRENDS

In Sweden and Europe today there are three principal types of food companies—local/regional companies, large national companies with some export business, and multinational companies (T. Ohlsson, personal communication). We believe that there will be in the future two types of food companies—local/regional and the multinational. Production is very heterogeneous with a wide variation of different systems, from small handicraft-based units to large production units that belong to the world's leading food companies.

We believe that in the future there will probably be two types of production systems in use—large-scale bulk production of, for example, sugar, margarine, or hard bread, and flexible systems that allow several products to be manufactured on the same production line. The food plant of tomorrow will either have to be highly flexible and produce a great variety of products or be capable of producing large volumes of high-quality (basic) bulk products. It is likely that bulk production will use high-quality raw material and manufacture high-quality products, and it will therefore require tools for defining quality. Such tools will measure the relevant properties of raw material and account for variations in them to adjust the production system so as to attain the required product quality after processing. Easy to use measurement techniques and sensors need to be developed for this purpose [5, 6].

These factors and trends will, in turn, compel company executives to take appropriate action and make the necessary decisions regarding the benefits of different approaches to production. Information technology (IT) will play a decisive role for spreading knowledge about the benefits of different approaches to production. However, computers are rarely used in food production today, which means that dedicated, user-friendly systems will have to be designed [5].

III. PROCESS DESCRIPTIONS IN COOKING
OPERATIONS

Cooking operations exist in various types of food plants, such as restaurants, and medium and large industries where different raw materials such as meat, bread, fish, and so forth are converted for consumption. The equipment differs, but the basic heat transfer media are roughly the same. At present, conventional heating techniques are most common in production units for cooking operations. Convection is used for boiling, baking, sterilization, pasteurization, and blanching. Water, fat, oil, or air are media used for heat transfer. Convection ovens, contact fryers, deep fat fryers, and autoclaves are the most common equipment for cooking solid foodstuffs. Scraped-surface heat exchangers (SSHE), tubular heat exchangers, and boiling pans or kettles are used to cook fluid products. In some cases microwaves or near infrared (NIR) techniques are also used [6].

The main intention of cooking operations is to reduce the activity of undesirable biological materials, for example, microorganisms and enzymes, or to change the physical properties, say, raw meat to a cooked product. At the same time, essential nutrients are destroyed, and other undesirable changes may take place.

Most cooking operations have been developed from trial-and-error tests over the years. They are often batch processes rather than continuous ones, and few of them are easily described by mathematical equations, computerized tools are therefore rare today, but they are under development. Simulators have, however, been designed for sterilization in autoclaves and in SSHE. Also, cooking and baking in ovens have been described by simple models [7].

In the first production unit described in the introduction, the processes use machinery that is easy to move and the process planning gives high flexibility. The manufacturer can pasteurize and blanch vegetables. The fillings can be sterilized, and dough and fillings are to be thawed. Some of the fillings are deep at fried, and some are boiled or oven cooked. As the production is based on frozen dough, mixing and proofing are done outside the actual production unit. There are possibilities to use microwaves for boiling, proofing, and thawing. Near infrared (NIR) techniques are also available for heating or baking. The staff is educated and can handle new techniques for heating. Sensors are installed, for example, temperature and water content devices, as well as sensors for measurement of aroma such as an electronic nose (see p. 480). In the production unit, computers are installed to make it possible to follow production. Data are collected in a logger and treated in accordance with so-called smart sensors. Production planning tools implemented on computers in the production hall are used by the operators, who can decide the best process for each product and can also decide which products should be prepared in accordance with customer orders. Parts of the processes are fully automated, where convenient.

In the second production unit described in the introduction, the process lines produce large volumes of soups for distribution in a large area, which could be, say, Europe. The machinery is not as flexible as in the production unit described in the first scenario. Heat exchangers are used for sterilizing or pasteurizing the products. The raw material is blanched continuously in a steam blancher. Conveyors or tubes with connected pumps distribute the product in the unit. Large packaging machines are used for poststerilization packing in plastic bags or paper packages.

The packages with prepared products are stored in room-temperature storage rooms. Raw material of high quality is used, and high-quality products are manufactured. Sensors measure relevant properties of the raw materials and take into account raw material variations to attain the required product quality after processing.

IV. DEVELOPMENTS IN COMPUTERIZED CONTROL SYSTEMS FOR COOKING OPERATIONS

The integration of computers and automation is one way of satisfying the need for increased flexibility. It must be possible for the food industry to introduce a higher degree of automation without whole plants or production lines becoming obsolete. The integration of computers in food production is complex, however, and involves more than just throwing hardware and software at the problem, as has been the case in the past [8].

Production that involves cooking of foods still depends fundamentally on the experience and knowledge of operators. There is more "art" involved than science. Few people can describe mathematically what happens when you cook a steak or sterilize a soup [3]. The operator himself acts as a human sensor: his nose, his feel in mouth, taste, fingers, "it is too soft," "it is too dark," "it looks good," "its feels thick."

Industrial cooking is often a very complex procedure, both in large-scale kitchens and in the food industry. The raw material changes continuously during the process, physically and chemically. Properties such as color, taste, flavor, and consistency are complex and subjective to measure with instruments, but controlling them is crucial to the food industry. The increasing demand for quality control and more information about the processes leads to a need to develop new tools and aids for process control and measurement of quality properties.

We anticipate a tremendous development during the next few years in the use of computers to control cooking operation. Computer software and hardware will gain new uses. Many future computer users basically are novices; computer systems must therefore be user-friendly so that the operator in the food production plant can handle them. Training and education has to be offered as development progresses, especially for staff in large-scale kitchens or restaurants [5].

The organization of work will be different in the future food production unit. The operators will take greater responsibility for planning the daily work. They will work in teams with a leader who is responsible for the development of the skills of each operator in the team. He/She will be the "coach." To realize this, computer aids will have to be developed and used as decision support for the planning, use of equipment, servicing, recipe composition, and the like. The technology and the working environment will be developed in parallel.

Development of computer software and hardware will be carried out by researchers, operators, and manufacturers as a team. The operator, being the expert on the process, will have an important role in this development work.

Information technology will play a decisive role for disseminating knowledge about the benefits of different approaches to production and also in optimizing the use of existing equipment and process systems. New technology will be introduced, but only after testing with simulators and optimization of the process system as a whole. The so-called minimal processing concept will be used more when this knowledge has been further developed. Kitchens can, for example, use sous vide systems and thus get better quality or more controlled quality.

A modified atmosphere can be used for packing of cooked meat that is minimally processed.

Knowledge of the interaction between the process and the product will be important in future food production. The interaction between raw material properties, process parameters, and final product quality is also a key to obtaining the best automated process where information systems and training are vital factors. A process identification is important to get the needed information.

In the following, some tools and computer aids that are being developed will be presented. Their advantages and limitations, what needs to be done, and possible difficulties on the way will also be discussed.

V. ON-LINE SYSTEMS

This section presents one method for systemizing computer aids. This could, of course, be done in various other ways.

A. Process Control and Sensors

It is evident from the literature that sensors in food processing is a popular field at the moment. Sensors act as the "feelers" and the "eyes" on a processing line, and one can measure things from temperature and flow rates to product and packaging properties. When it is discussed within the cooking industry what kind of properties need to be measured, the following list emerges:

Texture of meat or bread
Flavor/aroma of meat or bread
Color
Water content/water activity
Viscosity of soup
pH of soup

It is of course important always to measure the temperature both in the equipment and in the heated and the cooled product.

Typically the food industry has trailed other process industries in the use of microprocessor-based control systems. Other industries such as bulk chemicals, petroleum refining, and pulp and paper processing have narrowly controlled processes compared with the food industry. One reason for the lack of process control is that food processes are not easily automated because of the variability of food materials. Another reason is that adequate monitoring and sensing systems have not been developed (see Chapter 1 for more details).

In today's system, the sensors are coupled to instruments, and information technology tools convert the signal into useful data. Sensors that can measure, for instance, temperature, pH, and flow on-line have existed for some time. The market trends for sensors will show steady growth and will play a key role in the automation of food processes as well as for product development and process control. Noncontact instruments are appearing on the market, which is useful for the food industry. Low-cost sensors are being developed for various purposes.

New technologies will appear, such as

Laser techniques for measurement of moisture and flow
Sensors for measurement of crispness
Sensors for mouthfeel
Sensors that can measure water activity on-line
Sensors to measure moisture content on-line without calibration
Sensors for measurement of humidity in ovens
Sensors for measurement of freshness
Sensor fusion systems

The problems arise when the sensors are to be adapted to the real process environment. Many demands have to be fulfilled, such as nondestructive testing and no fouling. The signal, which is an objective measurand, has to be translated to a subjective evaluation of quality-related properties. The time for the response to the measurement has to be very short, less than 30 s, to allow changes to the process to be made and to correct deviations, both without delay. Existing sensors are often expensive and difficult to maintain. And finally, education and user-friendliness are important to take into account.

In a project carried out in Sweden, we studied industrial cooking and sterilization of fish balls, a typical Swedish dish [9]. Cod and haddock were the basis for the recipe. After mixing, the mixture is pumped to a mold machine and boiled before filling into cans and sealing. After filling, the cans are sterilized in an autoclave and stored. A hypothesis was the starting point of the project: "There is a relation between the instrumentally characterized property entity and the sensory property entity. A model may be designed to predict the effect of the process. By describing the static and dynamic process environments and their relation to product properties, quality properties can be controlled." In this way it is possible to measure an objective entity by an instrument that is related to sensory qualities judged by operators. Models were designed for the purpose, and today the system is used in the fish production unit.

The signals from sensors such as the electronic nose for baking or meat cooking are based on the relation to sensory analysis. The electronic nose is an instrument that comprises of an array of electronic chemical sensors with partial specificity and a pattern recognition system capable of recognizing simple or complex odors. Sensor fusion means that several different quality parameters such as water content and aroma consistency are measured by the same tool. The combined signal is related to sensory panels.

B. Flexible Manufacturing Systems

As a concept, the flexible manufacturing system (FMS) has been around for some years. It is directly related to the computer era and the technology that has evolved. The FMS is a computer-controlled array of semiindependent work stations and integrated materials handling systems designed to produce a family of related products with medium variety and medium production volumes of each [8].

If we recall the first example above, the food production system was very flexible in the local unit. Flexible systems allow several products to be manufactured on the same line. In the latter case, short changeover times will be possible for a variety of products in small volumes. The plant needs the ability to switch from one product to another or to vary the original recipe. Short start-up and stoppage times and minimum bottle-necks are required to achieve efficient production planning and control of the process.

At SIK, computer systems have currently been developed for a production line for meat cooking in which batch processing is combined with continuous stages, which may consist of different unit operations, conveyors, and packaging machines. Simulations can be made to find the optimum line for obtaining the optimum food product quality [3].

C. Food Plant Automation

The most basic and most difficult step in food plant automation is the development of a scientific understanding of the product manufacturing process. Traditionally, when a company commits to automation most of the effort is focused on selecting and implementing the control system. It is important to understand that without the development of adequate sensors and process understanding, application of any control system will produce little more than a sequential automation of equipment operation [8]. It will not produce the responsive process control necessary for the production of consistent, high-quality products. One of the biggest efforts when designing a modern production plant is to get the best out of the automation. It is often enough to get as much as possible out of the simplest changes, and it is not necessary to go as far as automating all product processes. There still are doubtless major advantages to be gained when parts of the production are automated.

Two basic components are required for fully automated process control: a means by which important process conditions such as temperature, pressure, and flow rate can be held constant at a predetermined setpoint, and a means by which the sequence of operations can be controlled [10].

In the second scenario of the introduction, the large-scale production unit of sterilized soup, most of the production is fully automated and only a few operators can handle the total production. Underpinning the automation are sensors that can measure the characteristics of the raw materials to relate them to the desired qualities of the product and knowledge of the process. The process is continuous, and the soup is manufactured on a large-scale basis. The quality characteristics of the manufactured soup are very well specified.

VI. OFF-LINE SYSTEMS

A. Process Simulators

To simulate means to copy or to pretend. A simulator is an apparatus that reproduces the conditions of a technical process—a flight simulator, for instance, consists of a cockpit and an electronic brain that simulates the flying of a real airplane. Simulation is now being used by the processing industry to reproduce complicated technical systems and processes. For food purposes, it has been used in food research for a long time, but in food production it is more rarely used [11]. Simulation is used for training of operators and also for decision support in production. The process is reproduced on the screen, and the graphic interface represents the real process. Changes in temperature, variations in the raw material, or 'changes to the recipe can be studied.

To develop a simulator, the process has to be described by a model, either in mathematical or statistical terms or on the basis of expert systems or a neural network. A poor model is different from reality and gives a poor simulation—no simulation can be better than its model. One of the problems with simulators and food processes is to find the proper input data on, for instance, the material properties, which differ from one material to another and change continuously during the process.

At SIK we have devoted a great deal of attention to the development of computer software for food processes. Such software is used in our own research both to learn about and understand different process operations and to minimize the number of trials and the work involved in the evaluation of the results. The software is being upgraded for use in consultancy work aimed at improving productivity and production efficiency as well as in the development of new processes and the study of what happens during processing [11]. Our programs are designed with different displays for input and output data. Input data are set with a mouse and keyboard by simply clicking on the button and entering a new value. We have put a lot of emphasis on user-friendliness.

At SIK we have been working on different cooking operations for many years. In the area of batch sterilization in autoclaves and continuous sterilization in aseptic processing, we have a broad knowledge base both from research and from practical development work in our pilot plant. Baking of different products, mainly bread, has been the focus of our research for 15 years. New techniques such as infrared radiation and microwaves have been applied to heat treatment of cereal products. These processes have been focused on since started our development work on simulators for food production.

In continuous cooking operations such as soup sterilization, it is impossible to measure the temperature inside the particles. So, outside of extensive trial-and-error procedures, you need some means of calculating the temperature.

ASEP is a simulation program for aseptic processes. This program has been developed at SIK on the basis of know-how and modeling of measurements of thermal processing. It was developed for continuous cooking in heat exchangers of fluid products containing particles. The simulations are done in real time. This gives rise to a number of problems, such as

One cannot afford too many operations in one time step.
One cannot afford too much memory.
One must handle different magnitudes in time and space.

Energy balances are used to calculate profiles in the liquid. These include the energy input from steam and energy loss from the liquid to particles moving at the same velocity as the liquid.

Within the SSHEs, the liquid is heated by steam in the jacket and loses energy to the particles. The energy balance is described over a volume element in the SSHE. In the holding tube the heated liquid will lose energy to the surrounding air via the tube wall and the insulation. Energy is lost from the fluid to the particle moving in the tube. The heat transfer situation of interest for simulation is a solid piece of food immersed in liquid. Food has relatively poor heat conductivity. For the convective heat transfer at the boundary, the derivatives of the solution are much steeper, and thus errors at the boundary dominate. If convective boundary conditions are properly handled, it will be possible to get along with few discretization nodes in the interior of the body. The F and C values (process time) are calculated separately at each time step for holding tube and the SSHEs.

The residence time of the particle in the holding tube is considered as the speed in relation to the average liquid speed. It is also possible to change the thermal properties of the product, the shape of the particle, the number of heat exchangers for heating and cooling, and so on. It is also possible to run the program for a product without particles. It is possible to change the parameters for the equipment so that the program can be used to simulate heating either in SSHEs or tubular heat exchangers. An overview of the entire process is available, making it possible to follow the temperature at all parts of the process simultaneously (Fig. 1).

Baking is one of the most energy-consuming processes in the food industry. Energy is needed to heat the oven and the dough and to form a crust and the crumb. The baking industry could save a lot of money if it could optimize the process, namely, know when the crumb is baked and control the crust formation in order for both parts of the bread to be ready at the same time.

In a multiclient project that we started a couple of years ago, and together with a number of experts from universities, equipment manufacturers, and the food industry, a simplified but relevant model was produced of the baking process in a tunnel oven. The model involves a combi-

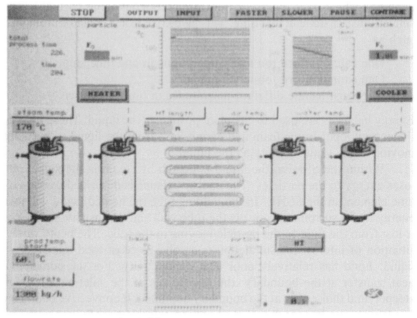

One of the graphs of the ASFP program

FIGURE 1 Overview of the aseptic process.

nation of radiative and convective heat transfer. Bread, pizza, and pie were the products studied. The influence of the heat transport on the quality and the water transport on the crust were taken into account in the model.

OVENSIM is a relatively new simulator in our collection. The model exists in two different PC-based platforms with graphic user interfaces—one for our research and one for industry (Fig. 2). These two simulators will be further developed by including mathematical models for the chemical and physical reactions that take place as a result of the heat and water transport. These reactions will give the final quality properties of the product. The dynamic heat transport will be combined with kinetics of the most important quality parameters (Fig. 3).

B. Decision Support and Quality Control

There is a lot of work being done on quality control at present, ISO 9000 and Hazard Analysis at Critical Control Points (HACCP) being two exam-

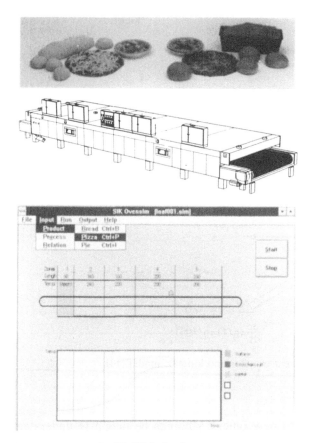

FIGURE 2 The OVENSIM simulator.

ples. Delegating specific responsibilities and authority to the lowest possible operating level requires that information systems and decision support be developed. These supports should be integrated into the production system and are an important part of the work. Computers and computer systems have a broad area of application in this context, and an increasing amount of information is being provided by computer systems. It is crucial, however, that correct data are available and that upgrading is done continuously at the production site. The information system should be tailored to the situation where it is used.

Knowledge-based systems are very well suited for these tools. The user of the system is the operator at the production line. The knowledge of the veteran operator can be documented in a readily accessible way

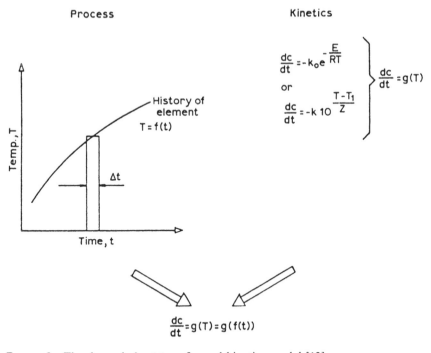

FIGURE 3 The dynamic heat transfer and kinetics model [12].

and can be passed on to newcomers, who can be taught to understand the total production process on the line.

It is important to develop software for user interfaces to communicate with the operator, but it is also important to design the hardware carefully. Shape, size, and color are potentially important factors in operator use of these supports. Researchers, operators, and manufacturers should put some effort into addressing these factors [5].

At one of Sweden's large production units of convenience foods, it was believed that it will be difficult to recruit personnel to the food industry in the future [7]. They are therefore attempting to create better job content and status. A new organization is being introduced with delegated responsibilities and authority, and a computer-based information system was developed as a decision-making tool for the operators on the lines. The crepe and pizza lines were used for pilot projects. New routines were introduced, and the line personnel were being trained so that they could get to know more about their company and their production and to learn both to plan the production on their own lines and to follow up costs and materials consumption. In addition, they are required to report certain results from the line to the computer, from which they may also obtain information.

C. Planning Systems

Against the background of increased flexibility and productivity, it is important to construct efficient tools for production planning. Computer systems for such purposes are rare in the food industry today. They have to be developed in consultation with the operators and tailored to each process. Terminals with data should be available to each operator on the line.

Many food processes traditionally consist of a combination of discrete batch functions and continuous unit operations or transport systems. These are coupled by buffer tanks, which are used for storage. A large amount of product may be locked up in the production line, resulting in an increase in investment costs. Planning systems to determine the optimum way of passing the product through the line are needed.

VII. FUTURE PERSPECTIVES

The development of computerized process control in industrial cooking operations is the key to developing the production systems for the scenarios described in the introduction. Researchers and manufacturers have to work together, and a number of pieces remain to be found in this jigsaw puzzle. Some of these are:

1. Development of software and hardware. User-friendly graphic interfaces have to be designed, and the hardware should look more like an instrument or a part of the equipment than it does today.
2. More knowledge about the interaction between the process and the product must be gained. The relationship between the raw material properties and how these are affected by the process in obtaining the desired properties in the end product should be studied further.
3. Characterization of the quality of the end product should be related to sensory data, such as taste, aroma, and so forth.
4. Materials data for the product are basically unknown today. These are required to understand the change of the raw materials into the desired end product. Examples of materials data are thermal properties and the diffusion properties of water.
5. Models based on different information technology tools, such

as expert systems, neural networks, and so on, should be developed for cooking processes.

6. Models for the influence of heat and mass transfer on nutrient and quality-related property kinetics are bases for further development of cooking processes as well as simulators.

7. Minimal processing will be used both in restaurant and industrial cooking operations.

At SIK we have developed a consultancy concept with the aim of increasing the efficiency and productivity of the food production systems. It is called the PEPP-talk concept (SIK's Programme for Efficient Processes and Productions). We can help the food industry analyze bottlenecks and give recommendations for action to increase productivity. It starts with discussions and analysis. The areas discussed are

Choice of equipment and production technology
Techniques for measurement of the product quality and control of the process
Process simulation in process development and control
Production planning
Environmental aspects and Life Cycle Analysis (LCA)
Work organization
Decision-making tools
Training

The purpose of this presentation was to discuss the development of computer aids for the future, to achieve increased productivity and efficiency in food production. Nobody knows exactly what the future will bring, but we have some ideas. We all have to work together on this. If we contribute with our respective knowledge and efforts, we will probably arrive at some of the tools described above.

We may be able to order the bread, pizza, or pie product directly from the local production plant, with quality properties that we have decided ourselves. The operators will know more about their processes, and the art will be mixed with scientific aspects.

REFERENCES

1. C. Skjöldebrand, Computer aids for efficient food production in the 90s, ACoFoP III International Symposium, Paris, SIK's Service Serie No. 904, October 1994.

2. N. Bengtsson, B. Hedlund, P. Olsson, Th. Ohlsson, and C. Skjöldebrand, Tekniska förutsättningar för småskalig livsmedelsproduktion, STU information (STU) [in Swedish], 604–1987 (1987).

3. C. Skjöldebrand, Moving from art to science, *Food Technol. International—Europ*: p. 115 (1991).

4. PA Consulting, *Information Technology: The Catalyst for Change*, Mercury Books, W. H. Allen, London, 1992.

5. C. Skjöldebrand and M. Schmidt Larsson, User friendly production systems for the food industry, *Trends in Food Sci. & Technol. 5*:89 (1994).

6. Th. Ohlsson, Boiling in kettles and scraped surface heat exchangers, Proceedings of Progress in Food Preparation Processes, Tylösand, Sweden, June 1986, pp. 71–105.

7. C. Skjöldebrand, Swedish food production for the 1990s, *Food Control* 4(4): 181 (1993).

8. J. T. Clayton, Flexible manufacturing systems for the food industry, *Food Technol. 41*(12):66 (1987).

9. C. Skjöldebrand, S. G. Eriksson, L. G. Vinsmo, and C. G. Andersson, Property characterization from direct measurements and operator observation, Food Processing Automation Conference III FPAC III, Orlando, Florida, SIK's Service Serie No. 881, February 1994.

10. A. Texeira and C. F. Shoemaker, *Computerized Food Processing Operations*, Van Nostrand Reinhold, New York, 1989.

11. C. Skjöldebrand, B. Sundström, H. Janestad, and C.-G. Andersson, Simulation of food processes: A tool for training operators in production, *Automation Control of Food and Biological Processes* (J. J. Bimbinet, G. Trystram, and E. Demoulin, eds.), Developments in Food Science 36, Elsevier, Amsterdam, 1994.

12. B. Hallström, C. Skjöldebrand, C.Trägårdh, *Heat Transfer and Food Products*, Elsevier Applied Science, London and New York, 1988.

18

Computerized Process Control for the Bakery/Cereal Industry

Gilles Trystram

High School of Food Science and Technology, National
Institute for Agronomical Research, Massy, France

I. INTRODUCTION

The transformation of cereal products from dough to biscuit, cracker, bread, or pastry is a very complicated process, involving many mechanisms to be carried out [1]. Many of these mechanisms are not well known, and their control appears to be difficult. From the consumer point of view, numerous properties are desired in the baked product. The taste, of course, is important, and the texture is a direct consequence of product preparation (mechanical treatment and formulation) and of the piloting of the baking itself. Color, both for the bottom or the top part of a product, is important, and other well-known properties are necessary, such as crispiness.

From an engineering point of view, some other properties seem to be needed. For example, to obtain a long time of conservation, the moisture content (or the water activity) of a biscuit is important, but for economic reasons, it is better to obtain the higher ratio of water content at the end of the process. Dimensions are related to texture, but even at desired textural properties problems may be observed; for example, when the thickness is too high for one biscuit, the packaging of several biscuits becomes a difficult task. The consequences of product density often involve the total weight, which must be under control for regulation purposes.

Depending on the product concerned, some of those properties become more important. Each application is a specific one. The only common ground is the large number of desired properties and the interaction among them during the transformation of dough to final product.

The process is mainly influenced by several steps. First, the formulation is important and permits one to improve some specificities from one product to another. Second, the mechanical treatment is crucial. It begins with mixing, but the viscosity influences the baking, and it is really related with mechanical processing of the dough. The baking itself is naturally an important part of the process, and one of the most complicated.

Control of such a plant appears to be difficult, and in fact there are only a small number of well-automated lines [2]. Some reasons may explain the situation. The understanding of mechanisms carried out during mixing, rolling mill, or baking is still poor. It is hard to control mechanisms that are not well understood. Even as more tools and explanations are proposed by research teams, the implementation of such results in the industry is not easy and takes time.

Another reason for the low level of automatic control of such processes is the diversity of the transformation mechanisms of baked products. Classical ways are encountered mainly, but new ways are now available for extrusion cooking. Independent of the type of process, it is observed that a mixing of batch and continuous operation is often encountered. The control of such a line is still difficult because the kinds of tools and solutions for automatic control are not the same for the two types of processes. There is no method available for the design of such an automated system.

The lack of sensors is, for all food processes [3], a limitation to the development of high-level, high-performance automation systems. For the baking of cereal products, the number and the diversity of desired properties imply difficulties in the design, the implementation, and the cost of sensors. Only a small number of properties are monitored on-line. An important part of the product evaluation is a subjective evaluation performed by the operators.

Finally, it must be highlighted that the design of automatic controllers requires information about the dynamics of the process [3]. There are not many studies of dynamics analysis of baking ovens, and so the design of appropriate controllers is difficult.

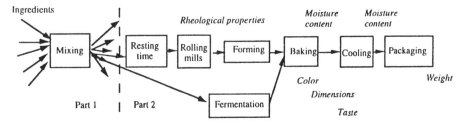

Figure 1 Principle of a line for baking a cereal product and illustration of some key properties.

The purpose of this chapter is to point out some industrial progress and to present results that can be applied on-line for the control of the baking of cereal products.

II. PRESENTATION OF THE PROCESS

Figure 1 presents the principle of a classical line used for transformation of dough into baked product [1]. The first step is always formulation and mixing, which is a multiple-input, multiple-output process. Often a resting time is used before rolling mills and forming of the product. In some cases a fermentation step (for bread, for example) is used before baking, cooling, and packaging.

Figure 1 illustrates some important factors along the line that may be controlled. For a specific process, control information can be obtained from one or another unit operation. The combination is important. As an example, the control of thickness of the product after baking is sometimes performed using the oven control, or in some other plant using the tuning of the roller mills. The state of knowledge is not sufficient, generally, to decide the best way for controlling the thickness. Most often, the control is manually performed by operators.

III. THE CIM APPROACH FOR BAKING PROCESSES

As presented in Chapter 20, the computer-integrated manufacturing (CIM) model is a way for analyzing and performing the implementation of computer-based functions for controlling a plant [4]. For the baking of cereal products, the Fig. 1 can be a guide for using a CIM approach in the plant. As indicated, two parts of the process are distinct. In the first one, a multiple-input, multiple-output process must be considered. For the mixing and formulation, numerous ingredients are used. Generally, the same mixing process is used for performing several different mixes.

On the other hand, after the mixing step, the process is a single-input, single-output plant. The kind of fluxes that are controlled is not the same. Thus CIM control of the plant can be analyzed in two phases: one specifically for the mixing process, and the other for the following line. In some cases the packaging part can be controlled through a specific solution.

If the implementation of a CIM solution is recognized as a way for improving the performance of the whole line, it is necessary to develop the intermediate levels. The first level concerns the implementation of sensors and actuators. The second is dedicated to the control itself, and the third is designing at the level of supervision and the operator help support system. As we will see, supervision is important for the case of baking.

IV. CONTROL OF THE CLASSICAL CEREAL PRODUCTS BAKING PROCESS

A. Classical Control of Mixing, Fermentation, Cooling, and Packaging

For a long time, some operations have been controlled using programmable logic controllers (PLCs). The concept is a simple on–off control, including sequence control, in which the main information used is the time (duration of the operation). Indeed, the measurement and specifically the on-line instrumentation is poor, and time appears to be the best way for controlling the sequence of action. For the mixing process, some improvements are implemented using temperature measurement. A threshold of temperature is then used as an event for the determination of the end of mixing. Time is still the most frequently used variable for rest period control. The same is true for fermentation, in combination with the simple control of ambient temperature, and for the cooling part which is often a noncontrolled sequence except through the velocity of the band.

As presented in the Chapter 3, dealing with modeling of food processes, the best way for analyzing and designing the control of simple on–off processes is the use of sequential function charts (SFCs). The power of PLCs permits implementation of more complicated functions. For the mixing process specifically, flexibility permits the introduction of a large number of product recipes. At this level of the process, the use of computer technology can improve the performance and the reproducibility of the process, including hygiene sequences such as cleaning-in-place, for example. The main limitation is still the number of available sensors for measuring the right properties of the transformed products.

The rolling mills are generally not controlled via product properties, but some control of strength is used to maintain them as constant as possible. Some experiments were performed for implementation of viscosity measurement of dough with associated feedback control of the rolls, but the measurement was not reliable, it is hard to implement, and the method for feedback design is not easy. Even if a solution can be developed in some cases, the generalization of such an approach is difficult.

B. Control of the Baking Oven

Before discussing automatic control of the baking process, we need to describe briefly the roles of each variable during baking, and with the consequences for controlling the oven.

Though different technologies are used, an oven is just a heat exchanger, in which heat is generated from the combustion of gas or the use of electrical energy [1]. An oven is a distributed parameter system [3]. Along the oven, a large number of actuators are distributed. The monitoring of the baking depends on the realization of different profiles along the space, inside the oven. The main process variables that influence the properties of the baked product, assuming a constant product entering the oven, are

1. The *air temperature*, the most commonly used variable, which is controlled both in roof and base of the oven through a classical proportional–integral (PI) controller (or on–off controller) from one temperature sensor per zone (Fig. 1).
2. The *air hygrometry*, which influences the rate of drying, and as a consequence the internal heat and mass transfer and reaction during baking. Air hygrometry is controlled through air exhaust at one or more points along the oven. Through most of the oven, there is no automatic control of ambient relative humidity. Only open loops are used.
3. The *baking time*, directly related to the velocity of the band.

And as a consequence of some operating conditions, air velocity, air pressure profiles, and heat flux profiles (for conduction, convection, and radiative fluxes) are important for baking (Fig. 2). Some results, as we will see, propose to control the baking through the control of heat fluxes rather than temperature profiles. Most of the variables are interrelated, making the process quite complicated. Figure 2 explains some influences of the

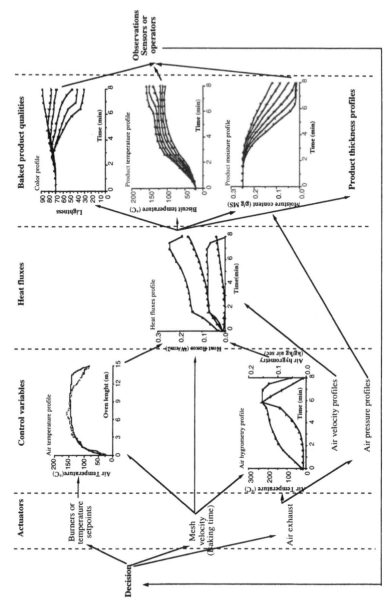

FIGURE 2 Principle of baking oven control. Influence of control variables on baked-product properties.

variables on the properties of a biscuit, and illustrates how the oven behavior is modified through automatic or manual controls. The main difficulty of the baking oven process control is then to obtain and to maintain the best profiles for the product along the duration of the baking.

Close relations are, of course, available between each profile. The main difficulty in control is to find a way to maintain all the profiles independently. The actuators are used for obtaining the profile of air temperature. This is the classical way to control the process, and most literature describes the baking curve, which is the air temperature curve. In fact, the modification of air exhaust is often used by operators. It permits modification, inside the oven, of the air velocity and, as a consequence, of convective heat fluxes. But another consequence is the evolution of the air pressure profile and the air relative humidity profile, which implies an evolution in the way the product is dried and baked. Recent results [5] establish that the main consequence of the evolution of these profiles is the evolution of the heat fluxes inside the oven. Direct relations are then observed between baked-product properties and heat fluxes. The consequences are modifications in the profiles of product properties as indicated in Fig. 2. The operator, or the sensor when it is implemented at the output of the oven, only sees the modification of the final value of the property. If the requirements are not satisfied, a modification is decided, by the operator or through an automatic device. Figure 2 illustrates the complexity of automatic control of baking ovens.

1. Sensors for Real-Time Quality Measurement During Baking

a. Sensors for Oven Control. Because the direct control of baked product properties is hard and not obvious, it is important to control, and to measure, some parameters not directly related to these properties. As indicated in Fig. 2, air temperature, hygrometry, and heat fluxes appear to be interesting variables.

The measurement of air temperature is classical. The main difficulty is the trying conditions inside the oven: high temperature (around 300°C), high humidity, and important fouling of the atmosphere due to fat and other baking by-products. Often the value indicated by air temperature sensor is false, with significant differences observed between reality and measurements. Nevertheless, an air temperature sensor is the most common sensor implemented inside a baking oven.

Recent progress in sensor technology allows high-temperature air hygrometers. Some applications are proposed, by oven designers, for including air hygrometers in oven instrumentation. Implementation problems are still hard to solve. The first one is the choice of the sensor location. The hygrometry profile is certainly important during baking, but the high cost of sensors prevents implementation of enough apparatus to control the whole profile. The location must take into account the most sensitive location. Such information is not available. The profiles presented in Fig. 2 illustrate some difficulties in the choice of sensor location. The number and repartition of air exhaust must be considered. Nevertheless, some technologies are available. Two ways are seen to take into account the main difficulties of environmental conditions.

The first way for hygrometer implementation is the use of a continuous sampling system, able to extract a small and representative flow rate of air and to ensure specific conditions before measurement. Mainly this consists of regulation of the air temperature of the extracted continuous sample so that the condensation on the sensor (or before it) is not possible. It is necessary to know the range of the air moisture content before the measurement [6]. A simple gold filter is used before the hygrometer to protect it from the volatile compounds coming from the baking. Such implementations are used with simple air hygrometers such as capacitive sensors (used for relative humidity measurements, some able to work up to 150°C), or cold mirror sensors, which can furnish directly the condensation temperature of the air. Psychrometers are used, too, specifically the high-temperature psychrometer proposed by Ultrakust [6], designed to work at high temperature and in a polluted atmosphere.

The second method involves implementation directly inside the oven. There is no such implementation, but it is important to notice the interest in gas-specific sensors for this purpose. Progress has been made in the design of selective gas sensors using semiconductors or polymers. One application is the measurement of water content or oxygen in gas. Applications are available for several industries in which the implementation conditions are difficult, as for baking. Future trends are easy to imagine. The domain of application is greater than the simple moisture content of the air. Other air compounds can be tracked as baking indicators. One such application was proposed for ethanol control during bread baking.

Other instrumentation for oven control concerns the heat fluxes performed by gas combustion or electrical heating of air. We discuss this later.

b. Sensors for the Real-Time Measurement of Baked-Product Properties. It is not possible to discuss all applications performed for specific cases. This section presents the main results known for the real-time control of some baked-product properties.

Moisture content. One of the most interesting properties is the moisture content of the product. Different studies were performed to develop sensors for moisture content [7, 8]. Two methods are available. Most applications are based on near infrared (NIR) absorption. The principle is simple. An NIR wavelength is partially absorbed at the surface of a product in relation with the number of OH bonds encountered. The comparison of a specific absorbed wavelength and a nonabsorbed one allows calibration of a sensor for moisture content measurement. This principle is put into practice on several apparatuses (Infrared Engineering, Quadra Beam, and other's), for on-line real-time measurements. Figure 3 presents a recording of moisture using the MM55G from Infrared Engineering [3]. Adjustments are made to account for the discrete presentation of the product under the sensor at the end of baking. The measurement is affected by surface variations, color, distance between sensor and top of the product, and so forth. As Fig. 3 shows, a numerical treatment of the signal (filtering and averaging) is necessary to obtain a reliable information. A classical 1% uncertainty performance is easy to obtain with these sensors.

Such sensors, as used now for crackers, biscuits, and other thin products, are not reliable for thick ones. Indeed, infrared absorption is only a surface measurement. The gradient inside a product is not measured. Another kind of application concerns the use of microwaves for the detection of the water content through the interaction of electromagnetic field with the dielectric and power dissipative properties of matter,

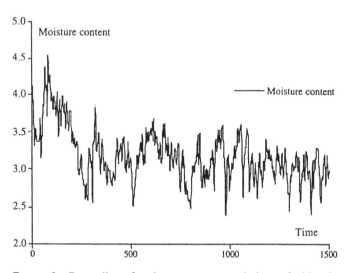

FIGURE 3 Recording of moisture content variations of a biscuit at the output of a indirect gas fired oven. Measurements were performed using an MM55G moisture sensor from Infrared Engineering. (From [3].)

specifically the water. Apparatus is available for the control of baking, drying, or extrusion. An accuracy of 0.5% is currently obtained, and the density can be determined at the same time as the moisture content [9].

Independent of the kind of sensor used, one must know the problems faced in moisture measurement. The main difficulty arises in calibration. Two steps are necessary. One is a static calibration; a set of samples is measured with the sensor and with a reference method, and the correction is determined. Then a dynamic calibration must be performed. Both steps are hard, and they must be checked frequently, because various factors can affect the sensor performance.

Color. Another important property in a baked product is color. Progress in sensor technology has produced new color sensors with improved performances. Two kinds are available. The first is illustrated by the sensor from Infrared Engineering (Colorex) or Hunterlab, for example. They are tristimulus color sensors, able to furnish a color measurement in an $La*b*$ space (or another color space related to this one when a normalized light, D65, for example, is used). The principle is the generation of a light (D65) and the presentation under filters in front of the surface of the product [3]. The resulting light is analyzed through a detector and transformed into an electrical signal. An illustration of the recording of such a signal is presented in Fig. 4. The signal is noisy because of the variation of color on the surface of the product. Calibration and comparison between laboratory measurement and sensor signal permit measurements with 2% uncertainty in the case of the Colorex. No strong variation was observed, and this kind of sensor can be used on-line for color quality

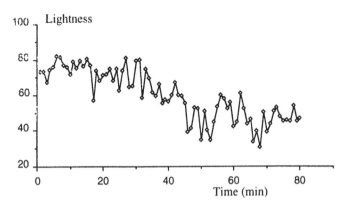

FIGURE 4 Example of recording color (lightness) of biscuit during baking using a sensor Colorex from Infrared Engineering. Measurement plotted after filtering of the original signal. (From [3].)

control purposes. MacFarlane [10] reports the principle and performance of a new apparatus from Hunterlab that can measure both the color and height of a biscuit.

Even if a triangulation system is used to cut the measurement when no product is under the sensor, one of the problems for implementation is to be sure that the sensor always provides the same information. If no travelling system is implemented with the sensor, the measurement is only related within one row.

Another way for detecting color evolution from baking is the use of color image processing systems. The cost and performance of color cameras allow their implementation into sensors. As an example [11], the comparison between laboratory colorimeter measurement and color image processing system gives a correlation coefficient of 0.98 for L, $a*$, and $b*$. The rapidity of the measurement is high and is compatible with real-time constraints. The main advantage is the ability to measure a complex or averaged value of the color. The choice of the right optics for the camera permits the measurement of more then one row, independently of the derive of the band.

An indirect way for measuring the color of a baked product is to measure the top-surface temperature of the product. Indeed, results are available in which good correlations are obtained between the surface temperature at the end of the oven and the color of the biscuit [12].

Dimensions. The control of the product dimension is essential for technological purposes and from the consumer point of view. The dimension, specifically the thickness, is greatly affected by baking. Several systems have been tested for thickness measurement. On-line, thickness can be measured by mechanical sensors that are in contact with the product and can follow the variations in the thickness. They are mostly used as indicators rather than as sensing systems. An ultrasound apparatus has been proposed, but at this time it is not in industry use. The same is true for the pneumatic sensor. The most powerful method seems to be the use of image analysis technology [11]. Several implementations have been proposed. The first uses a projected shadow from oriented lighting of the product [13]. A geometrical relation is easy to establish, but the on-line implementation and relation with the mesh are not accurate. Some other results propose the use of structured light, and the computation of the image permits volume evolution estimation.

Probably the most interesting system is the one presented in Fig. 5, where a laser diode is used. Using a known incidence angle for the laser, the trace on the product surface, compared with the trace on the mesh, is directly related to the thickness of the product. The same image analysis system allows one to measure the diameter or length and width of the

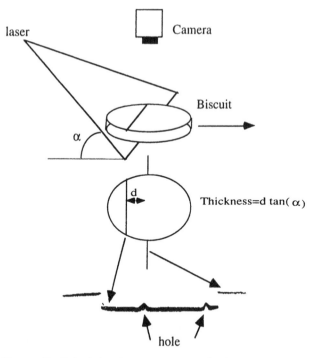

laser

Camera

Biscuit

α

d

Thickness=d tan(α)

hole

FIGURE 5 Principle of measurement of thickness using image analysis system, and illustration of the accuracy of the method.

product. An estimation of volume variation is then easy. An example of accuracy of the system is presented in Fig. 5, in which a hole at the biscuit surface is detected by the sensor. The accuracy of the measurement method is 0.05 mm (in practice, 0.1 mm is the effective accuracy), which is sufficient for the application. Shape variation can be detected too [11]. Figure 6 presents an example using image analysis on-line for the detection of length variations that are, for example, indications of broken products.

These results show how image analysis systems can be used to follow several properties of the baked product. Some complex applications are developed and implemented on-line in several plants throughout the world [14].

Heat fluxes.   Recent results obtained by different teams seem to be very interesting for the instrumentation of baking ovens [15, 16]. Heat flux analysis has established that most of the fluxes are radiative (between 50 and 75%), another part is convective fluxes, and conductive fluxes are around 10%. Direct relations are established between heat fluxes and

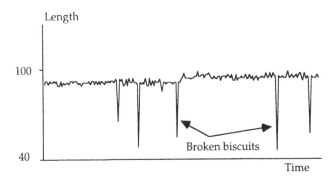

FIGURE 6 Recording of length variations of a biscuit after baking. Measurements are performed using on-line, real-time image analysis.

baked-product properties [5, 15–17]. As an example, Fig. 7 presents correlations between moisture content and lightness (L is used as color information) and total heat fluxes. The linear relationships observed are interesting results that permit control of properties through the control of heat fluxes. Some such sensors are under development at this time in several research laboratories and will probably be available soon.

Conclusion. Significant progress is being made, and more properties can be measured on-line. However, each case is a specific one, and engineers have to keep in mind that if the ability of a sensor is established and a calibration is obtained, during use, the calibration still must be verified and controlled. Most of the control systems give poor performance because of problems with sensor calibration. In the specific case of baked products, a result with one product is not easy to transfer to another one. Sensor location choice is still a difficult operation. See Chapter 2 for more information on sensors.

2. Automatic Control of Baking Ovens

Automatic control design is a two-step process. First, it is necessary to choose the variables that are to be controlled, and second, the quantitative tuning of the controller (including its choice) is done. For the first step, Fig. 2 illustrates the complexity of the choice. The purpose of this section is to illustrate trends for automatic control of baking ovens, work in progress, and some practical implementations.

a. Structure of Baking Oven Control Strategies. The kind of the oven is very important for the structure of the control strategy. Generally, three strategies are used: open-loop control, closed-loop indirect control,

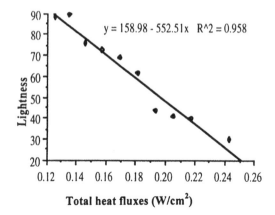

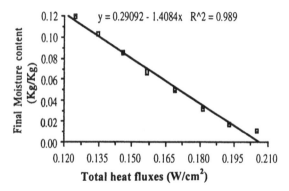

FIGURE 7 Relations between total heat fluxes and properties of a baked product (biscuit). Simulation results are obtained with a validated knowledge-based model. (From [5].)

and closed-loop direct control. Basics on the control strategies are given in Chapter 4.

Open-loop control. Because of the lack of proper sensors, most of the oven is not controlled. But without any control it is difficult for the operator to optimize the baking. Some classical controllers have been implemented. The first loop often realized is the PI control of the mesh velocity through a motor controller. It permits one to obtain a constant and reproducible baking duration. The same motor controllers are used in some applications for the automatic control of air exhaust. Depending on the number of air exhaust ducts, motors are used to tune as accurately as possible the rate of extraction. This allows one to obtain a fine air humidity profile inside the oven (and air velocity and pressure profiles, too; see Fig. 2). As indicated previously, air temperature controllers are implemented. Generally, several burners are grouped in a zone, both for sole and roof. Figure 8 explains one kind of control that is classically implemented. It is a low-level control, but it helps the operators, who have to tune a few setpoints and not a large number of on–off burners.

New technologies, such as the programmable logic controller (PLC), can produce better controllers than PI ones. For example, the number and the order of burners used for the control of air temperature in a zone is automatically decided by the controller. This kind of solution is a more flexible one, and safety is easily provided for (for gas combustion, for example). Another improvement is the linking of several adjacent zones together (Fig. 8). This permits one to obtain from one setpoint modification the coordinated modification of all the profiles of air temperature. Of course, such improvements need to be well applied by operators.

Another possibility, put into practice in some gas-fired baking ovens, concerns the total power available for combustion. Combustion effectively depends on the ratio of air and gas flow rates. It is possible to build a servo control capable of tuning the total power of the oven between lower and higher values. This improves flexibility, and the distribution of the power along the oven is better. A limitation arises from increased pressure losses if the pipes are too long [18].

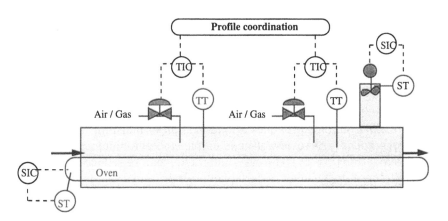

FIGURE 8 A classical strategy used for a baking oven control. Shown is case of a gas-fired oven with two zones and coordination of zones for getting temperature profiles.

Closed-loop indirect control. Although sensors are not available, some strategies are proposed to control indirectly some properties of the baked product. It is the case, for example, when product's top surface temperature is measured and used as a color measurement system (correlation has been established [12]. A feedback loop can be established using a cascade structure on a setpoint of one zone air temperature controller.

The duration of baking can be variable: depending on the product, between 3 and 30 min. For long baking times, disturbances are only detected after the duration of baking. For this, feed-forward controls are proposed. Easiest to implement are controls on the product presence in the oven. Indeed, if no product is on the mesh during some time, a hole is observed in the flow of the product. Then, a lot of product is burned because the use of heat fluxes is not good. It is necessary to reduce the heat fluxes beforehand to avoid product burning. A simple way is the installation of an electric cell at the input of the oven. The cell detects if a hole occurs in the flow of product, and a PLC is then able to reduce, in conjunction with the progression of the product, the heat flux profile. This is a simple example of a feed-forward control in anticipation of disturbances. Some trials are made for other disturbances such as variations of dough viscosity, but there is no general solution. It is important to note that the same control loop is able to take into account the transient behavior of the baking ovens. For example, start-up, shutdown, or production changes are crucial to control. Feed-forward control can minimize the amount of the product with lower quality.

Closed-loop direct control. When sensors are located on-line, it is possible to perform a closed-loop direct control of the oven. Applications are diverse. The most common situation is a simple single-variable control. For example, when the moisture content is the most important parameter, a simple controller is put into practice, acting on one setpoint of air temperature (generally on the roof, in the middle part of the oven). The same situation is encountered for color or thickness control. More complicated is the case where control of two parameters is needed. Often moisture content and color must both be controlled. There are interactions: a modification of color provokes a modification of moisture. Independent control is not easy. A review of some simple models for interaction is given by MacFarlane [19]. In some cases interaction can be reduced because of the difference in dynamics of each phenomenon, and a simple steady state interaction model is sufficient. In more complicated cases, the design of the controller can include interaction, but it needs a good and accurate analysis of oven dynamics.

Figure 9 is an illustration of a structure for automatic control of gas-fired baking ovens. Again, each application is a specific one and the figure only illustrates some possibilities and advantages of automatic control.

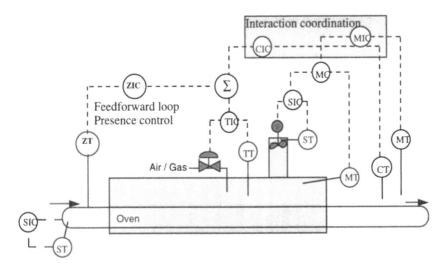

FIGURE 9 Advanced strategy for the control of gas-fired baking ovens. Feedback and feed-forward controls, including interaction of color and moisture content controls, are included.

b. Examples of Dynamic Analysis. Between qualitative analysis, necessary for the control structure design, and quantitative analysis for the tuning of the controller, studying the dynamics of the process is important. There are not many published studies and results concerning dynamics of the behavior of baking ovens. Of course, sensors are necessary to perform such a study. An example is presented in Table 1. It concerns the case of a small (15 m long), indirect gas fired oven, characterized by six zones for roof and base, independently controlled using classical PI control of on–off burners (34 burners). The oven is equipped with five air controlled exhausts. Experimental results are obtained when step responses are performed from a reference setpoint profile. Each setpoint of air temperature (for the six zones) is studied, and the step responses between air exhaust and baked-product properties are also studied. The table establishes several results. First, pure delays are observed for the lightness response versus air temperature. The design of industrial controllers takes this into account, and sometimes a PI with delay controller is used.

Second, all the step responses can be considered as first order. These are characterized by the rise time values (time to reach 90% of the steady state value) being highly variable with respect to the related control variable. This kind of result illustrates that oven baking is a complicated process with difficult dynamics. The choice of the best acting parameter on

TABLE 1   Example of a Dynamic Analysis of a Baking Oven Used for Biscuit Production

Property	Pure delay (s)	Static gain	Rise time (s)
Color: L			
Tz1	90	-0.01433	165
Tz3	30	-0.1737	250
Tz4	0	-0.09329	165
Tz5	0	-0.07992	100
Tz6	0	-0.02856	170
B1	0	-0.04552	335
B2	0	0.01744	35
B3	0	0.04156	55
B4	0	0.0172	45
B5	0	-0.04063	40
Moisture			
Tz1	0	-0.00758	250
Tz3	0	-0.01923	230
Tz4	0	-0.0183	500
Tz5	0	-0.00752	325
Tz6	0	-0.0096	180
B1	0	-0.0073	85
B2	0	0.00757	45
B3	0	0.0055	80
B4	0	0.0050	45
B5	0	-0.0029	40

this basis is not simple. It depends on the goal of the control. Such studies are necessary if accurate performances are needed for the control of the oven. An example of a resulting dynamic model of biscuit coloration is proposed in Chapter 3.

The result presented here illustrates that this analysis is possible. It takes time (and money), but it is necessary if we want to improve performance.

c. About the Controllers. Numerous controllers are available. Classically, PI and PI with delay controllers are used. Recent works propose implementation of fuzzy controls. Some advantages are related; the main ones are the simplicity of tuning and the good level of cooperation with operator needs. The performance is not better than with other controllers. The main difficulties concern interaction and cases in which importants pure delay is observed. Specific controllers must be designed, and the implementation on classical PLC is not so simple. It must be pointed out that good performances are obtained with simple strategies. A large number of industrial problems are solved at the upper level, when the operator can act independently on the plant. The relatively large time constant of the process permits this choice.

In some plants, the oven is never modified, and operators have to act on other parts of the process. This is a good reason to install a three-level CIM computer system for implementation of a supervision level able to help the operator.

V. OPERATOR DECISION SUPPORT SYSTEM

In the food industries, processes are often complicated because of nonlinearity, interactions, and the gap in knowledge concerning their control—lack of sensors, for example. Classical automatic control seems to be difficult, and many open loops are encountered in which the roles of the human operator become more and more important. Product properties that contribute to quality and process productivity depend mainly on the accuracy of operator reaction.

To decide on modifications to actuators, the human operator uses on-line measurements performed by sensors. But this is not the only source of information. Much is obtained from subjective evaluation. For baking of cereal products, color, shape, and decor are important product properties. Often there are no sensors located to measure such information in real time, on-line. The only way is human evaluation.

It is difficult to establish a link between human evaluation and sensor measurements. Nevertheless, the search of such relationships is an interesting way for progress in control applications. Because food quality is complex information, the use of sensor fusion is a possible technique for the integration of various information. Sensor fusion is an interesting approach, which consists of the definition of process indicators able to be processed on-line, in real time, from more than one measurement using smart data treatments, such as classification, modeling and so forth. Classification methods can likely perform such fusion tasks, especially because human evaluation is often more a classification than a continuous evaluation. Applications are available copying a human operator in the subjective evaluation of the color of biscuits from a baking oven.

Several studies have been concerned by classification and application; a recent work [20] introduces an interesting concept appropriate to color perception. A fuzzy sensor is used to evaluate human color perception from three numerical variables: R, G, B. This sensor is based on the concept of fuzzy subsets in multidimensional spaces. This notion is well adapted to baked-product color evaluation. Indeed, we do not need the use of fuzzy inferences because we are upstream of the decision. Therefore, we will use this method to link the colors of biscuits to the linguistic notions manipulated by the operators (Fig. 10).

The fuzzy multicomponent membership function concept [20] generalizes the classical notion of a fuzzy membership function. This idea is developed in fuzzy set theory [21]. The adjustment of the fuzzy multicomponent membership function is realized during a training stage (five points per class). Samples of cookies are taken from different ovens at an industrial site [22].

Figure 10 shows a recording obtained during an industrial test and then a comparison between operator judgment and fuzzy classifier, and also it shows the comparison with the results given by a Bayesian classifier. The results given by the two classifiers are comparable. Nevertheless, the fuzzy classifier is nearer to the gradual evaluation of the color of cookies by the operators than the Bayesian classifier. The results point out a global coherence between the two classifiers: Bayesian and fuzzy. Those results are good (more than 75% correct classification).

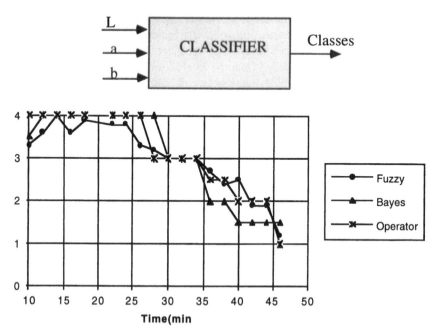

FIGURE 10 Recording of prediction using the fuzzy and Bayesian classifiers, and comparisons with operator classification.

The main interest of this study is the ability to reproduce the evaluation performed by an operator on-line, using a sensor and a data treatment. Such tools can help the operator make proper decisions. Similar applications are reported for verification of the conformity of baked products with quality standards [23].

To design an operator decision support system on the same principle, some expert system applications have been proposed. It is now well known how to design an expert system. The number of rules is generally large, and the time for obtaining the rules is important. The fuzzy expert system appears to be a good alternative that allows one to obtain a reliable decision support system in a reasonable time. Some applications are available in some plants. One of them, as an example, contained 61 rules, validated on the plant in 2 months [22]. The main constraints of such an approach are the existence, cooperation, and competence of the expert and the human problem of implementation and training of the operators for using these kinds of tools.

VI. CONCLUSION

The automatic control of the production of baked cereal products is complicated. If a part of the process is easily controlled, the lack of sensors and inadequate knowledge imply difficulties for building the right control structure. One part of the difficulty is that even when good solutions are implemented, the extrapolation from one operation to another need not be simple. Each case must be considered as a new one. Recent results allow one to take into account a part of the control problem as cooperation between operators and controllers. It is probably the best compromise for designing control of more reproducible production of baked cereal products. Automatic control of baking ovens is a work in progress, and more control loops in baking ovens will be implemented in the near future.

REFERENCES

1. P. Wade, *Biscuits, Cookies and Crackers* Vol. 1, Elsevier Applied Science, London, 1988, p. 200.
2. G. Trystram, F. Courtois, and M. Allache, Automatic control of the biscuit baking oven process, *Developments in Food Engineering* (T. Yano, R. Matsuno, and K. Nakamura, eds.), Blackie Academic and Professional, London, 1994, pp. 975–977.
3. G. Trystram and F. Courtois, Food processing control: Reality and problem, *Food Research Int. 27*:173 (1994).
4. G. Trystram, M. Danzart, R. Treillon, and B. O'Connor, Supervision of food processes, *Automatic Control of Food and Biological Processes* (J. J. Bimbenet, E. Dumoulin, and G. Trystram, eds.), Elsevier Science Publishers, Amsterdam, 1994, pp. 441–448.

5. D. Fahloul, Contribution à la maîtrise de la cuisson des biscuits, *Modélisation et simulation d'un four tunnel*, Doctoral thesis, ENSIA, France, 1994.
6. G. Trystram, P. Brunet, and B. Marchand, Humidity measurements in industrial baking ovens or dryers, 6th International Drying Symposium IDS88, Versailles, 1988.
7. K. Carr Brion, *Moisture Sensors in Process Control*, Elsevier Applied Science, London, 1986, p. 122.
8. E. Kress Rogers, *Instrumentation and Sensors for the Food Industry*, Butterworth Heinemann, London, 1993, p. 770.
9. J. King, On line moisture and density measurement of food using microwave sensors, Proceedings of the Food Processing Automation Conference IV, ASAE, 1995, pp. 127–138.
10. I. MacFarlane, On line combined color and height sensor for bakeries, Proceedings of the Processing Automation Conference IV, ASAE, 1995, pp. 34–39.
11. T. Moll, C. Guizard, and G. Rabatel, Off line colour machine vision for analyzing the biscuit baking process, Proceedings of the CAPPT'95 Conference, Ibra-bira, IFAC, Ostende, Belgium 1995.
12. S. Shibukawa, K. Sugiyama, and T. Yana, Effects of heat transfer by radiation and convection on browning of cookies at baking, *J. Food Sci. 54*:621 (1989).
13. G. W. Krutz and C. J. Precetti, *Automated Crackers Inspection Using Machine Vision,*, Paper No. 91–7535, ASAE, St. Joseph, Michigan, 1991.
14. M. Monnin, Machine vision gauging in a bakery, Proceedings of the Food Processing Automation Conference III, ASAE, 1993, pp. 62–70.
15. T. Fearn, R. Lawson, and D. Thacker, A heat-flux probe for baking ovens, *FMBRA Bull. 6*:258 (1986).
16. H. Sato, T. Matsumura, and S. Shibukawa, Apparent heat transfer in a forced convection oven and properties of baked food, *J. Food Sci. 52*:185 (1987).
17. U. De Vries, Products quality control strategies, Workshop in Baking Engineering, ENSIA, Massy, France, 1994.
18. P. Brunet, I. Savoye, F. Rapeau, and G. Trystram, Flexibilité d'un four de biscuiterie par régulation de la puissance thermique, Ind. Agr. Alim., 1990, pp. 537–546.
19. I. MacFarlane, *Automatic Control of Food and Manufacturing Processes*, Elsevier Applied Science, London, 1983, p. 320.
20. E. Benoit and L. Foulloy, Un exemple de capteur symbolique flou en reconnaissance des couleurs, *RGE* No. 3:22 (1993).
21. A. Zadeh, Fuzzy sets, *Information and Control* No. 8:338 (1965).
22. G. Trystram, N. Perrot, and F. Guely, Applications of fuzzy logic for the control of food processes, Proceedings of the Food Processing Automation Conference IV, ASAE, 1995, pp. 504–512.
23. R. K. MacConnel and H. H. Blau, Color classification of baked and roasted foods, Proceedings of the Food Processing Automation Conference IV, ASAE, 1995, pp. 40–46.

19

Computer-Based Controls in the Fish Processing Industry

Gour S. Choudhury and
C.G. Bublitz
University of Alaska–Fairbanks, Kodiak, Alaska

I. INTRODUCTION

The process of applying technological innovations to industrial applications tends to be sporadic. Technologies are most often applied significantly after development or are not implemented because their use generates other problems. The fish processing industry, probably more than any other food industry, exemplifies this practice. For several decades, the industry has been struggling with developing and implementing fundamental changes. Most recently, consumer demands for high-quality products and the need to stabilize production costs so as to compete effectively in international markets have been the primary objectives. Despite these imperatives the industry has not produced substantial changes in product types or mix, product quality has been inconsistent, and production costs have not been adequately controlled.

Development of technologies, particularly in electronic sensing and computer control systems, could prove central to eliminating existing production bottlenecks, increasing productivity, and creating opportunities for subsequent innovations. Fundamental changes of this nature, however, have been slow to develop, and regional variation continues to characterize and determine the course of innovation within the industry. The diversity and inherent variation of raw material, regional economic conditions, and seasonal characteristics of the labor force have all made the introduction of new technologies sporadic and variable.

Traditionally, the harvesting and processing of fish were conducted by family enterprises that relied exclusively on apprenticeship training. Customarily, this produced impediments to change, and new methods were viewed with extreme skepticism and implemented slowly. For 450 years, the fish processing industry produced mainly salted and dried fish products. Fish were generally in abundance, were caught close to shore, and were processed and marketed by the fishermen themselves. The introduction of canning, however, broke this trend. Because of the nature of its raw material, the fish processing industry paid close attention to developments in canning technology. In particular, the salmon industry needed a system to address the diverse and rapid influx of raw material. Annual salmon seasons were as short as 15 days, strictly limiting harvesting and processing time and forcing packers to produce their entire annual output within a brief period. The shortness of the season and unpredictability of the spawning runs placed a heavy emphasis on speed and efficiency in operations.

The introduction of canning imposed a pattern of technological implementation on the industry that to a large extent exists today. The complexities involved in mechanizing operations that manipulated fish, and the comparative ease with which machines could be used in operations that manipulated cans, produced a basic pattern of conjunctive but separate manual and mechanized unit operations. This pattern continues to exist, in that those unit operations which manipulate fish depend more on a human work force while supporting processes are more heavily mechanized.

Over the past few decades, however, some fundamental changes have occurred in the fish processing industry that have required the introduction of automated technologies. Adoption of freezing and refrigeration has resulted in development of an industry that currently produces primarily frozen products. The advent of the surimi industry has introduced more sophisticated processing technology into the industry. Surimi manufacture is the process whereby fish fillets are reduced to a fish paste, which is used to produce a variety of analog products that resemble other seafood, for example, imitation shrimp or crab legs. With the introduction of these product types, controlled product handling was required and more advanced and complex activities had to be performed. This has resulted in new technologies being adopted for harvesting and processing fish.

With these changes have also come major changes in the market areas being served. Fish processing is now very much integrated into the worldwide system of food production and distribution. Requirements imposed by these markets and products necessitate strict harvesting, handling, processing, and storage conditions on the producer. In many cases actual processing occurs at sea aboard highly advanced catcher–processor vessels. Because of size and living constraints, these vessels depend on modern technology to automate processes and minimize the hardships of the work environment. Smaller catcher vessels are increasingly required to control the harvest, handling, and storage to ensure premium quality fish are delivered to shore-based processing plants. To meet these demands the fish processing industry is looking for automated technologies.

The process whereby fish products are produced is a continuum of unit operations initiated by the capture of fish and cumulating with the production of the final foodstuff. Each step in this continuum affects succeeding steps and ultimately the type and quality of the final product. The interrelated nature of these steps, therefore, dictates that all component operations within this continuum be considered. It accomplishes little to automate one unit operation thereby creating a bottleneck in another. The degree to which computers are used to control and integrate unit operations, and improve overall efficiency during harvesting and processing of fish, is the subject addressed in this chapter.

II. FISH HARVESTING

Fish harvesting is a complex process entailing several distinct operations: encounter (or finding), capture, and onboard handling of the catch (Fig. 1). Factors controlling these operations can be divided into two major categories: those inherent to the fish and others intrinsic to the vessel. These factors determine the vessel's ability to find fish, the type of gear used, and the degree of handling and preprocessing that can be conducted. Of the onboard systems available, those designed to locate fish exhibit the highest degree of integration and use of microcomputers. In contrast, onboard handling of the catch features the least sophistication. To date, all preprocessing and onboard fish handling activities are accomplished manually, and no computer-controlled automated systems are available. This condition should change as existing and upcoming requirements for efficiency, economy, and human safety will make the development of automated handling systems an industry imperative.

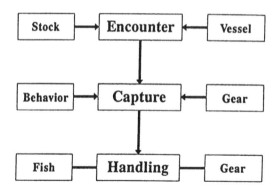

FIGURE 1   Fish harvesting operations and variables controlling those operations.

Development of auxiliary instrumentation and onboard electronics was initiated by the adaption of radar and sonar into navigation and depth finding systems for use aboard fishing vessels. The advent of these technologies permitted vessels to venture into deeper water and harvest new species in larger quantities. Until the early 1970s, navigation, fish finding, and communications were handled by single-function, stand-alone units. The advent of the Loran C (LOng RAnge Navigation) system, with latitude and longitude readings as well as time differences, demonstrated the advantage of interfacing information. The improvement in digital electronic circuitry that permitted marine applications was affordable, and had the necessary speed initiated this development.

Historically, onboard technological changes have occurred in an evolutionary manner, implementation occurring only when the technology was proven to be reliable and cost-effective. Today, this trend has been accelerated by huge demand, competitive markets, and processing advances mediated by rigid quotas applied to protect fishing stocks and national interest. These factors have all contributed to the need to find and adapt new technologies to ensure optimum productivity. Consequently, an increasing number of fishing vessels are relying on integrated microprocessor-controlled instrumentation to locate fish and for operation of the ship's control and monitoring systems.

A. Encounter

The first step in fish harvesting is to locate the desired species in sufficient abundance to warrant harvesting. Those factors that mediate the ability to locate fish are shown in Fig. 2. Fish availability is determined by the stock's distribution and abundance, the environment, and fishing intensity

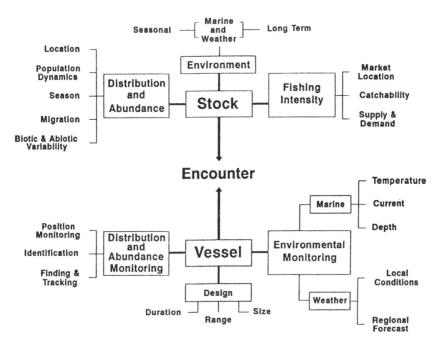

FIGURE 2   Encounter phase of harvesting operation, depicting parameters determining fish finding capabilities.

on the stocks. The size of the fishing vessel, its ability to monitor the environment, and techniques utilized to locate fish stocks are also factors in determining encounter. To accomplish the latter, a great deal of effort has been directed toward the development and automation of "fish finding" equipment. Today, most medium-sized fishing vessels are equipped with an array of navigation and hydroacoustic instrumentation dedicated to locating fish. This enables the vessel to precisely locate specific areas, permitting the operator to repeatedly track over productive grounds. Theory and operation details of many of these instruments are provided by Tetley and Calcutt [1]. The ability to locate fish is moderated by the vessel's range and trip duration. These factors are continuously evaluated and updated by monitoring the vessel's fuel consumption, speed, and engine operation. Weather conditions, which are monitored by a variety of onboard instruments, in turn mediate range and trip duration. Many of these systems rely on microprocessors or a central microcomputer for operation and information interfacing.

   An assortment of equipment is used to determine and evaluate vessel location and fishing activities relative to fish movements, depth, and abundance. Electronics such as radar, Loran C, satellite navigation, magnetic and gyro compass, autopilot, course alarm, course plotter, sonar, and echosounder not only help in navigation but are also utilized to determine the location and abundance of fish recources. Of the onboard electronic systems, the course plotter is probably the most sophisticated. This system stores and plots the vessel's previous course tracks and recalls and displays individual tracks on a CRT, permitting the operator to conduct fishing operations by retracing previously prolific hauls. To accomplish this, the course plotter system integrates marine chart information stored in computer memory with inputs from navigation instruments, including Loran, Decca Navigator, satellite navigation systems, and radar. Interfacing the course plotter with a system such as the Ritchie MagTronic Compass [2] permits the operaor to conduct fishing operations without manual plotting or piloting. The MagTronic Compass, which allows integration of the autopilot, Global Positioning System (GPS), Loran C, and radar, permits automatic steering—including course changes along a preplanned track.

Course plotter systems normally include alarms to alert the operator of way points or possible collision situations. The Simrad RS2800 [3] can be operated as a radar, a chart display/plotter, or a synchronized chart plotter with radar overlay. The radar image is projected onto a computer-generated chart to enable the operator to observe objects and/or hazards not incorporated on the stored chart. Other types of plotter systems such as the SEANAV 1050 [4] provide 3-D bathymetric plots instead of surface images. Software, such as Maptech [5] and WindPlot [6] is also available for personal computers; this provides similar plotting capabilities.

A meterological mapping software, Met Map, is available that receives live weather images from satellites and registers them onto computer-generated map projections [7]. This system also provides infrared images, which are calibrated for sea and land surface temperatures. National Marine Electronic Association (NMEA) data such as position, time, speed, water depth, barometric pressure, and other information can be displayed on the screen via an onboard GPS.

The trend for navigation and fish finding systems has been proprietary networking and integrating of multiple functions to a single display. Between manufacturers these systems have little, if any, intersystem communication and operation capabilities. In most cases, these are stand-alone machines with some interfacing capability. The first standardized interfacing format, NMEA 0180, was developed by the National Marine Electronics Association in 1980. This standardization was designed to interface a Loran C receiver and an autopilot and was upgraded in 1982,

NMEA 0182. More flexibility was provided by NMEA 0183, which permitted interfacing of Loran C and GPS with electronic chart displays, autopilots, radar, communication, and vessel instrumentation equipment [8]. These standards, however, have a major deficiency: although several pieces of equipment can simultaneously receive information, only one can transmit. NMEA is currently developing new network interfacing standards to enable multiple transmission units as well as multiple receiving units.

Onboard vessel monitoring systems (VMS) involving microcomputer technology focus on fuel conservation and consumption monitoring. A typical fuel monitoring system's output usually comprises both actual and average fuel consumption per mile, as well as the fuel flow in liters/hour, and the amount of consumed and remaining fuel. Additional information provides the vessel's range at the actual fuel consumption, the distance covered since the last resetting, and time and distance to destination. Computer monitoring of propulsion plant operations aboard fishing vessels has, to date, been extremely limited. Restrictions imposed on this application arise mainly from software limitations. These limitations were partially resolved by Bonnett [9]. He used an empirical approach to formulate prediction equations to analyze engine performance. Predicted values over a variety of operating conditions are stored in a computer and used as a baseline for determining future engine performance.

The goal of VMS is to automatically show long-term trends in system performance and to detect short-term anomalies in equipment behavior that may point to future problems or system decline. The term *incipient failure mode analysis* is sometimes used to describe this function. The average VMS typically monitors up to 150 sensors, which are of three basic types: (1) discrete switches, which have two states, (2) analog devices, which deliver a voltage or current proportional to a measured variable, and (3) rate devices, which deliver an electrical pulse proportional to a measured variable. An integral part of these automated packages is an alarm and monitoring system (AMS). The AMS is programmed to alert operators when equipment behavior warrants human attention. This alarm function is normally triggered when a sensor parameter exceeds a preset value programmed into the computer. The AMS typically looks at each sensor many times each second (100,000 measurements per minute) and performs simple averaging to determine a value [9]. These data are then displayed on gauges or monitor screens and are compared with preset alarm values. In keeping with traditional procedures, the measured parameters are periodically printed to produce hard copy logs. Logs can be reviewed to evaluate past performance and serve as a historical record.

B. Capture and Retention

The capture and subsequent retention of individuals of any given species are controlled by a variety of gear and behavior parameters (Fig. 3). A general review of fishing gear types is given by Regenstein and Regenstein [10]. Fish behavior is determined by both environmental and biological factors, some of which can be monitored and utilized to selectively harvest fish. For example, the mesh used in a specific gear design can provide either a high- or low-contrast image based on the visual acuity and contrast discrimination of the target species. The type of image presented will moderate the type of response by the fish and may lead either to escape from or retention by the gear.

Comparatively few systems have been developed for automating the capture and retention phase of fishing operations. Automated longlining, purse seining, and fillnetting systems have been developed but are exclusively based on mechanical controls. Systems utilizing computer controls are dedicated to trawling. The trawl systems that have been developed

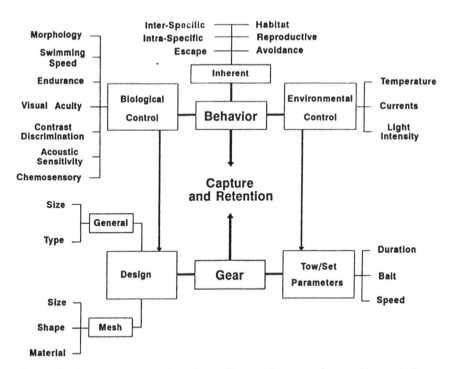

FIGURE 3 Capture and retention phase of harvesting operation, and interrelationship between controlling units.

are of two types: winch control systems and net monitoring systems. The former systems are designed to maintain the gear in the optimum fishing configuration during maneuvering or changing current conditions. Net monitoring systems survey the position of the gear relative to the fish and ocean bottom so that the gear can be maneuvered to optimize catch.

Muriaas [11] reported the development of a trawl control system designed to monitor the scope of the warps, tension in the warps, warp speed, and winch rpm. This information is automatically integrated and displayed by a microcomputer. The system incorporates computer-controlled mechanisms enabling it to shoot the gear to a preset warp scope, to haul the gear at the maximum possible speed without overloading the winch, and to keep tension in both warps equal at all times. System operation is based on the principle that oil pressure in the hydraulic motor is directly proportional to the load on the trawl warp. Two winch systems are connected by a balancing pipe, which maintains equal pressure in both motors. The computer control monitors all critical functions during trawling operations. Should any condition exceed preset limits, an audible alarm will sound and the cause will be displayed on the screen. Automatic shooting and hauling of the trawl to preset warp lengths, braking, and holding of the trawl in a constant position are accomplished by computer monitoring of pressure transducers on each trawl winch. Any increase or decrease in pressure in one winch results in the computer playing out or retrieving warp until the pressures are equal. Conversely, if the warp on one winch begins to play out, hydraulic pressure is increased to stop the winch. If the preset high pressure limit is exceeded, the winch will play out warp and a "hang-up" alarm will sound. Shooting and retrieval speed and warp length are monitored by rotation detectors mounted on each winch. High shooting speeds results in increased hydraulic drum pressure to decrease speed. Warp length is determined by the number of drum rotations during setting, with fine length adjustments achieved by balancing of hydraulic pressure after the preset warp length is reached. Output to the screen is in both digital and graphical format. All necessary drum speed and warp data (length, tension, etc.) are programmed through a keyboard.

Additional trawl information can be obtained using a multiparameter data telemetry system for simultaneous acquisition of parameters pertaining to net performance, environment, and vessel [12]. This system utilizes an onboard computer to process, integrate, and display data, and a set of remote sensors mounted on different parts of the trawl system and vessel. The sensors are connected to a multiplexer unit, which conveys signals to the onboard processor through a cable. Parameters such as depth, relative motion, mesh shape, catch, and door tilt are measured by variation in electrical impedance caused by compression and expansion of a coil or from motion of a pendulum. Other parameters measured include water flow, by electrical pulses from impeller rotation; salinity, using a platinum electrode cell; temperature, using a semiconductor encased in a metallic cover; solar radiation, through a photo diode; and net height, using an acoustic transducer.

Currently, most commercial trawl net monitoring systems only chart net opening and net position relative to the bottom. Scanmar [13] introduced ultrasonic telemetering type equipment for monitoring parameters such as depth, catch, temperature, and speed, along with other functional features. A very sophisticated system, "Net Nav" [14], has been introduced by Seametrix for monitoring headline height, warp tension at net, net geometry, depth, temperature, trawl board angle, net contents, and codend heading. Westmar [15] introduced a system utilizing a 32-bit microprocessor for signal processing and high-definition display. The Model TCS600E system includes forward-scanning sonar, vertical sounder, net profiling, pressure depth, water temperature, and catch sensors. The system includes a split screen option enabling the operator to view two parameters simultaneously. The sounder system incorporates two frequencies (60 and 160 kHz) to optimize use in varying fishing conditions.

C. Onboard Handling and Preprocessing

Once the catch is onboard, the fish are either bulk stored or preprocessed depending on species and market (Fig. 4). The nature of the handling, preprocessing, and storage is mainly determined by the volume of fish caught, target markets (e.g., the type of consumer product to be produced), and vessel size. Individual catch weights on medium-sized vessels can reach up to 50 tons, and on larger vessels up to 150 tons. Consequently, the technique used for handling and the speed at which preprocessing is accomplished are critical in maintaining overall quality. Olsen [16] outlined several techniques for the handling and preprocessing of fish. These systems all depend on manual operation, and as yet no computer-controlled automated system is available to assist in handling and preprocessing operations. Larger vessels, especially catcher–processor vessels, are more fully automated and employ systems similar to those utilized by shore-based processing plants, as discussed in Section III.

Occurring simultaneously and in proximity to the on-deck fish handling and preprocessing is the manual handling and maintenance of the fishing gear. A few automated systems for the shooting, retrieval, and baiting of gear have been developed. These systems are mainly stand-alone units that rely on mechanical controls for operation.

D. Integrated Systems

The next stage in the process of vessel automation will be development of computer integration of all the vessel's and gear's controls and instrumentation (Fig. 5). Olkhovskii et al. [17] presented theoretical aspects of integrated information systems for use on board fishing vessels. Ben-Yami [18] indicated that development of a fully integrated computer-controlled system would produce several operational divisions that could be further integrated into semirobotic systems. The functions he outlined included

1. *Control and Display*—monitor and control of instrumentation and machinery via a keyboard, and present data at one or more displays in verbal, digital, or video format.
2. *Warning and Summons*—correlate and analyze different sensor inputs, and issue warnings for potentially dangerous or hazard-

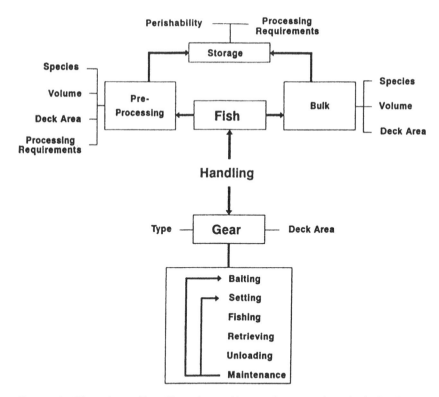

FIGURE 4   Flow chart of handling phase of harvesting operation, depicting factors determining gear and fish handling capabilities.

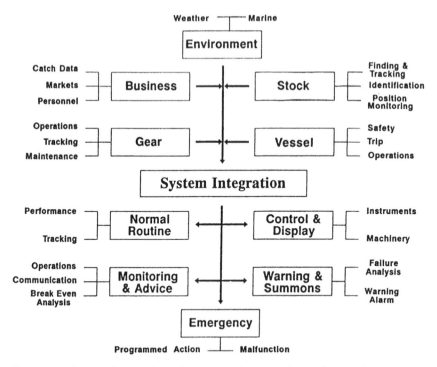

FIGURE 5 Schematic of onboard computer-integrated vessel operation system.

ous situations, system malfunctions, and operation parameters
such as variations in fish targets, overloaded codend, excess fuel
consumption, irregular engine performance, change in weather,
and so forth.

3. *Normal Routine*—execute programmed activities where imme-
    diate reaction is desirable, namely, steering vessel by isobath,
    chasing targets, steering vessel by compass course, maintaining
    cruising speed, and so forth. The latter can include steering the
    vessel and changing speeds and courses as preprogrammed or
    from input from auxiliary instrumentation.

4. *Emergencies*—provide instantaneous reaction to situations re-
    quiring multiple and simultaneous actions. Examples of this type
    of control include engine or fishing gear malfunctions requiring
    immediate reaction and evaluation of situations that endanger
    lives, such as monitoring and warning of vessel stability prob-
    lems (icing, leaking, or improper loading).

5. *Monitoring and Advice*—maintaining information on stores,

fuel, water, vessel range, trip duration, tides, currents, shipboard weather observations, species type and weight of catch, and so forth. Integration of this information with market data such as fish prices and landings through satellite communications would provide real-time information on the economic feasibility of ongoing operations.

Computer-based integration of all systems would simplify vessel operation and correspondingly make fishing operations safer. This would be the anticipated result of constant and precise monitoring of all integrated instruments, and the number and character of alarms and automatic reactions that could be preprogrammed or entered by the skipper. Normal operation, subject to technical software designs, could be preformed by single-function keys with preassigned functions, using menus, or by typing simple plain-language commands. This type of operation could extend to vessel, winch, and gear, as well as off-vessel components integrated into the system. Terminals located in strategic areas such as staterooms and the mess room would enable operators to monitor and react to situations from anywhere aboard the vessel. The skipper would also be able to react and take emergency action more rapidly using any of the terminals. Like several other industries and, as has happened in aircraft operation, computer-controlled instrument integration would improve the performance and safety of the crew, vessel, and gear.

III. FISH PROCESSING

The world's fish catch has steadily increased over the last 40 years from 21 million metric tons in 1950 to 97 million metric tons in 1990 [19, 20]. However, technological developments in the processing of fish have not been as extensive compared with those in other sectors of the food industry. Most fish processing is primary in nature with some automation or computer control being introduced at various stages. The advantages of computer control and process automation for improved performance and profit are increasingly being realized in the fish processing industry. The major hurdles for automation have been the complex geometry of fish, variable nature of the harvested catch, and nonavailability of adequate sensing and monitoring systems. The few unit operations where computer control has been successfully introduced are based on the physical attributes of fish. Most of these process control technologies are proprietary, and very little information is available in the public domain. The major fish processing operations where some automation or computer control has been introduced are fish fillet, roe, and milt production; fish canning; surimi production; curing and smoking of fish; and fish meal production.

A. Fish Fillet, Roe, and Milt Production

Fish fillet production is one of the major operations in the fish processing industry. The unit operations comprising the process are shown in Fig. 6. The sequence of operations may vary depending on the fish species processed and the finished product. The harvested fish are sorted and graded into two or more size groups, gutted, washed, headed, and filleted. The process of gutting, heading, and filleting is normally done by one machine (e.g., Baader filleting machine) or a series of integrated machines. For roe and milt production, the fish is first headed and then eviscerated for roe or milt removal. The headed and gutted fish are then subjected to filleting operation. Scale removal is done before filleting for fish species with scales. Each fillet is then skinned, trimmed, and inspected for detection and removal of parasites, wherever necessary. For some fish (e.g., red salmon), skin is left on the fillets for production of smoked products. The fillet is then weighed, packed, frozen, bulk packed, and kept in frozen storage before shipment.

The unit operations where automation or computer contort is used or available during the filleting process are sorting, grading, gutting, heading, filleting, and weighing. Sorting of fish is mostly a manual operation even though automatic sorting machines are available. The Fish Monitoring System (FMS), equipped with artificial intelligence, uses computer-aided machine vision to identify and quantify all fish by species, length, and weight [21]. The FMS employs computer controls to sort fish in real time in any combination of eight groups by species, length, or weight with a throughput of 120 units per minute. A herring sex sorting machine, HD1S (Neptune Dynamics Ltd., Vancouver, BC, Canada), is used in some fish processing plants to segregate mature herring by sex. Sorting is achieved by sensing and electronically analyzing the amount of energy transmitted when infrared light of variable intensity is directed through the belly of the fish containing roe or milt [22].

Automatic grading by length and/or thickness using mechanical devices and/or light emitting diodes, and by weight using dynamic scales is used in many fish processing plants [23]. The Scanvaegt Image System (ScanGrader 7100) can identify up to eight different product types using 14 different parameters at the rate of 240 items per minute per line. The image system is based on self-learning principles and a product profile can be automatically built by setting the computer to learning mode. The weighing table, equipped with a computer/photocell routine, ensures weighing of only one piece at the precise weighing time. The weight distribution parameters are constantly updated during production, providing very high precision batching [24].

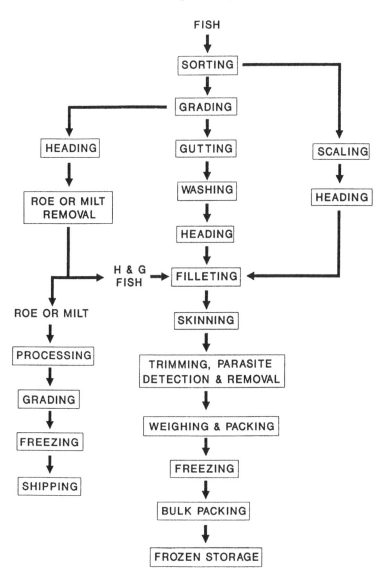

FIGURE 6 The unit operations comprising the process for production of fish fillet, roe, and milt.

The introduction of electronics has significantly increased the throughput of fish filleting machines [25]. The gutting, heading, and filleting operations are integrated in one machine that uses computer-controlled step motors to adjust the cutting tools. The size of the fish is electronically determined as the fish approaches the heading unit. The host computer uses this data to control the machine settings for each individual fish. A recent design (Baader 212) has incorporated an optional electronically controlled roe extraction unit, which removes roe and guts for subsequent manual separation. The headed and gutted (H & G) fish is either discharged or transferred to the conveying system of the filleting section, where fillets are produced by cutting tools controlled by the host computer. The machine has an electronically controlled pinbone cutter, moved by tandem stepping motors, for production of boneless fillets. The cutter is controlled by the host computer based on the physical parameters stored for each type of fish [26].

The operations following filleting are either mechanical or manual and tend to impede production. At present, parasite detection and removal is the rate-limiting step during fillet production [27]. Development of an automated system to assure parasite-free fish fillets is crucial for increasing throughput during fillet production. Various approaches that have been investigated are laser candling [28], ultraviolet light [29], X-ray [28], scanning laser acoustic microscope (SLAM), and pulse–echo technique [28, 30–33]. The major problem in application of these techniques is the inability to distinguish parasites from the surrounding flesh. The physical properties used to differentiate the nematode and surrounding fish tissue are very similar [28]. None of these methods proved successful, because of low sensitivity, poor resolution, and slow throughput. Choudhury and Bublitz [34] used electrical properties of fish muscle and associated parasites to develop an electromagnetic method to detect parasites in fish muscle. The fundamental change this method represents is simplification of the parasite detection problem by changing it from complex pattern recognition to simple current detection. This technique has the potential for automated industrial applications requiring high sensitivity, high resolution, easy defect recognition, and fast throughput.

B. Fish Canning

Canning is widely used in the fish processing industry as a method for preserving and shipping fish such as salmon, tuna, sardines, mackerel, herring, sprats, and pilchard [10, 35]. The chain of unit operations constituting the canning process is shown in Fig. 7. The fish preparation prior to filling varies depending on the type of fish being canned and the final

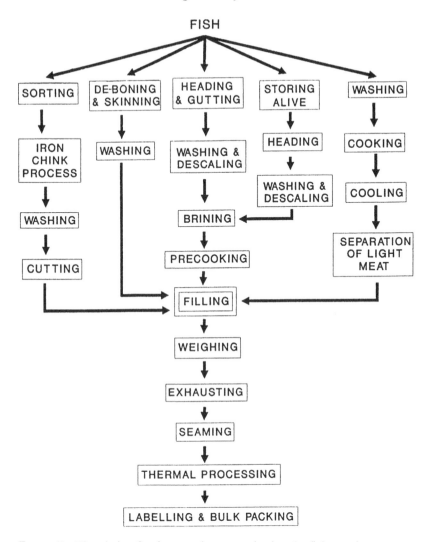

FIGURE 7 The chain of unit operations constituting the fish canning process.

product. The traditional salmon canning consists of processing the fish by an Iron Chink, which beheads and removes the roe, viscera, blood line, fins, and tail. Fish are then portioned before filling. For skinless and boneless salmon products, deboning, skinning, and washing are done prior to filling [10]. Pretreatment of herring, sprats, mackerel, and pilchards includes heading, gutting, washing, descaling, brining, and precooking.

In some cases, fish such as sprats are stored alive to empty the gut before heading [35]. Tuna are cooked and cooled before the light meat is separated for can filling [10]. Following the pretreatment, the filled cans are subjected to standard canning operations such as weighing, exhausting, seaming, thermal processing, labeling, and bulk packing.

Most of the canning unit operations are either mechanical or manual. However, a computer weight control and monitoring system (Digicon 2000) is used in some processing plants for statistical quality control (SQC) and filler adjustment for precise weight control [36]. The system consists of a host computer, checkweighers, and a motorized fill adjustment mechanism, which precisely maintains the average weight at a particular setpoint. An electronic checkweigher provides product weight to the Digicon 2000 system, which can control four filling stations simultaneously. The host computer processes the data, displays the results, and conducts statistical analyses. Current and cumulative production statistics are displayed on one computer monitor, while another shows the SQC chart. The system uses sophisticated statistical algorithms and digital stepping motors to adjust the filler to ensure precise weight control. Weight data, archived to disk, can be exported to standard spreadsheet or database software for further analysis and report generation.

C. Surimi Production

Surimi is washed and refined fish mince used as intermediate raw material for production of imitation seafood products. Alaskan pollock (*Theragra chalcogramma*) is primarily used for surimi manufacture. The technology for surimi production was developed in the 1950s and has since been improved with the introduction of new machineries. The process, described in Fig. 8, essentially consists of extracting muscle from fish, washing to eliminate soluble components, refining, and mixing with a cryoprotectant [37]. The steps before mincing can vary depending on type of fish and quality of the final product. In general, fish are subjected to washing, heading, gutting, filleting, and skinning before mincing. Alternatively, muscle can be extracted by mincing after heading and gutting. Small fish can be fed directly to a deboning machine after washing to obtain minced fish muscle. The minced fish are washed for two or three cycles to remove lipids, low molecular weight organic compounds, minerals, and soluble proteins. Washed mince is refined to eliminate residual bones, connective tissue, scales, and skin pieces. The refined mince is then dewatered in a screw press or decanter centrifuge and mixed with cryoprotectant and frozen for shipment to analog processing plants.

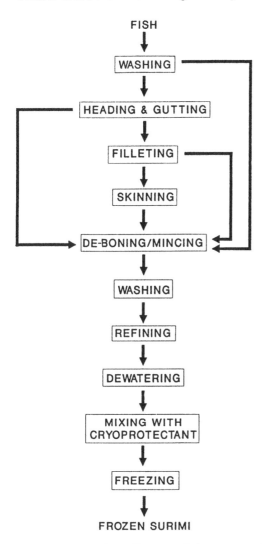

FIGURE 8   The flow diagram of the process for production of frozen surimi. (Adapted from [37].

The application of computer and electronic instrumentation to the surimi process has been shown to increase production, worker performance, and product recovery, and to improve quality consistency of the final product [38]. Surimi production can be fully automated. The operations preceding mincing are currently controlled by a computer, and sensors required to automate the remaining unit operations are available. In addition, mincing produces a consistent fish paste that can be pumped when mixed with water, enabling smooth flow of material from one unit operation to another. Computer control and process automation during surimi production has been achieved at various levels in different plants, and information on such systems is proprietary. The process is presumably controlled by a computer or a central processing unit that facilitates smooth transfer of material from one unit operation to another. Level control sensors, flow meters, and timers are used to control the ratio and amount of fish mince and water in different washing stages, and to control residence time. The availability of on-line moisture meters will help automate the dewatering stage.

D. Curing and Smoking of Fish

Fish used for manufacture of a variety of smoked products are salmon, herring, haddock, cod, whiting, mackerel, trout, sprats, eels, and oysters [35]. As shown in Fig. 9, the process essentially consists of fish preparation, brining or dry salting, and smoking. The fish preparation steps include washing, gutting, splitting, or filleting. The prepared fish are then brined or dry salted and smoked. Microprocessor-controlled systems are available to control the smoking operation [39]. The smoke environment (temperature and relative humidity) and the smoking time can be programmed to obtain the desired product. These parameters and the functions chosen and/or energized during the operation are displayed on a built-in panel. The system controls various oven phases using operating functions such as exhaust damper, fresh air intake, smoke inlet, steaming, liquid smoke, blower speed, and so forth. Each program can use segments for heating, drying, smoking, cooking, steaming, and showering. A preselected delay start can be included in each program or inserted manually. The self-diagnostic feature built into the microprocessor indicates common problem areas such as temperature high limit, blower failure, burner flame out, incomplete purge, motor overload, and inability to reach temperature and humidity setpoints within a reasonable time period.

E. Fish Meal Production

Depending on species, 30–80% by weight of the fish caught is not utilized for direct human consumption and is discarded as by-products or wastes. The fish processing by-products include trimmings, belly flaps, heads, frames, fins, skin, and viscera from fillet, surimi, and canning operations, and fish that are too small, soft, or too parasitized for fillet production.

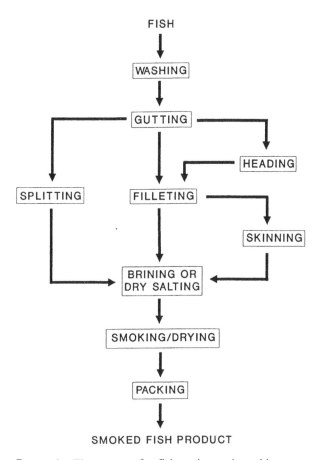

FIGURE 9   The process for fish curing and smoking.

The predominant method for processing fish by-products is production of fish meal and oil, described in Fig. 10.

The composite by-products are ground, cooked, and pressed to obtain wet meal and press liquor. The liquid fraction is separated in solid, oil, and water phases. Stickwater is concentrated in evaporators. The wet meal, solids from phase separation, and stickwater concentrate are mixed, and the mixture is dried. The dry meal is mixed with antioxidant, milled, and packed or sieved to separate bones before milling and packing. Alternatively, the dry meal can be sieved, mixed with antioxidant, milled, and then packed. The entire process is automated using programmable logic

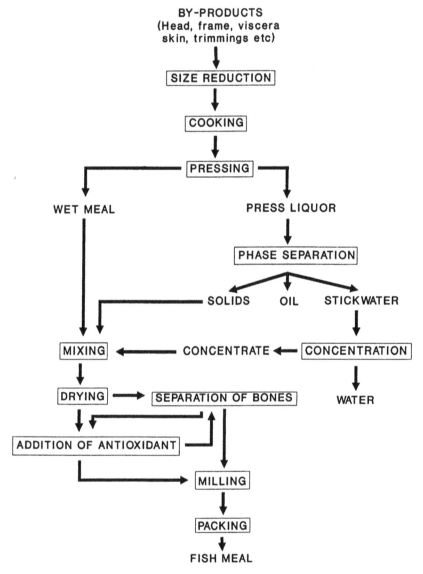

FIGURE 10 The flow diagram of the process for production of fish meal and oil.

control and is monitored from the control panel. The information on the use of computer controls in such systems is proprietary.

IV. SUMMARY

The use of computer-controlled automation in the fish processing industry is in the initial stage of development. Most onboard and in-plant computers are utilized exclusively for data gathering and signal processing applications. Only a few applications incorporate computer control of specific unit operations. Development of integrated systems for automation of the majority of unit operations, from finding and harvesting of the raw fish to packing of the finished product, will not occur in the near future. Fundamental research will be needed to develop sensors and determine relationships among the unique and highly variable parameters.

Development of sensors for effective monitoring of gear and fish–gear interactions will require extensive research. The current use of onboard computer-based systems is limited because of the lack of specialized application programs. The development of software, especially expert systems, would make the present use of computers on board fishing vessels more effective. Fishing-oriented software should be based on a user-friendly format that incorporates logbook, fishing grounds and weather information, environmental data, previous and current catch, market information, and vessel and gear monitoring system applications. Even with manual input and without special analytical facilities, this type of software would provide rapid and pertinent information necessary for decision making and fishing operations. Incorporating well-designed analytical functions and automatic information would increase the utility of onboard computer systems significantly.

Research will be needed to develop sensors and relationships among characteristics such as size, age, sex, maturity, seasonal variation, species, and the like before automation can occur. Sensor development should utilize physical and chemical attributes of fish as they relate to these variables. Collaborative research between industry and academic institutions will speed up efforts in these areas. The fish processing industry will invest energy and resources in development of secondary processing in the immediate future, and process automation will be an integral part of such development. The continuing growth of aquaculture will provide a high-quality raw material, eliminating some of the uncertainty associated with utilization of wild stocks. Aquaculture has the potential of providing a more consistent raw material harvested at the optimum maturity, and it will simplify automation of fish processing operations.

REFERENCES

1. L. Tetley and D. Calcutt, *Electronic Aids to Navigation*, Edward Arnold, Baltimore, 1986.
2. Anon., *National Fisherman*, August 1994, p. 21.
3. *Simrad Chart Plotter*, RS2800 promotional literature, Simrad Inc., 19210 33rd Ave. W., Lynwood, Washington 98036-4707.
4. Anon., *Pacific Fishing*, August 1994, p. 44.
5. Anon., *Pacific Fishing*, August 1994, p. 43.
6. Anon., *Pacific Fishing*, June 1994, p. 45.
7. Anon., *Alaska Fisherman's Journal*, August 1994, p. 36.
8. M. Crowley, Integrated electronics may be closer than you think, *National Fisherman*, August 1994, pp. 26–27.
9. D. E. Bonnett, Vessel vital signs monitoring, Proceedings of the World Symposium on Fishing Gear and Fishing Vessel Design, The Newfoundland and Labrador Institute of Fisheries and Marine Technology, St. John's Newfoundland, Canada 1988, pp. 380–384.
10. J. M. Regenstein and C. E. Regenstein, *Introduction to Fish Processing*, Van Nostrand Reinhold, New York, 1991.
11. R. Muriaas, Automatic trawl control system, Proceedings of the World Symposium on Fishing Gear and Fishing Vessel Design, The Newfoundland and Labrador Institute of Fisheries and Marine Technology, St. John's Newfoundland, Canada 1988, pp. 396–400.
12. T. K. Sivadas and K. Ramakrishnan, Performance evaluation of trawl system through multi-parameter data link, Proceedings of the World Symposium on Fishing Gear and Fishing Vessel Design, The Newfoundland and Labrador Institute of Fisheries and Marine Technology, St. John's, Newfoundland, Canada, 1988, pp. 401–404.
13. Anon., *Fishing News International*, October 1987, p. 29.
14. Anon., *Fishing News International*, October 1987, p. 24.
15. Anon., *Pacific Fishing*, October 1994, p. 48.
16. K. B. Olsen, European fish handling and holding methods, *New Directions in Fisheries Technology* (J. F. Roache, ed.), Department of Fisheries and Oceans, Ottawa, Canada, 1985, pp. 46–56.
17. V. E. Olkhovskii, M. N. Andrevv, A. A. Levin, and V. L. Yakovlev, *Automation of Navigation and Tactical Control in Fishing*, Israel Program for Scientific Translation, Jerusalem, 1972.
18. M. Ben-Yami, Integration of fishing vessel instrumentation under a single computer umbrella, Proceedings of the World Symposium on Fishing Gear and Fishing Vessel Design, The Newfoundland and Labrador Institute of Fisheries and Marine Technology, St. John's, Newfoundland, Canada 1988, pp. 392–395.
19. *Fisheries of the United States* (B. R. O'Bannon, ed.), National Marine Fisheries Service, National Oceanic and Atmospheric Administration, US Department of Commerce, Washington, DC, 1985, p. 32.

20. *Fisheries of the United States* (B. R. O'Bannon, ed.), National Marine Fisheries Service, National Oceanic and Atmospheric Administration, US Department of Commerce, Washington, DC, 1990, p. 33.

21. P. DeBourke, An outline of a fish monitoring system, Proceedings of the World Symposium of Fishing Gear and Fishing Vessel Design, The Newfoundland and Labrador Institute of Fisheries and Marine Technology, St. John's, Newfoundland, Canada 1988, p. 371.

22. N. J. C. Strachan and K. C. Murray, Image analysis in the fish and food industries, *Fish Quality Control by Computer Vision* (L. F. Pau and R. Olafsson, eds.), Marcel Dekker, New York, 1991, pp. 209–223.

23. J. Neilsen, J. H. Reines, and C. M. Jespersen, Quality assurance in the fishing industry with emphasis on the future use of vision techniques, *Fish Quality Control by Computer Vision* (L. F. Pau and R. Olafsson, eds.), Marcel Dekker, New York, 1991, pp. 3–20.

24. B. Amaral, Personal communication, Baader Food Processing Machinery, Baader North America Corporation, Fort Myers, Florida, 1995.

25. J. Graham, New technology for in-plant processing, *New Directions in Fisheries Technology* (J. F. Roache, ed.), Department of Fisheries and Oceans, Ottawa, Canada, 1985, pp. 105–110.

26. M. Trenovich, Personal communication, Baader Food Processing Machinery, Baader North America Corporation, Fort Myers, Florida, 1995.

27. C. G. Bublitz and G. S. Choudhury, Effect of light intensity and color on worker productivity and parasite detection efficiency during candling of cod fillets, *J. Aq. Food Prod. Technol. 1*(2):75 (1992).

28. D. L. Hawley, *Final Report: Fish Parasite Project* NOAA Grant No. NA: 85-ABH-00057, 1988, pp. 1–39.

29. J. H. C. Pippy, Use of ultraviolet light to find parasitic nematodes in situ, *J. Fish. Res. Bd. Canada 27*:963 (1970).

30. H. Hafsteinsson, K. Parker, R. Chivers, and S. S. H. Rizvi, Application of ultrasonic waves to detect sealworms in fish tissue, *J. Food Sci. 54*(2):244, 273 (1989).

31. H. Hafsteinsson and S. S. H. Rizvi, A review of the sealworm problem: Biology, implications and solutions, *J. Food Prot. 50*(1):70 (1987).

32. M. Freese, *Ultrasonic Inspection of Parasitized Whole Fish* Paper No. FE: FIC/69/0/12, 1969.

33. M. Freese, Distribution of *Triaenophorus crassus* parasites in whitefish flesh and its significance to automatic detection of the parasites with ultrasound, *J. Fish. Res. Bd. Canada 27*:271 (1970).

34. G. S. Choudhury and C. G. Bublitz, Electromagnetic method for detection of parasites in fish, *J. Aq. Food Prod. Technol. 3*(1):49 (1994).

35. R. McLay, Canning, *Fish Handling and Processing* (A. Aitken, I. M. Mackie, J. H. Merritt, and M. L. Vindsor, eds.), Ministry of Agriculture, Fisheries and Food, Tory Research Station, Edinburg, UK, 1982, pp. 115–125.

36. D. A. Yip, Personal communication, Digicon Engineering Incorporated, N. Vancouver, BC, Canada

37. Anon., *Introduction to Surimi Processing* (M. Okada and K. Tamoto, Supervisor), Overseas Fishery Cooperation Foundation, Akasaka Twin Tower, 17–22, Akasaka 2, Minato-Ku, Tokyo, Japan, 1986.

38. P. M. Nicklason, Applied technology in the high seas fishing industry, *J. Aq. Food Prod. Technol.* 2(2):113 (1993).

39. G. Martini, Personal communication, Enviro-Pak, Clackamas, Oregon

20

Computer-Integrated Manufacturing in the Food Industry

Bart M. Nicolaï
Katholieke Universiteit Leuven, Heverlee, Belgium

I. INTRODUCTION

In response to the steady increase of technology and general welfare after World War II, most industrial enterprises adopted an organizational structure based on the principles of Taylor. In this respect the enterprise was subdivided in several functional units corresponding to the several activities such as production, design, and inventory management. This organizational structure was motivated by the increasing complexity of manufacturing practice and the limits of human information processing capabilities [1]. Parallel to the exponential increase of computer performance, these functional units were automated, and concepts such as PICS, MRP, MRP2, DCS, and CAD/CAM were introduced and applied with varying success. As initially there was not much coordination between the automation efforts of the several units, this eventually led to an archipelago of *automation islands* with a multitude of hard- and software systems unable to communicate with each other. It became clear that there was need for a new automation paradigm in which the classical Taylorism would be replaced by a more integrated approach. This led to the introduction of the concept of "computer-integrated manufacturing (CIM)."

In a CIM system the different functional departments are linked to each other by means of a common database and appropriate software. In Fig. 1 the information flow in a typical food company where the CIM concept has been implemented is shown. The central database remains at the core of the system and is accessible to all users by means of the company computer network. This network may contain a heterogeneous mixture of computer types (e.g., PCs, Unix servers, mainframes) and even different subnetworks (e.g., Ethernet, Token Ring, SNA), operated by a variety of (network) operating systems. The different logistic departments are linked to the network, as well as to the design and manufacturing departments.

In this chapter several issues will be discussed related to the implementation of CIM in the food industries. Because of space limitations,

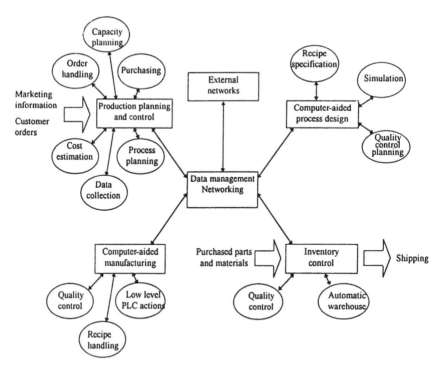

FIGURE 1 Information flow in a typical food processing plant where the CIM concept has been implemented. (Adapted from [1], by permission of the Springer-Verlag GmbH & Co.)

aspects such as financial accounting and inventory control will not be discussed here. For a full account of this subject, the reader is referred to standard textbooks on industrial management techniques. In Section II, the basic concepts of data and knowledge bases will be reviewed. Contrary to the situation in other industrial branches, computer-aided design techniques for food manufacturing processes are still in their infancy. However, some interesting developments recently appeared in the scientific literature, and these will be reviewed in Section III. Section IV is devoted to production planning and control, which will be focused toward the specific problems that arise in applying common techniques such as MRP, MRP2, OPT, and JIT to food manufacturing systems. In Section V, some issues related to computer-aided manufacturing will be discussed briefly. The practical implementation of the CIM concept in the food industries are discussed in Section VI, and some applications will be given in Section VII. The chapter will be summarized in Section VIII with some concluding remarks.

II. DATA AND KNOWLEDGE BASES

From Fig. 1 it is clear that the data is the glue that keeps all the components of the CIM framework together. It is therefore appropriate to have a closer look at the different approaches to store, retrieve, and manipulate data.

Data consists of facts acquired by means of human perception or physical devices. Examples of data are the temperature and pH in a vessel filled with fluid, the ingredients of a food product, and a purchase order by a customer. To achieve true integration, it is necessary that a unique, although possibly distributed, data storage facility exists that can be accessed by all functional units of the enterprise. This facility must offer a wide range of tools to introduce, update, delete, and consult data, while taking into account security issues.

Data may have a completely different semantic meaning depending on the person, environment, and situation. For example, the fact that the pH in a vessel is 8 might mean to the operator that the valve of the acid tank is blocked, to the food chemist that some vitamins are destroyed, and to the process scheduler that the batch has to be rescheduled. Data that is organized into a higher semantic level taking into account a particular environment is called *knowledge*.

Large amounts of data are most conveniently processed using appropriate database systems, and the corporate database constitutes the major information resource for the company. The sharp increase of research efforts toward artificial intelligence techniques since the mid-1980s has led to new and versatile knowledge representation techniques. Tools to build knowledge-based systems are now commercially available and are very useful for applications that need to process heuristic knowledge. Some successful applications of these novel data representation techniques will be outlined in the following sections.

A. File-Handling Systems

The first data-handling systems relied on the *file* concept [2]. A file is a collection of *records* (or plain text), and each record contains a number of *fields*. A field represents atomic data that is not further structured. For example, the file "Food   ingredients" may have a list of records such as "Cream," "Beef," and "Cashew nuts," and each record may have a list of fields such as "Protein content," "Fat content," "pH," or "Record   number." The contents of a field can be used for searching through the file and locating matching records. Note that the data stored in the fields may be alphanumeric or logical.

Files are often structured according to the application that will use the data, although this complicates maintenance and modification seriously. A major drawback of the file-handling systems is that the application programmer must be fully aware of the low-level physical file structure. This complicates the application software considerably. Also, no mechanism or language is available to interrogate, or *query*, the database, and the application programmer is forced to implement the required query operations. A further drawback is the absence of mechanisms to perform transactions (atomic operations on data). Further, although a file-handling system may be optimized for a specific application, it is most unlikely that it will be optimal for another application as well. As a consequence, several applications usually apply different file systems containing the same data but stored in a different file format.

File-handling systems are important from a historical perspective, but they are now replaced almost completely by database management systems.

B. Database Management Systems

1. Data Models

Database management systems (DBMSs) were soon introduced to overcome the problems associated with file-handling systems. The basic philosophy behind a database management system is that the data should be available to arbitrary application programs in a standardized way. Also, no redundant data should be in the database, to safeguard data consistency. To facilitate database management, modification of the database structure, and implementation of application programs, it is advantageous to separate the application software and the physical database from each other by means of a new *data model layer*. A data model is an abstraction that allows the data to be manipulated independently of the equipment used to store the data. An individual data item represents a property, or *attribute*, of some *object* to which a value is assigned. For example, "74.4%" is the value of the attribute "Water-content" of the object "Chicken meat, white."

Different data models can be distinguished [3], including

The hierarchical model
The network model
The relational model
The object-oriented model

2. The Hierarchical and the Network Models

The hierarchical model and the network model are both examples of *navigational databases*, in which the objects are stored as records and linked by means of pointers. The database is searched by jumping from one record to the other. While this structure allows very fast access, it is also rigid in the sense that the links between the objects are predetermined. Also, the application programmer must know the sequence of pointers to access specific data.

In the hierarchical model (e.g., the IMS system of IBM) the data is organized in a treelike structure. The objects and their attributes are ordered by parent–child relationships. While a parent can have more than one child, a child can have only one parent. This last property is precisely one of the major disadvantages of the hierarchical model. The network model, on the other hand, allows a child to have multiple parents. The network model was formalized by the CODASYL (Conference on Data Systems and Languages) initiative in the beginning of the 1960s. A CODASYL compliant database is network based, and it disposes of languages to describe and manipulate data, and to establish a link between the logical and physical structures of the database.

3. The Relational Model

The relational model [4] is based on the mathematical theory of sets. In this model, the data are represented in the form of tables. The example in Table 1 is used to illustrate the concepts. Assume that some food company produces foods for a variety of restaurants. The first subtable in Table 1 represents the set "Customer." Each customer is characterized by a list of attributes. A customer may place one or more orders. The set "Order"

TABLE 1 Example of the Relational Model for Data Representation; (a) Customer Table, (b) Order Table, (c) Food Table

Customer-ID	Customer-Name	Address	City	Country	
Customer 1	Il Travatore	Via Appia, 34	Rome	Italy	
Customer 2	The Golden Duck	Oxford Street, 66	London	United Kingdom	
Customer 3	Imperial Sushi House	Sotobori-Dori Av. 123	Tokyo	Japan	(a)

Order-ID	Order-date	Due-date	Food-ID	Quantity (kg)	Customer-ID	
Order 1	13 April	5 May	Food 3	5	Customer 2	
Order 2	24 April	7 May	Food 4	10	Customer 2	
Order 3	25 April	3 May	Food 2	5	Customer 1	
Order 4	29 April	10 May	Food 5	1.5	Customer 3	(b)

Food-ID	Food-name	Food-type	Price-per-kg ($)	
Food 1	Spaghetti Napolitana	Italian	4.0	
Food 2	Pizza Quattro Stagioni	Italian	5.0	
Food 3	Kidney Pie	British	4.5	
Food 4	Pork Sausage	British	3.5	
Food 5	Tempura	Japanese	6.7	
Food 6	Swordfish	Japanese	11.0	(c)

and its attributes are shown in the second subtable. The foods are characterized by a third subtable.

An important feature of commercially available relational database packages such as Oracle (Redwood Shores, California), Sybase (Emeryville, California), and Informix (Menlo Park, California) is the availability of a common *query language* to manipulate the data base. This language, SQL or *Structured Query Language*, is a high-level ("fourth generation") language as opposed to low-level ("third generation") procedural languages such as FORTRAN and C. It was originally developed by IBM [5] but has evolved toward a de facto industry standard. A typical SQL statement may look like

SELECT Food-name FROM Food WHERE Food-Type = 'British'

and has as its effect that it will display only the values of the attribute "Food-name" from the table "Food" and for which the attribute "Food-type" is equal to "British." The resulting subtable is shown in Table 2. Far more complicated queries can be constructed by using the set-theoretic

TABLE 2 Resulting Table
After SELECT Operation

Food-name
Kidney Pie
Pork Sausage

operations that are defined in the relational model. Suppose, for example, that we want to know the foods and their quantity that must be prepared before the 6th of May. This can be achieved by executing the following query:

SELECT Food-name, Quantity FROM Food, Order WHERE
    Order.Food-ID = Food.Food-ID AND Due-Date < 6 May

To process this query, the SQL engine will first create a new table by combining each row of the table "Order" with each row of the table "Food" using the algebraic JOIN operator. This table thus has in theory $4 \times 6 = 24$ rows and is subsequently searched for foods that satisfy the due date condition. The resulting subtable is displayed in Table 3.

Besides the JOIN operator, relational algebra provides a list of other fundamental operators, such as INTERSECTION, UNION, and DIFFERENCE. Details are beyond the scope of this chapter; the reader is referred to specialized works such as [3].

While this is a very simple example, it is easy to see that the JOIN operation may require considerable computer resources if large tables are involved. For example, joining a 1000-row table with a 500-row table (both reasonable dimensions) would result in a 500,000-row table, which could take up a considerable amount of memory space. Obviously, in commercial relational DBMS systems, appropriate mechanisms are implemented to avoid the actual assembly of the table resulting from a JOIN operation. Other drawbacks are the fact that several typical database applications,

TABLE 3 Table Resulting
from Complex Query

Food-name	Quantity
Pizza Quattro Stagioni	5 kg
Kidney Pie	4.5 kg

such as the BOM explosion mechanism in MRP (see Section IV.B), involve recursive JOIN operations, which also involve considerable computer resources. For speed-critical database applications, the network model is therefore still in use.

4. Object-Oriented Databases

The object-oriented paradigm for data representation is most easily explained based on an example. Assume that a food company produces prepared Italian food dishes such as pizza, lasagna, cannelloni, tagliatelli, and so forth. Different varieties exist for each dish, but all have some ingredients in common. For example, the following pizza varieties are produced: pizza margherita, pizza quattro stagioni, pizza Napolitana. All have a dough bottom covered with tomato sauce in common, but they differ in their toppings. Further, each variety may be labeled in several ways, depending on the customer.

It is clear that this data can be considered as objects related to each other in hierarchical way. The attributes of an object can have a value, for example, the attribute "Tomato-sauce-quantity" of the object "Pizza margherita" may contain the value "50 g." Also, some procedure, or *method*, can be associated with an attribute. For example, if the "Tomato-sauce–quantity" of an object becomes smaller than a well-defined threshold value, an alarm may be triggered to warn the operator. Several objects having common attributes can be represented by a generic data structure, or *class*. We could, for example, introduce the class "Pizza," with the attributes "Pizza-type," "Tomato-sauce-quantity," and "Dough-quantity." An object is defined as an *instance* of a class where values are assigned to the attributes.

A crucial feature of object-oriented data structures is the notion of *inheritance*. This implies that classes that are structured in a hierarchical way may inherit all attributes (with or without value) and methods from the parent class. For example, the class "Pizza margherita" inherits the attributes "Tomato-sauce-quantity" and "Dough-quantity" from its parent class "Pizza."

Different objects can communicate with each other by means of messages. An object-oriented program is actually a collection of a large number of objects that exchange messages, and that execute methods if some attributes are changed. In fact, object-orientation is the preferred paradigm for the construction of graphical user interfaces. Several object-oriented languages (C++) and environments (Smalltalk, Kappa, KEE—the two latter are trademarks of Intellicorp Inc., Mountain View, California) are now commercially available, and it can be foreseen that object-oriented programming will become the main data representation and programming paradigm. There is now also a fast growing number of object-oriented database companies (e.g., Object Design Inc., Burlington, Massachussetts; Objectivity Inc., Mountain View, California and many others), but a lack of standardization so far limits widespread use. Some of these vendors recently have formed a consortium (ODMG) that works on standards to allow portability of customer software across their products. An alternative to buying an object database is to add object capability to an existing relational database. The familiar SQL environment is hereby preserved at the expense of a considerable loss in performance [6].

5. Recent Developments

Database technology is a very active research topic. Recent developments such as semantic, hypermedia, and multimedia databases are discussed in [2], [3], and [7]. An important trend in database design is the use of *client–server* technology. This is essentially a logical, and in most cases also physical, separation of the application software and the database management system. In the case of a relational database, the application software (the "client"), which runs, for example, on a networked PC, typically generates SQL queries that are passed through the network to a fast computer (the "server"), on which the database resides and which executes the queries. The results are subsequently sent back to the application program and reported to the user.

C. Knowledge Bases

1. Representation of Knowledge

Earlier in this chapter, knowledge was defined as the organization of data taking into account their semantic meaning relative to their particular environment. As manufacturing operations involve a considerable amount of knowledge processing, it is clear that to automate this activity some means to represent and handle knowledge in a computer is required. This includes representing facts, combining existing facts to derive new facts, derive patterns from facts, and generalizing these patterns to theories. Several approaches, or *paradigms*, have been proposed in the literature for this purpose. Before discussing these, it is convenient to subdivide knowledge in two categories [3]:

1. *Declarative knowledge*, which consists of facts known or believed to be true, relationships among these facts, and constraints on allowable facts and relationships. The relationships are usually expressed in the form of rules.
2. *Procedural knowledge*, which consists of procedures to derive (*infer*) new facts using declarative knowledge.

The difference between declarative and procedural knowledge can be illustrated easily by means of an example. The fact "The temperature in the refrigerator room is 6°C" and the rule "If the temperature in the refrigerator is above 5°C, then the food will be spoiled" are both examples of declarative knowledge. The human mind possesses procedural knowledge that allows it to derive the following new fact by appropriately combining the fact and the rule: "The food is spoiled." This mechanism is called *logical inference*. If the number of facts and rules is limited, then both the rules and the inference mechanism can be programmed by using a conventional procedural programming language such as Pascal or C. The rules are then hardwired in the code in the form of IF/THEN statements. However, if a considerable number of rules and facts must be handled, it is more appropriate to make use of a language or programming environment (*"shell"*) that provides specialized knowledge representation and handling facilities. It is worthwhile to mention that modern (active/deductive) database management systems also have limited knowledge processing capabilities. The knowledge representation techniques to be described in this section, however, have possibilities that go far beyond the limited features found in database systems.

2. The Rule-Based Paradigm

The *rule-based paradigm* is the most popular paradigm for knowledge representation. A rule-based system typically consists of a database, a rule base, and an inference engine. The *database* contains the facts known to the system. The *rule base* contains all the relationships between the objects in the form of rules of the form

IF {conditions} THEN {actions}

Rule-based languages include an *inference engine*, which provides the facility to make logical inferences by evaluating the conditional part of the rule using the facts, and executing the action part of the rules. Two inferencing modes are possible. In *forward chaining*, the system starts from the initial data of the problem and executes the rule set. This will most probably alter the database content, and the rules will be executed again using the freshly generated facts. This is repeated until some value can be assigned to a predefined goal variable. For example, in the previous example the goal variable could be the logical variable "Food-is-spoiled," to which after the first inferencing cycle the value "true" is assigned. In *backward chaining*, a value is assigned to the goal variables, and the rules are extracted for which the action-part matches the goal state. The conditional part of these rules are subsequently considered as the next goal to be matched, and the cycle is repeated until a fact is found that satisfies the conditional part of one of the extracted rules. Both chaining mechanisms have their advantages and disadvantages. Integrated environments are now available (e.g., the Kappa environment, Intellicorp) that simplify the construction of a rule base.

3. The Logic-Based Paradigm

The *logic-based paradigm* for knowledge representation relies on the logical deduction of *propositions* from known facts. A proposition is a simple statement of a fact, such as

Tank 1 is empty

Bottle filling station 1 is broken

These propositions represent simple facts and are called *atomic*. Atomic propositions can be combined to *compound* propositions using the logical operators or *connectives* AND, OR, NOT, and IMPLIES, for example,

Tank 1 is empty AND Valve 3 is open

Bottle filling station 1 is broken OR Tank 1 is empty IMPLIES Order 5 will be late

Atomic propositions that are combined using permitted connectives are called *well-formed formulas* and can be true or false. A well-formed formula that is true is called an *assertion*. A set of assertions (*axioms*) that are assumed to be true in the real world is called a *theory*. A well-formed formula that can be proven to be true for the given theory is called a *theorem*. Proposition calculus provides a means to prove a theorem given a set of axioms. For example, assume the following set of assertions:

1. Bottle filling station 1 is broken.
2. Tank 1 is empty.
3. Valve 3 is open.
4. Bottle filling station 1 is broken OR Tank 1 is empty IMPLIES Order 5 will be late
5. NOT Order 5 will be late IMPLIES Customer 6 will be satisfied

and assume that we want to prove the theorem "Customer 6 will be satisfied." By combining the atomic propositions (1) and (2) (which we know to be true) and assertion (4), we can deduce that the proposition "Order 5 will be late" is true, and is, hence, an assertion. The inference rule that we used here is called *modus ponens*. Further, from the latter assertion and assertion (5) we can deduce that the proposition "Customer 6 will be satisfied" is false. Although correct, this method is highly intuitive and difficult to automate. Therefore alternative methods, which rely on the notion of a *clausal form*, are more suitable [3].

Predicate logic was introduced to generalize propositions. A *predicate* is a property of an object and can be considered as an alternative way to express propositions. For example, the propositions

Tank 1 is empty

Bottle filling station is broken

can be expressed in predicate logic as

empty(Tank 1)

broken(Bottle filling station)

The power of predicate logic lies in the fact that some kind of *parameterized* predicates can be defined, such as

empty(x)

where x is a parameter that is not defined but which may be substituted by some value that is relevant to a given situation. The Prolog language is essentially based on predicate logic and has proven to be a powerful means to process knowledge.

III. COMPUTER-AIDED PROCESS DESIGN

Often the term *computer-aided design* is used to denote computer-aided drafting activities, which are aimed at replacing the traditional paper drawings by computer generated drawings. Symbol libraries relevant to food manufacturing plants are available for several major computer-aided drafting packages. Examples are the TankCAD system for the project planning of ice cream factories [8] and the system developed by Lacted A/S (Silkeborg, Denmark) to produce flow diagrams for dairy and food processing lines [9]. Related activities are computer-aided design of food packages. Guise [10] describes a system installed at Saint-Gobain Embellage (France) to design wine bottles. A bottle shape is designed on the computer to meet weight and capacity requirements. The model can be visualized in three-dimensional space, giving a very good idea of the final bottle shape. The system installed at United Glass (U.K.) can even transmit modeling and tooling information direct to the CNC (*Computer Numeric Control*) machines in the manufacturing department. Computer-aided design of novelty gum and candy product packages is described in [11]. Koletnik [12] describes the computer-aided design of novel ice cream cornet shapes, and baking plates for use in the manufacture of hollow wafer products. While these applications of computer technology are important, they are not specific to foods and will not be discussed here further. In the remainder of this section we will concentrate on computer-aided *process design* issues.

Although computers have been used for quite a while in the chemical process industry for process design purposes, their use in the food industry has been limited by several factors. In the chemical process industry, most processes are conducted in *continuous* mode and involve large product quantities. The reactions that take place are typically relatively simple, involving only a few components at the same time. Moreover, the reaction kinetics are usually well understood and the physical parameters of the products involved are known. In the food industry, on the contrary, the *batch* mode is the prime mode of production (65% of the food and beverage industry [13]). The products involved are often complex food materials and involve a large number of ingredients. The reaction kinetics are very complicated and often unknown, and the physical parameters are very variable between different batches and are usually unknown. As a consequence, the existing software for the computer-aided design of chemical processes is not well suited for the design of food processes and must be modified to be useful.

A. Recipe Specification

Batch processes can be described by means of *recipes*. According to the ISA SP-88 standard on Batch Control Systems Models and Terminology, a recipe is "a complete set of information that specifies the control requirements for manufacturing a batch of a particular product." Four types of recipes are distinguished: general, site, master, and control recipes. A *general recipe* is generic and transportable between different plants. A *site recipe* is site-specific and is compliant to site-specific constraints. A *master recipe* takes into account specific equipment requirements and configurations and is as such the recipe that provides the information for producing a batch or a series of batches of the same product. A *control recipe* is the lowest recipe level and contains the batch ID and characteristics of a single executed batch, including operator- or system-generated information.

A recipe consists of a header, the equipment requirements, the formula of the product, and the procedure. The *header* provides information about several features of the recipe such as the author, the identification of the recipe and product, the version, and other useful information. The *equipment requirements* specify the type and size of equipment to be used, construction materials, and so forth. The *formula* is essentially a list of ingredients and their quantities, and setpoints of process variables and their duration. The *procedure* consists of a list of actions and their sequence to produce the required product, which is specified in the header as described in the formula using the listed equipment.

The design of a recipe is a creative action traditionally performed by a cook, whose main concern is to develop a food that is visually attractive and that smells and tastes good. However, as a consequence of the industrialization of food production processes, other concerns have emerged, such as the sensitivity of the food with respect to oxidation, the stability of emulsified foods, and issues of microbial safety. Consequently, the industrial design of foods has evolved toward a highly specialized field based on diverse branches of science, including biochemistry, process engineering, microbiology, heat and mass transfer, and mathematics. Several attempts have been made in the past toward the development of software to simplify this complex design task. Some important advances in the area of thermal food processes will be described here.

The design of thermal sterilization process is well established and is based on the analysis of the heat penetration in the sterilized food and the kinetics of thermal inactivation of microorganisms (see Chapter 9). For conduction-heated foods, it has been suggested to use a mathematical model (the Fourier equation) to predict the temperature inside the can as a flexible alternative to actual temperature measurements. Unfortunately, analytical solutions are only known for relatively simple problems involving generic product geometries (slab, sphere, cylinder) and temperature-independent thermophysical properties. Further, complicated procedures (the Duhamel theorem) must be applied to take into account time-varying boundary conditions. For this reason, it was suggested [14] to numerically solve the Fourier equation by means of a finite difference method, and to use the computed center temperature as an input for the calculation of the process lethality by numerical integration. This was implemented into a program that is one of the first examples of computer-aided food process design software. As a further improvement, the use of time-varying retort temperature profiles was considered [15] to maximize the retention of thiamine while safeguarding the required process value. This eventually lead to the STERILMATE software package for computer-aided design of sterilization processes [16]. More elaborate computer-aided optimization procedures have been described in [17] and [18].

To overcome the limitations of finite difference methods with respect to the geometry of the food, the finite element method has been suggested in the literature by a number of researchers for the design of thermal processes. In Fig. 2, a finite element grid is shown for a glass jar filled with baby food [19].

A major development in computer-aided process design was the CookSim package [20]. This package is essentially a knowledge-based

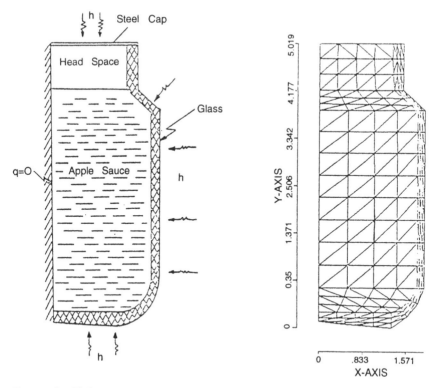

FIGURE 2 Finite element grid for a glass jar filled with baby food. (From [19], reprinted by permission of the Institute of Food Technologists.)

system that guides the user toward a safe thermal process design by automatically solving the mathematical models underlying the heat transfer process and the associated microbial kinetics. The CookSim package consists of four components. The *knowledge base decision support component* forms the central part of the system. It provides facilities to browse and edit data, and tools to display simulation results and compare processes and bacterial destruction rates. The knowledge base is built on top of the rule-based expert system shell ART. Rules for simulation model selection and thermal process optimization are included. The object-oriented *database component* contains the recipe, consisting of the food, the package, and the process in the CookSim terminology. The food is a particularly complicated object, which consists of ingredients in different proportions. Each ingredient has various parameters such as its physical state, and microbial and nutritional target parameters. The inheritance mechanism is effectively used to avoid unnecessary duplication of data.

The *simulation component* includes a finite difference procedure for conduction heat transfer analysis, and algorithms to calculate the thermal inactivation. The *user interface component* is graphically oriented and provides an easy way for the user to interact with the software. The authors also envisaged several techniques for process optimization, including neural nets, genetic algorithms, heuristics, and simulated annealing.

A related approach was followed in the development of the ChefCad package for computer-aided design of complicated recipes consisting of consecutive heating/cooling steps [21–23]. The global architecture of the package is shown in Fig. 3. The package is implemented on top of the C-based proKappa expert system shell, which provides rule-based reasoning and object-oriented data structures, and which runs on a Unix workstation. The *data and knowledge base* contains the declarative and procedural knowledge of the system. The declarative knowledge encompasses all the data in the system, including the current recipe, a list of food ingredients (the complete food table is in the system), species of microorganisms and the parameters of their growth/inactivation models, equipment types such as ovens and refrigerators, and so forth. The procedural knowledge base contains finite element routines for the numerical solution

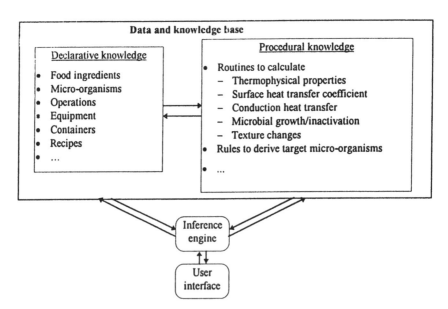

FIGURE 3 General architecture of the ChefCad package.

of 2-D heat conduction problems, an automatic finite element grid generator, routines to calculate the thermophysical properties from the chemical composition of the food, routines to calculate the surface heat transfer coefficient of the heating/cooling fluid, and differential equation solvers for the microbial growth/inactivation and texture changes. The *inference engine* is the core of the system. It is a part of the programming environment and contains procedural knowledge for making logical inferences. It is not immediately accessible to the programmer. The inference engine processes the user requests that arrive through the user interface. The necessary declarative data are fetched from the data and knowledge base, and they are passed to the calculation routines, which are then fired. The calculation results are then transferred back to the user interface for visualization. Also, a microbial safety diagnosis of the recipe is made by inferencing appropriate rules. The *user interface* is obviously the most visible part of the system. It is graphically oriented (X-Windows/Motif), and the user can interact with the system by means of graphical widgets such as push-buttons and pull-down menus. The time course of important process variables such as the food center temperature, the microbial load, and the texture can be visualized easily. The main window of the package is shown in Fig. 4.

B. Flowsheeting

Flowsheeting packages originated in the late 1950s in the petrochemical industries. Their main purpose is to calculate steady state material and energy balances during the operation of complicated chemical processes. The process engineer can then optimize the process by evaluating the effect of changing some process conditions on the product yield, waste production, and energy consumption of the process. Most flowsheeting packages actually consist of a language interpreter or compiler, a material properties database, models for different unit operations such as distillation columns, heat exchangers, and so forth, a calculation routine, optimization algorithms, and reporting facilities. The two main frameworks applied in simulators are the equation-oriented strategy and the sequential modular approach. In the former, a system of nonlinear algebraic equations is derived from the flowsheet and is subsequently solved using an iterative algorithm. In the latter, the flowsheet is broken down into modular blocks representing the unit operations in the process, which are then solved sequentially [24].

The use of flowsheeting in the food industry has been limited because of several reasons. First, most process flowsheeting packages are oriented toward continuous processes, which are conveniently described by alge-

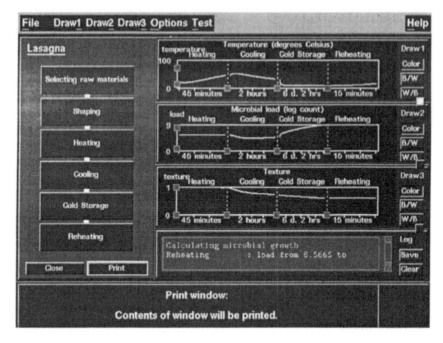

FIGURE 4 Main window of the ChefCad package. (Reprinted by permission of the American Society of Agriculture Engineers.)

braic equations. Food processes are, however, often conducted in batch and must therefore be modeled by means of differential equations. Most flowsheeting packages do not offer the ability to handle differential equations, so that less convenient general-purpose dynamic simulation packages (e.g., CSSL [25]) and related packages such as CSMP [26] or ACSL [27] or even low-level FORTRAN or C coding must be used. Further, most flowsheeting packages do not offer the ability to handle solids (with the exception of the ASPEN [AspenTech, Cambridge, Massachussetts] and GEMS [28] packages), whereas foods are often solid. Also, the kinetics of many subprocesses that take place during food processing remain largely unknown, as do the physical parameters of foods.

Nevertheless, some interesting case studies have been reported in the literature. Several examples of flowsheeting using the GEMS package were described [29]. The GEMS package [28] was originally developed for the paper and pulp industry. It incorporates a variety of blocks describing unit operations for the paper and pulp industry, and an executive program to connect the blocks and keep appropriate records. In a first example, the recovery of peel-oil from oranges during juice extraction was simulated. Alternative schemes were compared to minimize the operating problems that were encountered in the original setup. In a second example, the performance of a potato blanching process was optimized for controlling sugar content while minimizing energy consumption. In a third example, single- and multiple-zone drying systems were optimized by minimizing the energy consumption, increasing production capacity, and improving product uniformity.

Another batch flowsheeting package, PPDPAK, was developed at Purdue University [30] for the analysis of whey utilization, and later modified [31] to include all types of fluid food products. The package incorporates equipment and costing subroutines, a physical property prediction package, utility cost subroutines, and an optimization routine. The operating library of PPDPAK contains 14 unit operations related to food processing. It allows easy addition of new unit operation models, both steady and transient state. The package was used to investigate the most economical way to handle the whey waste stream from a cheese manufacturing process by producing whey protein concentrate by ultrafiltration. The ultrafiltrate permeate would then be fermented to ethanol. The option of preconcentrating the permeate feed using reverse osmosis before fermentation with the intention of reducing the fermentation and recovery equipment size was investigated using PPDPAK. The two process flow diagrams are shown in Fig. 5. The simulations indicated no payoff in applying reverse osmosis preconcentration.

The BOSS simulation package was developed for both single- and multiproduct batch processes [33]. The BATCHES package (Batch Process Technologies, Inc., West Lafayette, Indiana) evolved from the BOSS simulation package and includes a physical properties prediction module and interactive data input/output processor [32]. A BATCHES simulation model consists of three building blocks: equipment network, recipe network, and sequencing information. In the *equipment network* the equipment and their connectivity is specified. Scheduled maintenance and random events such as equipment failure can be included as well. The *recipe network* describes the recipe of the manufactured good as a network of tasks and subtasks. A recipe library is available. Subtask parameters such as the subtask duration, material input and output requirements, resource requirements, and state-dependent logic execution must be defined. The *sequencing information* includes the sequence of the operations to manufacture the products, the initial process status, and the materials arrival schedule. A number of tools such as icons, zooming, block copying, defaults, and on-line help are available to simplify the model definition. The process can then be simulated to calculate mass balances, equipment and

Process without reverse osmosis   Process with reverse osmosis

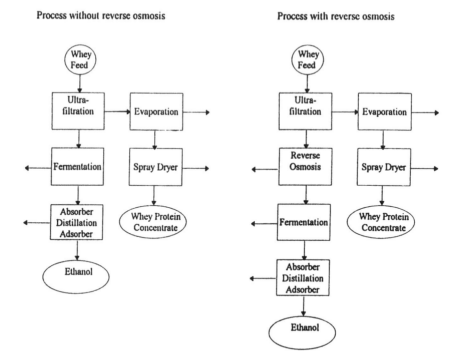

FIGURE 5 Flowsheet for conversion of whey into whey protein concentrate and ethanol. (From [32], reprinted by permission of Kluwer Academic Publishers.)

resource utilization statistics, cycle time, and waiting time information. Several graphical representation facilities (line graphs, Gantt and pie charts, animations) are provided.

Several applications of BATCHES to food process operations have been described in the literature. The BATCHES package was applied [32] for the simulation of a fluid dairy plant producing a mix of milk, ice cream, and cottage cheese with subcategories of each of these products. A schematic diagram of the process equipment in their simulation study is shown in Fig. 6. Raw milk is pasteurized in two semicontinuous pasteurizers, and stored as intermediate product in one of the surge tanks. Six intermediate products differing in composition are produced (buttermilk, 2% fat milk, and so on). The intermediate products are packaged in the package line into 36 different types of finished products, which are differentiated according to their package size. Some of the key operating features of the dairy plant were

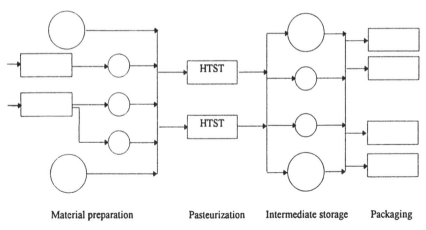

Material preparation   Pasteurization   Intermediate storage   Packaging

Figure 6 Schematic diagram of process equipment. (From [32], reprinted by permission of Kluwer Academic Publishers.)

The plant operates each day for approximately 20 h, in which a product mix as determined by the product planning department must be produced.

The amount of cleaning of the pasteurizers is to be minimized by appropriate scheduling.

The packagers stop frequently because of mechanical breakdowns or when cartons get jammed in the machines.

The pasteurization of some intermediates must take place in a specific pasteurizer.

The BATCHES model was applied to determine which solution was the best from an economic point of view in the case of a product demand increase of 64%: (1) adding an additional packager, (2) adding an additional packager and pasteurizer, or (3) adding an additional packager and creating a buffer stock. It was shown that with the first solution the required throughput could not be achieved. The second and third solutions both lead to the required throughput, but the third solution was $150,000 cheaper than the second. This indicates the financial gain that can be realized using process simulation. Pedrosa et al. [34] extended the analysis to 11 intermediate, 35 finished products, and 1 subproduct (cream) and employed statistical techniques to investigate the equipment utilization. They were able to show that most of the equipment was oversized, with an average utilization of below 60%, and that an expansion of 20% of the fluid milk production could be accomodated in a 24-hour cycle. Also, the simulations suggested some ways to improve the plant performance by reducing the volume of the surge tanks and optimizing electricity consumption.

Another application of BATCHES to food processes was described in [35]. They investigated the utilization rates of the pieces of equipment of the brewhouse of an industrial brewery by means of a BATCHES simulation and found that these rates were generally quite low and that some pieces of equipment were largely oversized. Based on these findings they proposed two new configurations for the brewhouses, thereby reducing by up to one-third the number of equipment items. In a recent publication they established an optimal recipe-dependent production schedule that minimizes the steam consumption peaks in the brewhouses.

Parallel with the evolution from text-based toward graphically oriented computer operating systems (Unix/X Windows, MS-Windows products, OS/2, MacOS), there is also a tendency toward graphically oriented simulation environments. One of the most successful is the WITNESS package, distributed by AT&T Istel (Redditch, U.K.). A production line can be assembled intuitively on the computer screen by pointing and clicking. A mixture of discrete and continuous elements can be used, and automatic performance reporting is provided. Although most applications are situated in the automotive, electronics, and pharmaceutical industries, WITNESS has been used successfully in the food industry, for example, to optimize ice cream manufacturing processes. Graphical programming environments such as EXTEND, LabVision, and others are also well suited for simulation processes. An example of the use of EXTEND to simulate water use in food processing plants was presented in [36].

Dynamic simulation of chemical processes is a very active research area. Current activities concentrate on simulators for processes described by a large number of differential and algebraic equations (e.g., SpeedUp [37]), which allow the engineer to define the plant model by connecting library unit operations, while additional unit operations can be defined, for example, those relevant to food processes. Combined discrete/continuous processes are also the subject of many research efforts (see [38] for a literature overview).

C. Quality Control Planning

The everlasting quest for quality is an essential feature of modern enterprises. The quality control activities in the food processing industry differ from those of most other industrial branches, as the quality of the produced items may change as a function of time. As a consequence, most food items have a limited shelf life and the manufacturer must be concerned not only with the present quality of the foods, but also with their future quality, even at a moment when they are downstream in the distribution chain.

Of most concern is the microbial quality of the food, because microbiologically spoiled foods are health concerns. Much effort has been devoted toward increasing the microbial quality of foods, and this has been achieved to some extent by means of software tools described in Section III.A. A qualitative approach that has gained much popularity lately is the HACCP methodology. HACCP (Hazard Analysis of Critical Control Points) is a preventive and systematic approach toward quality management of food manufacturing processes. It encompasses the identification of hazards associated with the production, the distribution, and the particular use of each food product, and the assessment of their severity and risk by well-documented and verifiable means. It also prescribes the determination of the actions that must be performed to control identified hazards at critical control points (CCPs), the monitoring of the criteria that indicate whether the CCPs are under control, the preparation of corrective actions if control is lost, and final verification to ensure that the HACCP system is efficient [39]. The HACCP procedure is compulsory in the new Food Hygiene Directive of the European Union (93/43/EEC).

The application of the HACCP methodology to a given food process requires much specific knowledge and involves considerable administrative effort. For these reasons it is suited to automation. A knowledge base [40], named Mirfak, was described, incorporating the knowledge related to bacterial contamination in a food production plant. This knowledge has been described using the HACCP formalism and is implemented as a *hypertext* system under the HyperCard environment. A hypertext document is essentially a document that consists of a large collection of pieces of information (text, graphics, sound, video) that are connected to each other by means of *links*. The user navigates through the document by clicking on active areas (specially formatted text, icons, buttons, etc.) that represent the links between the documents. The help tool of many software packages is typical example of hypertext, as well as World Wide Web documents. The Mirfak system contains a series of files, or *stacks* in the HyperCard terminology. The records of the stacks are called *reference cards*. Stacks or reference cards concerning microbiological analyses, foodstuff processing, and explanation of the HACCP system are implemented. There is also a stack of rules and standards for a good hygienic design, one containing miscellaneous information such as the characteristics of growth and destruction of the main microorganisms, and a final *object* stack describing the generic components (e.g., production and storage apparatus, premises) of a sandwich production factory. Several browsing tools (scanning, exploration, search, navigation memory, and a directory) are available. The provision of inferencing procedures is envisaged for future releases.

FIST-HACCP (TNO Voeding, Zeist, the Netherlands) is a MS-Windows-based software package for computerized HACCP analysis. It offers support in recording, processing, and reporting the complete HACCP protocol in an easy step-by-step approach. *Company data* such as personnel, process steps, and production documents are defined first. These data are common for all subsequent HACCP analyses. Next, the actual HACCP analysis can be accomplished. To this end the HACCP data must be entered, including the people involved in a particular process (the management committee and the HACCP team); the documents that describe the terms of reference, the product description, and the intended use of the product; and the production process. The latter can be represented visually in the form of flow charts. Once the production process is specified, the user can enter for each process step the potential hazards and the preventive actions to control these hazards. A decision tree then assists the user to determine whether a given process step is a critical control point. If so, the corresponding critical control point data can be entered and will be displayed in the flow chart. The continuous monitoring of the HACCP implementation is guaranteed by defining appropriate verification procedures. A wide range of reporting facilities is provided as well.

IV. PRODUCTION PLANNING AND CONTROL

The production planning and control activity is concerned with the detailed planning of the manufacturing activities of the plant. It involves the following tasks:

> Planning of manufacturing requirements and purchasing of raw materials
> Capacity planning
> Detailed production scheduling
> Recipe management
> Production monitoring and, if necessary, rescheduling

Several levels of planning can be distinguished, depending on their scope.

Strategic planning involves the long-range planning of the activities of the company. It is concerned with translating the business plans and sales plans into long-range production schedules. It enables the planning of resource requirements and material requirements for materials with very long delivery times. It involves high-capital decisions, such as purchasing new equipment and hiring new personnel. Decisions are taken on the basis of long-range demand forecasts.

The *tactical plan* usually covers a period of several months and involves the establishment of the *master production schedule* (MPS). The MPS is an overall schedule involving the production of goods, the capacity (availability of processing equipment), and the purchasing of raw materials.

Operational planning involves the establishment of a detailed production schedule and covers short time periods, usually on the order of weeks. The operational plan includes scheduling of batch operations, physical control of materials such as receiving, shipping, and stock management, inventory adaptation to short-term order changes, maintenance planning, purchasing and vendor control, and so forth.

In this section some approaches to production planning and control will be described, as well as the software tools available on the market to automate these activities. For a more elaborate treatment, the reader is referred to the literature.

A. Independent Demand Inventory Systems

For some products the customer demand is relatively constant, and the inventory level typically behaves as shown in Fig. 7. The inventory level decreases almost linearly, and when the *order point P* is exceeded, an order for a fixed quantity Q of the product is placed at the supplier. After a *lead time L*, the product arrives at the plant and the inventory is replen-

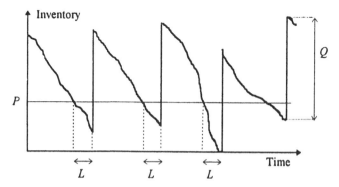

FIGURE 7 Typical course of the inventory level of an independent demand product.

ished. This type of inventory system is called an *independent demand inventory system*.

Alternatives such as periodic review systems are also in use. The independent demand inventory system can be automated readily. Most logistics packages on the market support this inventory system, offering features such as record updating, reporting, automatic reordering of items that are below their order point, generating exception reports, and recomputing decision parameters such as P, Q, and L [41].

An obvious drawback of the independent demand inventory system is clear from Fig. 7, where the inventory drops more sharply than expected after the third point crossing, thereby causing a brief shortage of inventory. For this reason forecasting procedures have been incorporated in many software packages, but it is clear that the success of an independent demand inventory system still relies on the predictability of the demand.

The fundamental drawback of the system, however, lies in the fact that it implies the existence of buffer stocks, which can be costly and are not always possible, particularly in the case of perishable products.

B. Material Requirements Planning (MRP) and Manufacturing Resources Planning (MRP2)

In practice many foods are manufactured from food components, which in turn may consist of subcomponents, and so on. This is shown in Fig. 8 for lasagna, which is assembled from pasta sheets, tomato sauce, and cheese. Tomato sauce consists of canned tomatoes, baked minced beef, cream, and an herbs mix. This can be continued down to the ingredient

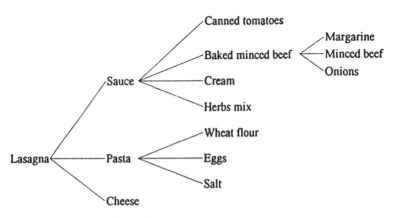

Figure 8 Composition of lasagna.

level, where ingredients are defined for this purpose as the food components that are purchased by an external supplier. In the discrete manufacturing processes, for example, an automobile assembly line, the term *Bill of Materials* (BOM) is often used to list the components and subassemblies that make up a product.

The independent demand inventory system introduced in the previous section can no longer deal with manufacturing processes involving a hierarchy of subcomponents, as it would imply an unrealistically large amount of inventory. The concept of MRP (*Material Requirements Planning*) was introduced in the late 1960s to aid in managing *dependent demand* inventory systems. The main principle was to reduce the magnitude of the buffers of intermediate products and to reduce lead times.

The MRP algorithm actually relies on a procedure known as *"explosion of the master production schedule."* The procedure will be explained based on the example outlined in Fig. 9, which is adapted from [2]. Suppose that a lasagna order schedule for a period of eight days is known (Fig. 9a), as well as the lasagna in the form of stock or unallocated manufacturing orders (Fig. 9b). The net lasagna requirements for the eight-day period are then determined from the gross requirements and the available lasagna as shown in Fig. 9c). For example, at day 0, the available lasagna (200 + 100 pieces) is sufficient to fulfill the requirements (100 pieces), and there is a remainder of 200 pieces. At day 1, the available lasagna (200 pieces + the 200 remaining pieces from day 0) is still larger than the required amount (300 pieces), and a stock of 100 pieces remains. At day 2, 200 pieces are required, so that the net lasagna requirement is equal to 100 pieces. Because the batch size is 200, one batch must be made available at day 2 (Fig. 9d), and because one batch lasts approximately one day, the production is scheduled for day 1. This in turn implies that the necessary components (sauce, pasta, and cheese) must be available on day 1 (Fig. 9e). For each of the components, the foregoing procedure is now performed recursively, until the starting dates for the manufacturing of the last subcomponents that are still prepared in the plant are known. Taking into account the supplier lead time, the order dates for the ingredients can now be calculated.

An important drawback of MRP is that it does not take into account the finite capacity of the plant. This was overcome in the 1980s by applying the MRP concept to determine labor, machine, capital, purchasing, marketing, and shipping requirements. As this implicated the existence of a common database, it can be considered as a significant step forward in integration of the administrative functions of the company. The updated approach is referred to as MRP2 (manufacturing resources planning) to distinguish it from the original MRP.

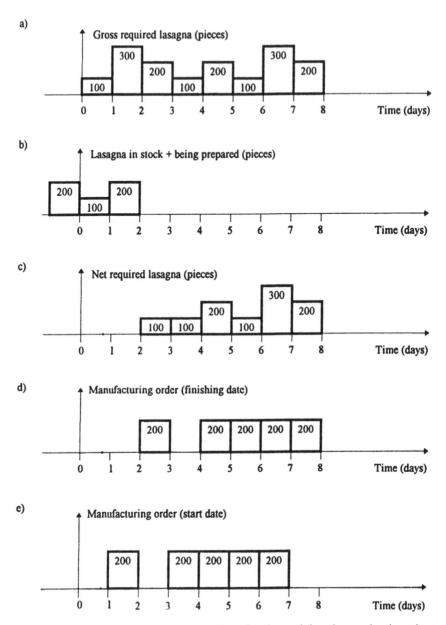

FIGURE 9 Example of the MRP procedure for determining the production plan of lasagna. (Adapted from [2], by permission of John Wiley and Sons.)

A large variety of packages with MRP2 functionality is now available on the market. Many of them have special features to deal with the particularities of the food industry such as the production of multiple end products from single raw materials (e.g., legs, wings, breast, and liver from chickens), limited storage times of foods, parallel production of components in batches, seasonal dependency of raw materials, and so forth [42].

The PRISM package is distributed by the MARCAM corporation (Newton, Massachusetts) and is designed specifically for the process industries. It consists of 28 modules that cover the total production chain. Advanced graphical user interfaces can be built. The package is based on the *"production model"* concept. A production model combines formula or recipe requirements with operational information. PRISM uses these models for collecting detailed product costs, for planning capacity and material requirements, and for defining production rates. Recipes and formulas can be created, evaluated, and modified by means of a PC-based formula management application. The effect of changes to formulas on the physical and chemical component properties and cost can be simulated. PRISM can exchange scheduling information with plant floor process control systems from major suppliers such as IBM, Hewlett-Packard, and Digital Equipment Corporation. Statistical data, costing data, and material availability information are up to date and readily accessible. Tools are available for quality management. Finite-capacity scheduling algorithms are incorporated. PRISM runs on IBM AS/400 computers, and a client–server implementation on all major Unix platforms is available. Some PRISM implementations in the food industry are described in [43] and [44].

C. Scheduling

The scheduling activity involves determining the sequence of operations to take place on each piece of equipment with due regard for capacity and with the objective of fulfilling the due date constraints specified in the production plan. As for continuous processes this is relatively straightforward, we will focus on batch processes in the remainder of this section.

The following factors are responsible for the difficulties encountered in process planning in the chemical industries [13]:

Cleaning operations required between the manufacture of different products

Limited raw material resources and production capacity

Operating preferences for making certain products using certain items of equipment

Equipment constraints such as size and materials of construction

Finding the appropriate balance between inventory level so as to be able to respond rapidly to customer orders against the storage capacity available and the cost of storage

Because of these factors the scheduling problem rapidly becomes a combinatorial problem of high dimension, and there is a definite need for computer assistance in this area. Classical scheduling approaches are based on linear integer programming techniques. However, the limited ability to represent a realistic situation, the required computer time, and lack of flexibility have restricted successful applications in the food industry and the chemical industry in general. Also, the human scheduler uses a large amount of expert knowledge that is difficult to represent in a numerical format. For these reasons alternative approaches based on artificial intelligence techniques seem to be more appropriate, and scheduling is currently the area of manufacturing management that is being most actively researched from a knowledge-based perspective [3]. Most MRP(2) packages now include a scheduling utility as well or can be interfaced with scheduling packages from other suppliers.

The SuperBatch system [45] is based on a continuous-time process representation and produces a schedule that satisfies operational feasibility constraints in detail, such as complex simultaneous transfers, continuous operations, tasks involving several processing vessels and utilities, and so forth. The plant equipment (vessels and connections), operating procedures, the production requirements, and operating constraints (such as equipment availability) must be defined in a high-level language. The scheduling algorithm requires the intervention of the human scheduler for some decisions, such as the order of batches, but takes care of the actual scheduling itself. SuperBatch was successfully applied in a number of industrial case studies, including a study at one of the largest dairy plants in the U.K.

GRIP (FYGIR logistic information systems, Rijswijk, The Netherlands; Potters Bar, U.K.; Burlington, Massachusetts) is an example of a commercial scheduling package developed for the semiprocess industry, and it offers many features, including planning of recipes with multiple actions, tank and silo planning, sequence-dependent changeover times, and simulation facilities. The observation that a scheduling procedure consists of a quantitative task that is accomplished by the computer, and a more intuitive task that must be performed by the human scheduler, lies at the basis of the software. Special efforts were made to enhance the interaction between the human scheduler and GRIP. The human scheduler can introduce decisions about the production plan by manipulation of graphical symbols using a mouse pointer. The GRIP system comprises the following constituent parts [46]:

A relational database, containing all data on the production process and planning.

Scheduling heuristics for batch creation, sequencing, and scheduling.

An interactive planning board based on a *Gantt chart*, showing batches on a horizontal time axis and a vertical resource axis (Fig. 10).

An event-driven constraint calculation and checking mechanism to deal with constraint violations such as overlap of batches, batches that start later than scheduled, and so forth.

Dynamic charts for inventory levels and personnel requirements (Fig. 10).

Graphical and text reporting facilities.

On-line interfaces to logistical systems, spreadsheets, and optional algorithms. SQL queries are supported.

GRIP was successfully installed in leading food and chemical companies.

The PROSPEX package [47] uses a combination of conventional simulation, modeling, and scheduling algorithms to represent all finite-capacity constraints, together with an English language rule system to express the scheduler's preference and operating strategies. The software is rule based and object and is now being implemented commercially in a flour-milling plant and in a cake and biscuit manufacturing company. An interesting feature of the package is its ability to show the reasons ("excuses") that an order fails to be executed.

Recent advances in scheduling software include real-time scheduling systems that can take into account process deviations having implications for the validity of the schedule (e.g., DSP, a G2-based software distributed by Gensym Corporation, Cambridge, Massachusetts).

D. Optimized Production Technology (OPT)

The OPT system (Optimized Production Technology) is currently marketed intensively as an alternative to the MRP system [3]. It concentrates on the efficient use of bottleneck resources. To this end the latter are scheduled using a proprietary scheduling algorithm. The noncritical orders are fitted afterward into the critical order schedule. The OPT system relies on a clear definition of the bottleneck resource, which is not always possible in practice. It is also sensitive to process deviations.

E. Just-in-Time (JIT) Production Systems

In the 1950s until the beginning of the 1980s, companies successfully applied the concept of "economics of scale" to produce large quantities of

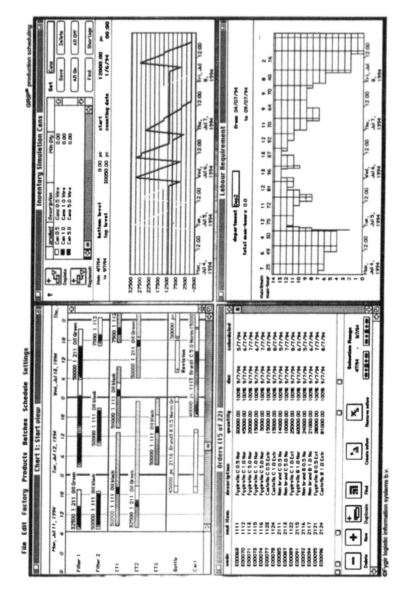

Figure 10 Screen display with simultaneous display of selected reports and charts. (© Fygir logistic information systems b. v.)

a relatively small range of products. The rationale behind this concept is the assumption that manufacture of a new good requires a lead time and involves start-up costs. The consequence of this approach was the necessity of large product inventory levels and intermediate product stocks. However, during the last decades consumer demands have evolved toward a very variable demand for a large variety of products. This requires a small response time of the production facilities, and also a diversification of the product line. These attributes are no longer compatible with the habit of keeping large stocks of goods. As described above, MRP and MRP2 were introduced to reduce buffer stocks, but unfortunately they are particularly prone to process disturbances that occur frequently in the food industry because of the varying quality and availability of raw materials. The software companies have tried to overcome this problem by increasing the buffer amounts or by adding "slack" time. However, this is directly opposite to the main principle of MRP. The *Just-in-Time* (JIT) principle was successfully introduced primarily by Japanese factories as an alternative production system.

The JIT approach is simple and consists of producing the right products in the quantities required by the downstream processes (or customers) at the right time. This is accomplished by elimination of activities and equipment that contribute no added value to the product such as lead times, transit times, and intermediate and end product stocks. While in a conventional "*push*" production system, the production is governed by intermediates that are waiting to be processed, in a "*pull*" system such as JIT the production is regulated by the consumption of goods at the end of the production chain. The introduction of a JIT system involves all aspects of the production system. Some of the key ingredients are optimization of product flow and factory layout to reduce transportation time, coordination of operations, capacity increase by purchasing additional equipment if necessary, reduction of the number of machine breakdowns by introducing preventive maintenance, increase of personnel flexibility, just-in-time deliveries by suppliers, and enhancement of overall quality. Note that the JIT principle is highly compatible with food production systems, as the decrease of stocks and lead times automatically enhances the microbial, nutritional, and visual quality of the foods.

The *Kanban system* is a popular implementation of the JIT principle. This system is based on cards (Kanban is Japanese for card) that are associated to each of the intermediates of a product. The cards are in fact documents that are attached to the part or container and that contain information about the contents of the batch. Assume, for example, a cookie baking process. To each pallet of packaged cookies corresponds a card specifying the amount of cookies. If the customer purchases a pallet, the corresponding card is detached and returned to the *Kanban post* (a card holder). Likewise, the empty pallet is returned to the pallet stack. As long as there are cards in the post, the cookie packager takes the oldest of these cards and packages a pallet of cookies. But, therefore, it needs unpacked cookies from the cookie bakery line. The same procedure is applied here, and in each subsequent production step. In a modern production facility the cards are obviously replaced by a paperless system. The Kanban system is very simple but implies a regular and known consumption trend. If not, safety stocks must be provided, which make the system less attractive. Therefore in many modern plants the production planning system is a combination of the Kanban system and an MRP system.

V. COMPUTER-AIDED MANUFACTURING

Computer-aided manufacturing (CAM) deals with automation at the factory floor level. The topic of process control with issues such as recipe handling and low-level PLC control is well covered by previous chapters and will not be covered in great detail here. Here, some issues relevant to the integration of all subsystems will be discussed.

A. DCS, SCADA, and Knowledge-Based Systems

As outlined in Chapter 4, a *distributed control system* (DCS) consists of a group of *programmable logic controllers* (PLCs) connected to each other and to host computers by means of a network. Each controller can be programmed to work relatively independently, and the host computers are responsible for coordination of the programmable controllers. The operator interface is typically a collection of computer screens in the control room and around the plant floor. A DCS is typically used for control purposes, with special emphasis on analog control [48]. A wide variety of control algorithms, ranging from simple PID (proportional–integral–derivative) control up to advanced adaptive control, is typically provided, and the control engineer simply has to define the tuning parameters. A Unix/X-Windows computing environment supported by a suitable relational database management system is usually the software platform of choice.

Off-the-shelf SCADA (Sequence Control And Data Acquisition) packages are aimed at gathering, logging, and reporting information. Libraries for interfacing with PLCs are usually available. Modern SCADA packages provide a broad functionality, including communication with databases, PLCs, and operators, recipe handling, and reporting, and often run under an OS/2 or MS-DOS/MS-Windows environment.

Knowledge-based systems have been suggested in the literature for decision support to process operators (see [49] for an overview). A complex expert system has been developed by the Campbell Soup Company for solving problems in the thermal sterilization of canned food products [50]. The system contains specialized expertise of retiring senior scientists. An expert system, called SOYEX, was presented [51] to correct the problem of high residual flake fats in the soybean crushing industry. The system is based on expert knowledge from experienced process engineers. An expert system, called LETTMAN, for management decision making on handling lettuce in the fresh produce department of a retail supermarket was described in [52].

B. Quality Analysis and Control

On-line quality analysis and control has become a crucial topic in modern food manufacturing plants. This is particularly the case for a JIT environment, where the aim is to minimize product defects.

Statistical process control (SPC) is a methodology aimed at *preventing* rather than *correcting* product quality deficiencies. In SPC the natural, random variation of product properties is considered to be normal. Changes in the statistical characteristics of the product properties such as mean, variability, or probability distribution shape are assumed to result from assignable causes, which should be corrected. Even if a process is under statistical control, it is possible that the variability is outside acceptable limits. The process is then said to be no longer *capable* of producing the product within acceptable specification limits. The principles of SPC can be found in many textbooks (e.g., [53]) and will not be discussed here further.

Many integrated CAM packages provide statistical process control facilities. For example, the MONITROL/UX environment (Hilco Technologies, Inc., Earth City, Montana) provides a full range of charts and diagrams for presenting and analyzing the manufacturing process. Current process variables can be displayed using trend charts, control charts, histograms, and scatter plots. A wide variety of standard charts is provided. Charts and diagrams can be grouped and automatically sent to the quality department's screens. An important feature is that out-of-statistical-control conditions can be defined to trigger predefined procedures or alarm conditions.

VI. IMPLEMENTATION ISSUES

In a CIM oriented company, all the different automated modules must eventually be tied into a global information system that rests on the common company database. Obviously, the introduction of CIM touches all aspects of the company, and consequently, every layer of the company management hierarchy should be represented in the CIM implementation team.

A typical CIM implementation consists of three phases:

In the *definition phase*, the concept is defined and the areas that are expected to gain the most are identified. An important and difficult task this phase is to achieve the commitment of all management layers. To this end the management should be familiarized with the CIM concept. A clear feasibility study is essential to this end. An inventory of the current state of automation must be made, to be able to recycle as much as possible of existing hardware, code, and databases. A list of candidate automation projects is established, and priorities are set in functions of the strategic plan of the company. Because of the complexity of the task, it is often wise to include external consultants who are experienced in CIM implementations.

In the *implementation phase*, important issues such as the database system (codasyl, relational) and architecture, the hardware (computers, PLCs, field bus/dedicated bus), the operating system(s), and network software are to be determined by the joint CIM team. Most of these items are related to each other. For example, some integrated software packages are only available on specific computer hardware, support only well-defined relational database management systems, and can communicate with specific PLC types. The actual implementation is carried out according to a *bottom-up* strategy, where the different departments of the company are first automated before incorporating them into the global CIM structure. Information system design methodologies such as MERISE or SADT [2] can be applied advantageously. For the development of knowledge-based systems, the KADS system (Knowledge-Based Systems Analysis and Design Support) has been suggested [2].

In the *feedback phase*, the CIM implementation is evaluated and a feedback to the definition or implementation phase is established. It is important to realize that because of the current growth of information technology some parts of the system will become outdated before the project is fully implemented. A CIM implementation therefore will be a continuous activity and is by no means an endpoint.

An important decision is the amount of in-house programming versus standard software packages. Verstraeten (personal communication) compared 17 standard software packages for integral logistics planning. He concluded that about 80% of the specific logistics functionality requirements of a typical company in the semiprocess industries can be fulfilled by standard logistics software packages. Further, most of the standard packages provide a 3GL or 4GL programming environment to add company-specific functionality. The use of standard packages should be considered carefully, as it can substantially reduce implementation time.

Several software packages now on the market that offer an integrated environment for process management. The IPROM package (Akzo Systems Nederland b.v., Ede, the Netherlands) is an environment to fulfill all needs of the process industries, and it consists of two modules. The PROLIS (Process Logistics Information System) module combines logistics functions such as MRP (Material Requirements Planning), CAP (Capacity Allocation Planning), and DRP (Distribution Requirements Planning), and order and inventory administration. The module PICA2000 (Process Information and Control Application) supports registration and reporting of batch and (semi)continuous production processes and allows for process optimization based on the process information. CIMPLICITY (GE Fanuc Automation, North America, Inc.) is a distributed system consisting of a large number of modules and is available for a large number of software/hardware environments (network, bus, operating system, PLC). The base system provides an interface between PLCs and users and offers a broad range of functionality including graphic status monitoring, management and reporting of point data, alarm management, and logging and reporting facilities. Optional modules include on-line SPC, database logging/reporting, maintenance management, fault diagnostics, MRP, trending, automatic control and recipes management, and many more. APIs (Application Program Interfaces) are provided to seamlessly integrate specific software into CIMPLICITY. If required, the package can be implemented on top of a single relational DBMS system.

The implementation of CIM typically involves one or more years. StarKist Caribe, a tuna processor, spent almost two years to implement an MRP2 system [54]. This involved replacing the old IBM mid-range System 36 by IBM AS/400 machines running the PRISM package. The following PRISM modules were implemented: production analysis and planning, resource processing, planning and management, financial support, customer order management, foundation, and purchasing and production analysis. Afterwards StarKist added quality, warehouse, and activity costing modules. A considerable effort went into porting a maintenance management package from the System 36 to the AS/400 environment. PCs are connected to the AS/400 network to provide spreadsheet and word processing facilities. The payroll system also runs on a PC network that exchanges files with the AS/400 MRP2 modules. The system is estimated to save the company over $1 million a year. The system is planned to be extended to include computer-aided manufacturing facilities as well.

This example indicates that substantial financial returns can be expected from a CIM implementation. However, in general it is not easy to estimate the financial return. For a more detailed discussion of these aspects, the reader is referred to [2].

VII. APPLICATIONS

Few plants are actually automated according to the full CIM concept as outlined in Fig. 1. Usually only those aspects that are expected to yield the highest profit by automation are implemented. In a survey on the use of CIM in food processing companies [55], it was found that financial accounting tasks were automated in almost all companies; 77.4% companies applied computers for production planning issues such as inventory control (100%), MRP (66.7%), and MRP2 (29.2%). Less than half of the companies reported use of computers in distribution management (45.2%) and computer-aided manufacturing (41.9%). Computers were also used in quality control (35%), materials handling (16.6%), maintenance scheduling (6.5%), and computer-aided design (6.5%). While the exact figures are probably somewhat outdated because of the rapid evolution of information technology, the trend is likely still the same. However, a literature survey (Table 4) shows that CIM is (at least partially) implemented in all sectors of the food industry.

Much attention has gone to the implementation of CIM in the meat industry. A general framework for the application of CIM to slaughterhouses was outlined in [56]. Weinberg [57] discussed a CIM implementation in a meat processing plant that is based on the BISON software. Special efforts have been made to assist the meat cutting. The *cutting method module* describes a stage of cutting or a cutting process, taking

TABLE 4 Applications of CIM in the Food Industry

Product type	References
Aseptic processing of particulate foods	[80]
Meat industry	[57]–[65]
Beer breweries	[66]–[73]
Chocolate production	[81]
Beverages	[73], [84]
Dairy, ice cream manufacturing	[75]–[79]
Pasta	[82]
Oils and fats industry	[83]
Snack foods, prepared meals	[74], [76]

into account the animal species. The way the initial material is subdivided into individual parts is represented on a video screen with printer. A cutting method library is included, which can easily be expanded or modified. Other aspects such as automatic tracking of meat and meat products as they are moved around the plant, financial issues, production planning, recipe development, material acquisition, weighing of finished products, and operation of storage facilities are discussed in [58]–[65].

Asahi et al. [66] discussed a CIM implementation in a 5.4 million hl/ year brewery. The monthly production plan is determined based on middle- and long-term demand forecasts. From the monthly production plan, the material and capacity requirements plan is established. The planning software runs on a host computer connected to a first LAN (Local Area Network). The quality control system runs on a computer connected to the first LAN and supervises the production cycle, the maintenance activities, and the monitoring, forecasting, and logging of the quality of raw materials, intermediary products, and end products. It also serves as a gateway to a production floor level LAN. The storage and transport system resides on a computer connected to the first LAN and takes care of the packaging units and the inventory and shipping tasks. It is connected to a third factory floor LAN. Similar CIM implementations are described in [67]–[73].

Martens [74] described the implementation of JIT and CIM in a production plant for "sous vide" type prepared meals. The ChefCad package (see above) is applied to investigate changes in product formulation and process parameters on the microbial load and texture of the food. An MRP package has been selected and adapted. Proprietary scheduling software was developed to optimize the sequence of batches in order to maximize the utilization rate of the packaging machine, which is the bottleneck of the production. A batch control software package (FERRANTI PMS) has been selected and adapted to automate recipe handling. The package runs on a process computer that can communicate with the low-level PLC platform on the one hand, and the inventory package, which runs on a Unix system. A completely new plant was built to implement JIT requirements such as zero defects, uniform production load, and reduction of setup times. Many difficulties were encountered in the implementation process, mainly due to biological variability of ingredients, the wide variety of products, the impossibility to automate tasks such as tasting, and the lack of standardization of hardware and software.

A relatively complete CIM implementation in the dairy industry is described in [75]. The software system consists of three components: the PRISM package, which is used to manage manufacturing and customer order functions; the POMS (Plant Operations Management System); and

DMACS, a distribution process monitoring and control system, which are combined to form an operational system for day-to-day running of the plant. The use of DCS systems in an integrated dairy processing plant and related issues are discussed in [76]–[79].

Other application areas of CIM include aseptic and retort technology [80], chocolate production [81], pasta manufacturing [82], oil and fat processing [83], dry packaged snack foods [76], and beverages [84]. It is very likely that many companies have now implemented CIM to some extent but do not publish their experiences because of strategic reasons.

VIII. CONCLUSIONS

The implementation of CIM is an important strategic decision that involves commitment of all hierarchical levels in the company management. The underlying technology in hardware and software is now available to make a comprehensive CIM implementation feasible. Although the food industry has been somewhat reluctant so far, this is changing very quickly, and it can be expected that those companies willing to implement the CIM concept in the near future will gain a substantial competitive advantage.

ACKNOWLEDGMENTS

The author wishes to thank Josse De Baerdemaeker, Toon Martens, Henk Olivié, and Eric Duval (Katholieke Universiteit Leuven, Belgium) for valuable discussions. Egon de Waart and Maxim Brouwer (FYGIR logistic information systems, Rijswijk, the Netherlands) are gratefully acknowledged for providing screendumps of the GRIP scheduling package. Peter Verstraeten (Serneels, Verstraeten and Partners, Gent, Belgium) is acknowledged for providing information on logistics packages, and Jaak Grobben (ALMA Universiteitsrestaurants, Leuven, Belgium) for advice on SQL programming. The author is Postdoctoral Fellow of the Belgian National Fund for Scientific Research (N.F.W.O.).

REFERENCES

1. A.-W. Scheer, *CIM: Computer Steered Industry*, Springer-Verlag, Berlin, Heidelberg, 1988.
2. J.-B. Waldner, *CIM: Principles of Computer-Integrated Manufacturing*, John Wiley and Sons, New York, 1992.
3. R. Kerr, *Knowledge-Based Manufacturing Management*, Addison-Wesley, Reading, Massachusetts, 1991.
4. E. F. Codd, A relational model of data for large shared data banks, *Commun. ACM 13*:377 (1970).

5. M. M. Astrahan and D. D. Chamberlin, Implementation of a structured English query language, *Commun. ACM 18*:580 (1975).
6. C. Tristram, Do you really need object databases? *Open Computing 12*(9): 72 (1995).
7. K. Parsaye, M. Chignell, S. Khoshafian, and H. Wong, *Intelligent Databases: Object-Oriented, Deductive Hypermedia Technologies*, John Wiley and Sons, New York, 1989.
8. Anon., The TankCAD system for the project planning of ice cream factories, *North Eur. Food Dairy J. 54*:298 (1989).
9. R. Hansen, Technical expertise as a selling item, *North Eur. Dairy J. 53*:25 (1987).
10. B. Guise, The glamour of glass, *Food Process. 58*(4):51 (1989).
11. R. Lingle, Streamlining package design through computers, *Prep. Foods 160*(7):86 (1991).
12. H. Koletnik, Spezialprodukte im Waffel- und Eistuetenbereich, *Zucker- und Süsswarenwirtsch. 47*:114 (1994).
13. P. Sawyer, *Computer-Controlled Batch Processing*, Institution of Chemical Engineers, Rugby, U. K., 1993.
14. A. A. Teixeira, J. R. Dixon, J. W. Zahradnik, and G. E. Zinsmeister, Computer optimization of nutrient retention in thermal processing of conduction-heated foods, *Food Technol. 23*(6):137 (1969).
15. A. A. Teixeira, G. E. Zinsmeister, and J. W. Zahradnik, Computer simulation of variable retort control and container geometry as a possible means of improving thiamine retention in thermally processed foods, *J. Food Sci. 40*:656 (1975).
16. K. H. Kim, A. A. Teixeira, J. Bichier, and M. Tavares, *STERILMATE: Software for Designing and Evaluating Thermal Sterilization Processes*, Paper no. 93–4051, ASAE, St. Joseph, Michigan, 1993.
17. J. R. Banga, J. M. Perez-Martin, J. M. Gallardo, and J. J. Casares, Optimization of the thermal processing of conduction—heated canned foods: Study of several objective functions, *J. Food Engg. 14*:25 (1991).
18. C. M. Silva, M. Hendrickx, F. Oliveira, and P. Tobback, Critical evaluation of commonly used objective functions to optimize overall quality and nutrient retention of heat-preserved foods, *J. Food Engg. 17*:241 (1992).
19. D. Naveh, *Analysis of transient conduction heat transfer in the thermal processing of foods using the finite element method*, PhD thesis, University of Minnesota, 1982.
20. P. R. Race and M. J. Povey, CookSim: A knowledge based system for the thermal processing of food, *Expert Systems and Their Applications*, Avignon, France, 1990, p. 115.
21. M. Schellekens, T. Martens, T. A. Roberts, B. M. Mackey, B. M. Nicolaï, J. F. Van Impe, and J. De Baerdemaeker, Computer aided microbial safety design of food processes, *Int. J. Food Microbiol. 24*:1 (1994).
22. B. M. Nicolaï, J. F. Van Impe, and M. Schellekens, Application of expert systems technology to the preparation of minimally processed foods: A case study, *J. A Benelux Q. J. Automatic Control 35*:50 (1994).

23. B. M. Nicolaï, W. Obbels, M. Schellekens, B. Verlinden, T. Martens, and J. De Baerdemaeker, Computational aspects of a computer aided design package for the preparation of cook–chill foods, Proceedings of the Food Processing and Automation Conference III, Orlando, Florida ASAE, St. Joseph, Michigan, 1994, p. 190.

24. J. N. Petersen and D. C. Drown, Computer aided design in the food processing industry. I. Overview of process flowsheeting, *J. Food Technol. 20*:397 (1985).

25. J. C. Strauss, The Sci continuous system simulation language (CSSL), *Simulation 9*:281 (1967).

26. H. Speckhart and W. H. Green, *A Guide to Using CSMP*, Prentice-Hall, Englewood Cliffs, N.J. 1976.

27. E. E. Mitchell and J. S. Gauthier, Advanced continuous simulation language (ACSL), *Simulation 26*:72 (1976).

28. L. Edwards and R. Baldus, *GEMS—General Energy and Material Balance System: A Modular Computer System for Pulp and Paper Applications*, Idaho Research Foundation Inc., 1972 (cited by [29]).

29. D. C. Drown and J. N. Petersen, Computer aided design in the food processing industry. II. Applications, *J. Food Technol. 20*:407 (1985).

30. S.-Y. Hsu, *A multilevel approach for preliminary process development and a demonstration on developing whey processing systems*, PhD thesis, Purdue University, W. Lafayette, Indiana, 1984.

31. P. S. Moyer, *Computer Aided Food Process Design*, M. S. Thesis, Purdue University, W. Lafayette, Indiana, 1987 (cited by [32]).

32. S. L. Havlik, L. Deer, and M. R. Okos, Computer-aided engineering in the food industry, *Food Properties and Computer-Aided Engineering of Food Processing Systems* (R. P. Singh and A. G. Medina, eds.), Kluwer Academic Publishers, Boston, 1988, p. 507.

33. G. Joglecar and G. V. Reklaitis, A simulator for batch and semi-continuous processes, *Comput. Chem. Eng. 8*:315 (1984).

34. A. C. Pedrosa, M. R. Okos, and G. V. Reklaitis, Simulation of non-continuous food processes, Proceedings of the Food Processing and Automation Conference II, Lexington, Kentucky, ASAE, St. Joseph, Michigan 1992, p. 375.

35. D. Mignon and J. Hermia, Energy integration of brewery operations, *Automatic Control of Food and Biological Processes* (J. J. Bimbenet, E. Dumoulin, and G. Trystram, eds.), Elsevier, New York, 1994, p. 401.

36. J. I. Maté and R. P. Singh, Simulation of water use in food processing plants, Proceedings of the Food *Processing and Automation Conference II*, Lexington, Kentucky, ASAE, St. Joseph, Michigan, 1992, p. 316.

37. C. C. Pantelides, SpeedUp—recent advances in process simulation, *Comput. Chem. Eng. 12*:745 (1988).

38. P. I. Barton, *The Modelling and Simulation of Combined Discrete/Continuous Processes*, PhD Thesis, Imperial College of Science, Technology and Medicine, London, 1992.

39. ICMSF, *Micro-Organisms in Foods. 4: Application of the Hazard Analysis Critical and Control Point (HACCP) System to Ensure Microbiological Safety and Quality*, Blackwell Scientific Publications, Oxford, 1988.

40. E. Laporte, G. Muratet, O. Cerf, and P. Bourseau, Development of a knowledge base for the improvement of hygiene in the food industry, Proceedings of the AIFA Conference on Artificial Intelligence for Agriculture and Food, Nîmes, France, October 27–29, 1993, p. 79.

41. L. J. Krajewski and L. P. Ritzman, *Operations Management—Strategy and Analysis*, Addison-Wesley, New York, 1990.

42. S. Serneels, Produktieplanning in de semi-proces. Schemerzone tussen twee werelden. *Logistiek Manage. 12*:7 (1993).

43. L. Stevens, MRP II comes to process manufacturing, *3X/400 Information Management 3*:56 (1991).

44. A. Laplante, Executive report—gaining a pricing edge. Finding the leeway to price effectively. *Computerworld 29*(10):113 (1990).

45. S. Machietto, Automation research in a food processing pilot plant, *Food Engineering in a Computer Climate*, IChemE Symposium Series No. 126 Preprints, IChemE, Rugby, U. K, 1992, p. 179.

46. E. de Waart, GRIP: Interactive software for production scheduling in batch process industries, *J. A. Benelux Q. J. Automatic Control 34*(3):18 (1993).

47. J. C. Taunton and C. M. Ready, Intelligent dynamic production scheduling, *Food Res. Int. 27*:111 (1994).

48. D. Mack, Computers in control, *Automation in the Food Industry* (C. A. Moore, ed.), Blackie, Glasgow, U. K., 1991, p. 75.

49. A. A. Teixeira, Computer technology in the future of food manufacturing, *Changing Food Technology. 3. Food Technology: A View of the Future*, Technomic Publishing, Lancaster, Pennsylvania, 1989, p. 123.

50. R. A. Herrod, Industrial applications of expert systems and the role of the knowledge engineer, *Food Technol. 43*(5):130 (1989).

51. L. A. Deer, D. D. Jones, and M. Okos, An expert system for soybean oil extraction, Paper No. 88–544, ASAE, St. Joseph, Michigan, 1988.

52. R. S. Muttiah, C. N. Thai, S. E. Prussia, R. L. Shewfelt, and J. L. Jordan, An expert system for lettuce handling at a retail store, *Trans. ASAE 31*:622 (1988).

53. G. B. Wetherill and D. W. Brown, *Statistical Process Control: Theory and Practice*, Chapman and Hall, London, 1991.

54. B. Sperber, Computer integrated enterprise. StarKist scales down to steer through high-tech waters, *Food Process. 62*(3) (1993).

55. N. A. Aly, A survey on the use of computer-integrated manufacturing in food processing companies, *Food Technol. 43*(3):82 (1989).

56. G. Lorenz, Strategic variables, processes and technologies in the meat production system. Current trends and research requirements, *Fleischwirtsch. 71*:175 (1991).

57. H. Weinberg, Computer-integrated weighing and data systems for cutting rooms, *Fleichwirtsch. 71*(5):555 (1991).

58. D. Antritter, EDV für die Fleischverarbeitung, *Fleischerei 39*:890 (1988).
59. T. Bishop and D. Smith, Should the meat industry digest CIM? Proceedings of the International Congress of Meat Science and Technology 35, vol. 1, 1989, p. 93.
60. P. Schimitzek, Beschaffungs- und Produktionsplanung in der CIM. Die betriebswirtschaftliche Bedeutung der Rezepturoptimierung, *Fleischerei 40*:604 (1989).
61. P. Schimitzek, Computerintegrierte Production in der Fleischwirtschaft: CIM in production, *Fleischerei 42*:30 (1991).
62. P. Schimitzek, CIM in der Fleischwirtschaft: Identifikationssystem als Voraussetzung zur Verfolgung der Warenbewegungen, *Fleischerei 42*:941 (1991).
63. P. Schimitzek, CIM-Anwendung im Schlacht- und Zerlegebetrieb, *Fleischwirtsch. 72*:861 (1992).
64. P. Schimitzek and M. Studer, CIM-Anwendung in der Fleischwarenindustrie: Erfahrungsbericht anhand eines Beispieles aus der Praxis, *Fleischerei 44*:349 (1993).
65. L. C. Nielsen, Tulip–Vejle Nord Plant, Proceedings of the International Congress of Meat Science and Technology 35, vol. 1, 1989, p. 190.
66. K. Asahi, M. Kitaguchi, and M. Kinoshita, Die neue Asahi Brauerei in Ibaragi, *Brauwelt 131*:2273 (1991).
67. Anon., Heineken on the move, *Food Eng. Int. 11*(4):21 1986.
68. P. Lange, Prozess-Datenerfassung und Anlagendokumentation. Betriebserfahrungen mit einem integrierten Rechnersystem zur Prozess-Datenerfassung (Monitoring) und Anlagendokumentation (Computer Aided Design), *Brauwelt 127*(15/16):648 (1987).
69. K. Koide, S. Nishie, and T. Nakatani, Neue Braustaette der Kirin-Brewery in Yokohama, *Brauwelt 130*:1375 (1990).
70. L. Ngan, Tiger hi-tech brewery, *Ferment 3*:252 (1990).
71. H. Weisser, Betriebsdatenerfassung in Brauereien, *Brauwelt 131*:864 (1991).
72. T. S. Kong, Computer integrated stock administration, *Ferment 5*:345 (1992).
73. G. Mailänder and G. Staab, E-Bilanz—Ein Energie-Bilanzierungssystem im Rahmen von CIM, *Brauwelt 132*:2178 (1992).
74. T. Martens, JIT and CIM concepts, applied to a large scale catering production plant, *Developments in Food Engineering* (T. Yano, R. Matsuno, and K. Nakamura, eds.), Blackie Academic and Professional, London UK 1994, p. 990.
75. Anon., First CIM site operational at NZ Milk factory, *Food Technol. New Zealand 26*:8 (1991).
76. M. Royer, Is there a DCS in your future? *Food Eng. Int. 14*:45 (1989).
77. C. Honer, Mid Valley Dairy: America's most automated milk plant, *Prep. Foods 157*:74 (1988).
78. C. Honer, Automated plant produces 17,000 gal of ice cream per hour, *Prep. Foods 156*:106 (1987).
79. C. Honer, Dairy plants get smarter, *Dairy Foods 90*:87 (1989).
80. C. E. Morris, Teaming up to advance aseptic and retort technology, *Food Engg. 65*(1):111 (1994).

81. Anon., Sweet running for automated chocolate production, *Confectionery Prod. 58*:646 (1992).
82. L. Lirici, Computer-integrated manufacture in the modern pasta factory [in Italian] *Tec. Molitoria 39*:1025 (1988).
83. Anon., Integrated oils and fats processing, *Oils Fats Int. 5*:36 (1989).
84. J. Busby, How ACCOS supervises the blending of Pepsi, *Brew. & Distilling Int. 20*:22 (1989).

Index